Professional Pilot Career Guide

Professional Pilot Career Guide

Robert P. Mark

McGraw-Hill
New York   San Francisco   Washington, D.C.   Auckland   Bogotá
Caracas   Lisbon   London   Madrid   Mexico City   Milan
Montreal   New Delhi   San Juan   Singapore
Sydney   Tokyo   Toronto

Library of Congress Cataloging-in-Publication Data

Mark, Robert P.
Professional pilot career guide / Robert P. Mark.
p. cm.
Includes index.
ISBN 0-07-134691-0

000000.0000 1999
000'.0—dc21   00-00000
                                                                CIP

McGraw-Hill

A Division of The McGraw·Hill Companies

9 0   FGR/FGR   0 9 8 7 6 5

ISBN 0-07-134691-0

The sponsoring editor for this book was Shelley Carr, the editing supervisor was Sally Glover, and the production supervisor was Sherri Souffrance. It was set in Palatino per the BSF design by Kim Sheran and Joanne Morbit of McGraw-Hill's Professional Group Composition Unit, in Hightstown, N.J.

Printed and bound by Quebecor World / Fairfield.

McGraw-Hill books are available at special quality discounts to use as premiums and sales promotions, or for use in corporate training programs. For more information, please write to the Director of Special Sales, McGraw-Hill, Professional Publishing, Two Penn Plaza, New York, NY 10121-2298. Or contact your local bookstore.

This book is printed on recycled, acid-free paper containing a minimum of 50% recycled, de-inked fiber.

For Nancy and Abigail.

Contents

Foreword

CONGRATULATIONS! You hold in your hand a book that could be the starting point for a very exciting, professionally rewarding, and challenging career. Becoming a professional pilot takes a lot of work, but the rewards, both personal and financial, are well worth the investment.

Our astronauts talked about having the "right stuff." What is the right stuff to become a professional pilot? I've spent 17 years heavily involved in the hiring process for a major airline. To me, what makes a professional pilot can be summed up in one word—attitude.

Before you launch your career as a professional pilot, I suggest that you assess your attitude toward work and life in general. Are you committed to excellence in everything you do? If you can honestly say you do strive to be the best—in everything from your academic preparation to your performance in the cockpit—and that you take every opportunity to improve and learn and perfect your skills, then, in my opinion, you're on the right track.

People often measure excellence in terms of flight hours. I don't believe that's an accurate gauge. I've hired pilots with 350 hours and pilots with 20,000 hours. I believe these pilots can perform equally well if they first have the proper attitude.

After 35 years of professional flying, I still personally critique every flight I make, whether I'm flying a Boeing 747 or a single-engine aircraft. I write down everything I can think of to improve the flight next time, and I review those comments from time to time. I'm still waiting to see a perfect flight.

Always strive to be the best as a pilot, and good luck. There's a great opportunity for those who are willing to make the commitment.

William Traub
Captain, B-727, 737, 747, DC-6, DC-7, DC-8

Acknowledgments

This book could not have been written without the sincere and honest input of a number of aviation professionals. I've tried to keep a detailed list of all who offered help in the writing or critique of the text before my editor Shelley Carr at McGraw-Hill ever saw it. If I've left anyone's name out, I sincerely apologize.

My thanks to Bill Traub, Kit Darby, Greg Brown, Scott Spangler, Jan Barden, Michael Thelander, Paul Berliner, Liz Clark, Gil Wolin, Cody DieKroeger, Louis Smith, Nick Verdea, Karen Kahn, Scott Randell, Sean Reilly, David Jones, Nancy Molitor, Derek Martin, David Manning, Sark Boyajian, John Bauserman, Craig Washka, Diane Powell, Jim North, Neal Schwartz, Pete Beckmeyer, Steve Mayer, Greg Watts, Joe Siok, Warren Cleveland, and Sandy Anderson.

My thanks also to David Jones and Tom Bossard for allowing me to reprint the following articles: Charter Flying: The Good, The Bad and the Ugly; Chicago Express—The ATA Connection; CommutAir—Keeping Its Pilots Busy; Your Logbook: More Than Just Numbers; How to Prepare for a Prehire Simulator Check; Unions: Can Pilots Exist Without Them? These articles were reprinted by permission (1996-1998), Bossard Publications Inc./*Flying Careers* magazine, 800-492-1881

The following articles were reprinted by permission from *Air Inc.*, Atlanta, Georgia—800-AIR-APPS: Landing a Flying Job Outside the United States; Will a Type Rating Get You Hired?; Flight Schools are Not all Created Equal; Finding a Flying Job After Age 50; The Upgrade Decision; Work Rules: How They Really Work. My thanks to Kit Darby.

Thanks to Scott Spangler for his editorial on the CFI shortage, reprinted with permission, (*Flight Training* magazine, 1998.)

Introduction

"The flight was short, but it was nevertheless the first in the history of the world in which a machine carrying a man had raised itself by its own power into the air in full flight, had sailed forward without reduction of speed and had finally landed at a point as high as that from which it started."

Orville Wright in a statement to the press December, 1903

When I was a kid finishing high school, I wanted to fly more than anything else in the world. I spent much of my free time riding my motorcycle out to places like Chicago's O'Hare International Airport and Midway airport, as well as some of the general aviation fields around Chicagoland, just to spend hours parked outside the fence watching the aircraft come and go. This was the mid '60s, so there were still plenty of opportunities to see some of the famous propeller-driven airplanes like DC-3s and DC-7s, as well as the turbo-prop Lockheed Electras that have since marched into history. Then there were the jets, DC-8s, Boeing 707s, DC-9s, and the Convair 880s and 990s. Twin-engine corporate turbo-props were just appearing at the general aviation fields. I knew all those names as well.

I spent hours thinking about how I'd feel behind the controls of some airliner or corporate jet. I took pictures, too . . . hundreds of them, most of which survive to this day in albums stashed in the closet. I think those pictures helped me stay motivated and focused on my goal. As I parked my old motorcycle by the airport fences, I knew that I wanted in, somehow, anyway I could manage it. I was 17 years old.

I wish I had known someone involved in the aviation industry then— a mentor, someone who could have shown me the ropes, who could have helped me avoid many of the pitfalls. If I'd had someone like that to teach me, I never would have spent so much time moving in so many different directions over the years, some directions that actually seemed to lead me further away from my goal rather than closer to it. I didn't wear my first airline uniform until I was in my 30s.

The '60s was a time, too, when airline pilots were hard to find. Believe it or not, I once saw a newspaper ad for airline pilots for a major carrier. The only major requirement was a private pilot's license. The airline paid for the rest of your training if only you'd be good enough to come and work for them. Many pilots did.

And now, as we approach the twenty-first century, with an economy surging ahead at a breakneck pace and airlines reporting record profits, there is again a pilot shortage—of sorts—as the airlines maneuver through one of the greatest expansions in aviation history. With a solid economic footing, more and more people are flying, creating a demand never before seen. In addition to new demand for air travel that always calls for increased numbers of pilots, the airlines have also begun retiring vast numbers of ex-military pilots who began flying during the early days after the Viet Nam War. This expansion, according to leading aviation number crunchers, means that approximately 19,145 pilots will need to be replaced in the next decade. Between retirements and expansions, the airlines will be hiring something over 12,000 pilots in 1999. While the numbers will decrease some after the turn of the century, the totals are still expected to be nearly 8000–9000 pilots per year until the year 2007. These jobs are spread out among the major, national, and regional carriers around the United States.

What that means to you as a potential applicant is that your chances of finding a cockpit position are better than ever. And when pilots are hired at the airlines, there is a trickle-down effect on the rest of the industry. As many new airline pilots are drawn into the system through ab initio programs or are moving out of the regionals or corporate positions, they leave jobs to be filled. This means more opportunities for new flight instructors, charter and corporate pilots, and regional and national airline pilots as well.

As a fellow pilot said to me recently, "Pilots are finally in the driver's seat." In 1997 and 1998, I met dozens of pilots who received multiple job offers at the same time. Pilots were sometimes jumping ship quickly, too, meaning they'd sign on at a regional, stay for a year, and move on to a major carrier. On the corporate side, pilots would often stay long enough to pick up a type rating or two and depart for greener pastures, their major motivation being that they could.

But a word of caution here, too. As little as five years ago, I was writing about what a slump the airlines—and hence the rest of the aviation industry—was mired in. During a two-year period from about 1992-1994, the airlines combined lost more money than all the airlines had made in profit since airlines began flying in the 1920s. This industry is cyclical. You might pick up this volume when times are on the downturn. But don't fret. All the techniques you'll learn here still apply. You might just have to work a little harder and be a bit more patient.

Despite the unprecedented need for more pilots than the system would seem to have the ability to currently produce, the competition is still incredibly stiff. One airline hires about 6 pilots for every 75 they interview. While this might seem to be contrary to the laws of supply and demand, understand that the problem stems from the fact that the shortage I alluded to earlier is based upon a shortage of qualified pilots. Although many carriers will interview people with 1500 hours of total time and a multiengine and instrument rating, the pilots they're actually hiring have considerably more time than that, on the order often of 3000 to 4000 hours or more at the majors. But the numbers of available pilots with these credentials is decreasing.

Echoing Bill Traub's words in the Foreword you just read, the attitude you approach your job search with is extremely important. In addition to Capt. Traub's focus on excellence, I'd like to add my own thoughts that resourcefulness and persistence will be critical as well. You'll need to know what tools are available to you that can make your job search and the process of getting hired easier. These range from an understanding of how to write and deliver the letter that explains your initial interest in a company, to getting your foot in the door for the interview, what to do when you have the interview scheduled, and what to expect until you complete your training and begin flying revenue trips.

One of the most important things you should do as you climb the ladder of success in aviation is to look back to where you came from. You're sure to see other pilots on the way up. Give them a hand. Share the information that you've gained along the way. That's what this book is all about. Sharing resources and information with others.

There may be times when, in frustration over trying to land your dream job, you'll feel like giving up. You might even think that a well-paying office job is where you should go. Don't do it. Take some advice from someone (me!) who fell into that trap but escaped to fly again. The money in an office job may appear to be more to start with than many aviation jobs offer, but after a few months or years of sitting around an office looking out the window at the sky, you'll kick yourself for having sold out.

You may learn that the path to your coveted flying job might not be as direct as you had at first expected. Some of the diversions might include a move out of state to gain experience in a particular segment of flying—like Part 135 or Part 121—or even in a particular position in a particular airplane, such as PIC (Pilot-in-Command) turbine on a turboprop, rather than SIC (Second-in-Command) on a jet. You might even realize at some point that you need to modify your goals from time to time. But if you keep your written goals in sight—why not tape them to your refrigerator?—and if you're willing to be realistic about the competition, as well as your own skill level, a career as a professional pilot can be in your future.

What you hold in your hands is not simply a book. It's a guide. But the word guide is not simply a part of the title. It is also a significant component of the philosophy that went into the writing of the text. This is a guide to resources; some are people, experts in the aviation industry, some are locations that are a storehouse of information. This guide will drive you in the directions you need to reach your goal of becoming a professional pilot.

I won't pull any punches with you. I'll also answer some thought-provoking questions such as "What exactly is a professional pilot anyway?" You'll learn some practical tips on training and hear my personal perspective on some of the institutions that many in the flight training and airline world hold so dear. I'll also share some of the potential traps you'll want to know about as you make your way toward the left seat.

In this text, you'll find a number of relevant articles of mine—as well as interviews and remembrances of pilots within the industry—that share valuable insights into the hiring game. But what makes this text so valuable—and completely different from most other aviation career texts on the market—is that its scope is beyond airline hiring. As we spoke of earlier, major airline hiring leaves a terrible void in other areas of aviation. But hiring on with a charter company or a corporate operator is not the same as hiring on with an airline. It's not necessarily more or less difficult, but it is different. But before you can even talk about a new flying job, we'll take you through much of the decision process and the inherent soul searching you'll need to do to decide which area of aviation fits your lifestyle best.

Just as my goal when I was a full time flight instructor was not simply to spoon feed students the answers, realize that my goal in writing this book is to teach you to think differently by the time you finish reading it. Hopefully you'll agree that this is the best investment you've made in your career. And since this book is updated on a regular basis, please feel free to share your insights about the industry and your own personal hiring process with me via e-mail at: rob@mark-comm.com.

The Web site for this book is also located at: www.ufly.com/pilotcareer. Feel free to visit from time to time and update me on your professional progress.

Finally, I'd like to add this disclaimer—isn't there always one? Unless otherwise stated, the opinions and ideas expressed in this text are mine and mine alone. I encourage you not to simply accept them as gospel and the only method of moving from point A to point B in your career. Instead, allow the experiences I relate to do their job, which is to help shape the way in which you think and thereby approach the obstacles you'll face in your search for the perfect flying job.

Good Luck.

Robert Mark

1

Your Career Starts Here

Word of Mouse— Networking and Technology

There is only one way to locate a job in aviation—networking. Whether you're exploring the outer reaches of the Internet, wandering around an airport with a fist full of resumes to hand out, or using information from an online or print publication, it's all networking.

If networking is the method to find a job today, the computer and the telephone are your primary tools. A pilot today—aspiring or experienced—is lost unless he or she owns a computer, has access to the Internet with an e-mail address and understands the intricacies of the Web, as well as has the ability to target a resume to the audience they're after. The information revolution has made it much easier for valuable content to be placed in locations for pilots to use as well. The trick sometimes is to learn where the information is located and how much of it is valuable—and credible.

During most of the interviews for this book, it became clear that when a pilot wanted a position, they all began by gathering information. But how they gathered that information, as well as how much information they put together before they began the application process, is what made the difference between success and failure. It is because of this need to gather information that companies like Air Inc, Flying Careers, the Berliner-Schafer Group, and fltops.com have emerged. They offer some of the answers to a pilot's need to find the information. But as we'll see later on, each serves a different market and delivers their products and services in a slightly different manner.

In the airline world, as well as in the corporate or charter marketplace, the story is much the same as you've been hearing throughout your life. That getting hired is not always a function of what you know, as much as who you know. The reason for this is simple. Hiring new employees is a difficult task. And people want the job being offered, but for a variety of reasons. Some want a career, some want to pay the rent and put food on the table, and still others want a place to hang out until something more lucrative appears. It's the employers' tasks to try to decipher the information they receive in answer to the questions they ask and tests they administer and synthesize them into an answer—is this the person I want to hire or not? Will this person really work as well with other pilots and ground personnel as they seem to believe they will? Will this pilot stay around only long enough to pick up a type rating and head for greener pastures?

Most employers then have their work cut out for them, but so do pilot applicants. So what is the easiest way to deliver the right information to an employer during an interview? Simple. Find out what they want and then give them what they want. Sounds a bit mercenary I know, but it really is not.

Understand that the most valuable part of networking is not just to you, but to a potential employer as well. You want a job and they want to hire a pilot, but the right one. Anything that you can do, from a personal visit to a phone call to dropping the name of someone the chief pilot knows to get your foot in the door will reduce that potential employers anxiety about hiring YOU!

That's the name of the game.

Throughout the remainder of this book, we'll talk about Web sites and information resources on the Web and techniques you can use to make you and your resume stand out to a potential employer. We'll also talk about how you can use one of the Internet's most valuable resources—e-mail—to learn more about the company you're thinking about hiring on with, by talking directly to people who work there now or—sometimes even more valuable—talking to someone who recently left a company.

Greg Brown, pilot, aviation author and system operator for two aviation message boards on America Online offered his perspective on the reasoning behind most networking efforts. "It all boils down to personal credibility. As an aircraft operator, you want the best and sharpest people working with you. That's why networking is so important. The more people who know and respect you in the industry, the faster you'll move up the career ladder. The other reason networking is so important is that there may not be a job available today. Your mission is to be remembered when they do have an opening and need to fill it in a hurry. A specific amount of your time should be dedicated to getting to know people, not simply logging flight hours."

And what do you do if you don't know someone at the company you want to work for? Easy. As one pilot told me, when he decided he wanted to work for Continental and knew no one, he started hanging out at the terminal and just walked up to Continental pilots he saw hanging out before a flight. He introduced himself as a pilot applicant to be and offered to buy coffee for a little information. A number of relationships developed because he made the effort to stay in touch with these pilots. One even turned into a letter of recommendation. This applicant is currently attending ground school at Continental.

Finally, it is important to understand that the definition of networking does not in any way imply that this is a method for only you to learn what you need to find a job. Networking is an interactive experience. As you'll hear a number of my comrades through this book say, you can't just take—you have to put something back into the system that keeps the information about our industry flowing, including job information.

As you begin your search for a new job, pass on the tips you've discovered to others who are also searching. Once you're online, this is easy enough to do through the message boards on CompuServe, AOL, or even those found on some of the aviation Web sites you'll hear discussed during the book.

Stay tuned.

Where do you see yourself next year, or perhaps in five years? Can you imagine flying airplanes for a living? Thousands of people can imagine it, but not nearly enough of them will actually become pilots. Why? Because some people believe that learning to fly is simply beyond them. Sure— transforming yourself from a fledgling student into a commercial and eventually into an Airline Transport Pilot involves a lot of work, considerable sacrifice, and not a small sum of cash. What is worthwhile having in life that doesn't tax you some? But by taking advantage of a wealth of resources—both written and in the form of personal relationships—you can do it. And believe me, the work is worth the reward.

I can't imagine a better profession than flying 30,000 feet above the earth and looking out from a vantage point that few others have, finding yourself in one of the greatest theaters in the world as you watch thousands of miles of wondrously changing clouds and landscape pass beneath. It's a treat that's tough to surpass. In addition to enjoying the beauty and majesty of this perspective on the world, I've spent the better part of my life making money at some form of flying . . . and so can you.

Flying airplanes for a living is a proud profession that dates back about 85 years to an era when flying meant a very serious risk to your life, not

Figure 1-1. Early days of commercial flying.

to mention the lives of any passengers you were carrying. (See Fig. 1-1.) In the days of the barnstormers, aircraft engines were simple but unreliable, so total engine failures were commonplace. The fuselage and wings of the first commercial aircraft were nothing more than cut pieces of aircraft spruce that were nailed and glued together to form the frame. Then, the fuselage and wings were covered with fabric and painted with many coats of aircraft dope for strength. A wood and fabric airplane is a concept that probably sounds a bit scary to people watching shiny aluminum 747s, 757s, and MD-11s fly overhead. But, for the record, that doped fabric and wood airframe was incredibly strong, much stronger than the engines.

The first commercial aircraft carried names like Ryan, Douglas, Curtiss, and Boeing (Fig. 1-2). They ranged from simple, single-engine aircraft capable of carrying just one pilot and a few pouches of mail to more sophisticated types that could carry 6 to 8 passengers with a little room left over for baggage. As aircraft grew more sophisticated, the title of commercial pilot began to take on the meaning of someone with, if not yet a respectable job, at least a regular salary. Notice, though, that no one said anything about longevity because, in the '20s and '30s, it was still considered quite risky to be flying for a living. Although pilots were always viewed as somewhat crazy, they were also looked on as super-men and superwomen of sorts: brave individuals, adventurers who defied the odds and cheated death every time they took to the skies. It's

Figure 1-2. Early United Airlines Boeing 247.

because of these men and women that modern aviation evolved into what it is today.

Over the years, aircraft reliability increased to where passengers were no longer considered completely crazy to fly. But pilots have still, to this day, been looked on by most people through somewhat heroic eyes (Fig. 1-3). Walk into a social situation and watch people's faces turn impressively toward the man or woman who tells the crowd they haul people around the sky in shiny silver airplanes. If you haven't seen that, there are only two reasons. One, you don't hang out with the right people yet. And two, you haven't been watching the faces of people at parties very well. You might think I'm bringing up some rather egocentric topics here, and I must admit that I am, but with good reason. Pilots tend to have rather outgoing, assertive personalities. This nature of theirs is a large part of what makes a pilot good at his or her job, but it's certainly not a requirement. A flying career will bring you years of pleasure, job satisfaction, and adventure, not to mention some rather superb financial rewards. So, if you're looking for a profession that will turn people's heads, as well as your own, I might add, there's nothing better than flying.

Figure 1-3. Pilots have always seemed somewhat heroic to passengers.

What Is a Professional Pilot?

Many people might initially identify a professional pilot as an airline pilot, but there are many more kinds of professional pilots around. There are corporate pilots shuttling the chief executive officers of some Fortune 500 companies. All over the world, there are pilots flying aircraft of all shapes and sizes, loaded to the gills with boxes and mail destined for just about every other point in the world . . . and someone needs to fly those airplanes. Freighters run the gamut from the sophisticated, Boeing 747s (Fig. 1-4) that cargo carriers like Fed Ex and UPS fly, to older commercial jets like the DC-8s that you'll find flying their tails off for smaller freight carriers like Detroit-based Kitty Hawk. Small freight operators still fly a few of the famous old airliners, the Douglas DC-3, as well as old pewter-colored Beech 18s.

At the local general aviation airport, you'll see another group of professional pilots—the flight instructors. These men and women will turn the sometimes confusing chapters of your journey from private to professional pilot into knowledge designed to make you a better professional

Figure 1-4. Northwest Airlines Boeing 747.

aviator. Teaching is a noble profession, not one to be taken lightly. Have you ever passed by a farmer's field and watched a small aircraft swoop low over a soybean field and spout a cloud of mist from behind? These agriculture pilots earn their living every day saving fields of produce from the destruction of pests and disease. If you want to talk about thrills, imagine flying along at 20 feet off the ground at 100 mph and making some wild-looking gyrations at the opposite ends of the field to maneuver the airplane back for the next pass.

There are also air ambulance pilots. They're on call for a 24-hour period, just in case there's an emergency that requires them to transport a severely ill or injured patient from one place to another. In this kind of flying, the hours are never the same, and the destinations are seldom repeated.

And I'm not finished yet. There are banner tow pilots who fly low over a field to grab a long banner, trail it behind the airplane, and display an advertising message to thousands of people below. You don't cover much territory, usually, but the flying is a real challenge.

Still hungry for professional pilot jobs? How about a forest fire tanker pilot? Obviously, they don't operate in as many locations as other forms of flight operations, but tanker flying can be a real adventure. For Hollywood's look at this side of the profession, try renting the video "Always," starring Richard Dreyfuss.

Consider the heroes of the FAR Part 135 Air Taxi rules, known as on-demand charter pilots. They could be called out to fly a quick load of freight from a parts plant in central Illinois to an automaker in Detroit to protect an assembly line from a needless shutdown. On-demand could involve last-minute plans by a dignitary or the local appearance of a rock-and-roll group. Not long ago, I saw a crew load up a Lear jet for a trip carrying only a single box the size of an egg crate. It contained a human heart headed for an anxious recipient 900 miles away. Don't forget the pilots who fly canceled checks or who report highway traffic from inside a cockpit each day.

Pilots are needed to fly sleek helicopters from corporate heliports on top of large buildings, as well as from the local airport. (Fig. 1-5.) Helicopters are special-use flying machines that can be stopped virtually on a dime and can vertically descend into a tight space to accomplish their

Figure 1-5. Commercial helicopter at work.

mission. Besides their use as a corporate shuttle vehicle, helicopters, by virtue of their agility, are one of the prime pieces of transportation used by Emergency Medical Services (EMS) for roadside evacuation of accident victims. In some locations, helicopters often compete with aircraft for aerial spraying and traffic reporting jobs.

But even after talking about some of the flying jobs available from time to time, I still haven't really put a finger on what makes a pilot a professional. Some people believe you're a professional only if you fly large aircraft. Some think that the moment you begin receiving your first paycheck, your status is transformed from trainee to pro.

A Partial List of Pilot Jobs

- Airline pilot—major, national, regional
- Corporate pilot
- Freight pilot
- Fractional ownership pilot
- Charter pilot
- Certified flight instructor

- Traffic patrol pilot
- Forestry pilot
- U.S. Customs Service pilot
- FBI pilot
- Pipeline Patrol pilot
- Banner tow pilot
- Law enforcement pilot
- Ferry pilot
- Sight-seeing pilot
- FAA Airspace checkpilot

Actually, I don't believe any of these terms define a professional pilot. They simply define a job. Professional—anything—is a philosophy, a way of life. I think professionalism really begins with the person (Fig. 1-6), not the job. If you measure your self-worth and your position in the world by

Figure 1-6. Professionalism really begins with the person.

either the aircraft you fly or the numbers on your paycheck, you're involved in the wrong profession, whether it's flying or brain surgery. I've known flight instructors who flew around in four-place Cherokees in a shirt and tie who were more professional than some airline pilots I've known who had become truly jaded by their aircraft type and their six-figure income. Make no mistake about it, a six-figure income is a paycheck we'd all like to have. But if all you pick up from flying is a paycheck every two weeks, you're missing some of the best that this career and life itself have to offer. Fly because you love it. Fly because you can think of no greater job in the world. If you begin your career like this, then nothing that happens to you will dim that flame of enthusiasm.

But, like any profession with a good salary and benefits, these flying jobs will not just fall into your lap. Don't be surprised if, after your initial training is complete, airlines, corporations, or the military don't fall all over themselves lining up for your services.

If you've spent time reading the history of aviation since airline deregulation in the late 1970s, you'll see that what began as unbelievable expansion of flying activities turned quickly to not just a preoccupation with the companies' bottom line, but also a rather myopic concentration on the dollar as the only consideration in running a flight department. What that has left for you potential flyers today is a route that can easily be planned but not as easily executed. But no matter what, FAA forecasts of flying activities believe that there will be growth within the industry, and so do I. How much and where are another story, but, according to 1998 FAA forecasts, the total pilot population is expected to be 652,500 in 1999, with 126,600 commercial pilots and 134,800 Airline Transport Pilots. These figures reflect a significant increase after many years of decline in pilot numbers. By 2009, the FAA believes the total pilot population will hover near 791,000, with 134,300 commercial pilots and nearly 157,000 holding ATP certificates. (Figs. 1-7, 1-8.)

But a moment of skepticism is certainly required here. The government could be wrong, and the figures could be worse . . . or they could be better. Many forecasts expect more growth in the regional industry than in the majors, as the majors relieve themselves of unprofitable short routes and pass them out to their affiliated regional carriers.

Pilot hiring is cyclical, as are so many other industries in the United States. You'll have to face the fact that, if you're 25 right now, you might not be able to quickly and easily jump into an airline or corporate cockpit for a while. But how important is that, really? Don't let your solid career plan be tarnished by the '80s desire to "have it all . . . right now." If you need a bit of consolation as you accept your first flying job, consider those pilots over the age of say 45 or 50. While a young pilot can wait around for a few years for the industry to turn around, those older

Figure 1-7. The captain of this A320 would hold an ATP certificate. *Courtesy Airbus Industrie.*

pilots will be well into their 50s, which makes employment for them very difficult. So, if you're 25, just beginning your flying career, or employed in a flying job that you don't want to make a career out of, stay calm. Look at the job as a means to get to your career goal of becoming a professional pilot. The industry is cyclical. Pilots will grow older, and they'll retire, making way for more people to move up the ladder.

Attitude: How's Yours?

Your job is to become a professional pilot. If you're going to be successful, you must adopt a professionally correct attitude. Without the right attitude, attaining your goal of becoming a professional pilot could just be the toughest plateau you've ever set up for yourself. You must mentally prepare yourself for the challenge of not just learning to fly, but also for the even greater challenge of finding the right job after you've learned to fly and gained the experience necessary for attaining your goal.

But please don't confuse my dose of aviation reality with the thought that the career is not worth pursuing. Quite the contrary. Flying professionally

Figure 1-8. A340 cockpit. *Courtesy Airbus Industrie.*

is worth the work it takes to get there, but finding the right job in aviation is going to be work. Just like a doctor or a carpenter, though, owning the basic skills doesn't guarantee an instant job.

What's the attitude, the right stuff, so to speak, for getting ahead in aviation? Rest assured it's not just maintaining a positive mental attitude. A positive attitude is only a part of becoming successful. Think about it; don't you know people who are very bright, people who know just what they should do to become successful, and yet never really succeed? No matter how bright they seem to be on the outside, they just never seem to get the right break or be in the right place at the right time or sell themselves in the best way possible when the time is right. So, just positive thinking isn't going to make you a professional pilot. I believe the reason most people don't succeed at the flying game is not because they don't want to succeed, but because they're afraid of failure.

Fear of failure probably squelches more careers than just about anything else. You must believe you can accomplish your goal, but even more than that, your plan must include a highway towards that goal, a path that's designed to help you leap past those points where the rejections arrive. Your plan must include a method for you to take action on that plan.

One young lady I interviewed for a story a few months back told me she had been asked to interview three different times by United Airlines before she was finally hired in 1992. Most people either give up after being rejected once or don't bother to interview at all because they're afraid of not being accepted. I'm no psychologist, so I'm only going to give you my thoughts on this—what I've heard pilots tell me over the years. You can't fail if you don't apply . . . some say. Others believe they were rejected because the people at the airline or company specifically didn't like them. Don't give up; keep trying, no matter what. This is the only attitude for success.

Sure, there'll be times when you might not give a good interview, but ask yourself, and answer yourself truthfully: Were you really prepared? Did you possibly set yourself up a bit for failure? So what about the woman hired at United that I mentioned a moment ago. How did she cope with so much rejection from such a large organization? She knew her goal was to fly for United, and she wouldn't let anything turn her from that path . . . nothing. If they rejected her, she didn't take it personally. She took a deep breath and asked herself which way to turn around this obstacle in her path.

Do you believe that most medical residents enjoy working 36-hour shifts at hospitals that often take them away from their family and friends as well as away from much-needed rest? I doubt it. But what these people do realize is that it's useless to have a plan that only talks about goals, with no way of dealing with the hurdles life is going to toss at you along the way. The people who make it are the ones who understand that, while they can't control life, while they're unable to predict what kinds of roadblocks will be tossed in their paths, they can control the way they react to events. You're in control of your own life. If there's such a thing as a positive attitude, I believe you must be positive about just what you can really control.

The Plan

Too often, pilots—young and old—believe that you aren't a pilot unless you're an airline pilot because that's where all the status, as well as all the money, is. Personally, I don't agree. If you really believe that all there is to aviation is sitting in the cockpit of a 747 on a long international route watching the islands and ice packs flow by beneath you, then I think you're headed into the wrong field. I really believe what an old aviator friend who now happens to pilot a Lockheed L-1011 said, "Do what you like, and the money will follow." In other words, if you chase only the money, you're setting yourself up for unhappiness in the long run. Avoid the "I want top dollar and I want it right now" disease.

The first words at the top of your list should be your goal. The gurus who research employment as well as personal goals tell us that the most important item you can accomplish in your plan to become a pilot is to write that goal down on a piece of paper. Writing the goal down in some concrete fashion makes the goal take on a life of its own. This is the first step toward making your goal become reality.

Your plan is only a road map (Fig. 1-9), a route to eventually arrive at your goal of becoming a professional pilot. But, like any route taken on the ground, your career plan shouldn't be so rigid that it's incapable of allowing you to take a detour where necessary. There might be situations that force you to change your plan slightly, such as an unplanned opportunity to fly an aircraft in a state 1,000 miles away on a short-term contract for six months. Do you ignore the chance just because you didn't have it written down in a plan you produced a year ago? I certainly hope not. Certainly, too, a time might come when a roadblock to your career will appear. Consider all your alternatives before you make any decisions. If part of your plan was to spend the summer months dusting crops or towing banners and the job you'd hoped for disappears because the company went out of business or perhaps hired someone else, what are you going to do? You must have some alternative course to steer your plan out of the muck and onto high ground.

The best method for keeping your career aloft rather than waiting at the terminal is to keep your education dynamic. Currently, I read seven aviation magazines each month. One of these is published on a weekly basis, so that's like reading eleven. Very often, though, you'll notice the same story more than once and consider reading it a waste. But if you're constantly in touch with your industry, always aware of what companies are producing what products, what aircraft are the newest off the line with what sorts of systems, you have a great chance of knowing just what to talk about during an interview. The other advantage of staying in touch through many magazines is the variety of perspectives you'll receive.

Another benefit, too, is the awareness of the job situation you'll gain. I recently read a story in one of the trade magazines about a company producing new aircraft in Wisconsin, and I wondered whether or not there might be a place for me, even in a small operation such as that. I had extensive experience with the aircraft they were producing, and I thought

Figure 1-9. A pilot's position on a regional Jetstream 41 requires a plan to achieve.

it was worth a chance to make contact with the firm and offer my services as a pilot. Unfortunately, this particular opportunity didn't turn out favorably. But it might have. My motto is that I have absolutely nothing to lose by asking; the worst that could occur is the company would say no, and no one ever died from being told no. Whatever your methods, maintain contact with the industry.

The plan you produce for yourself should be realistic. If you expect to fly right-seat on a 757 after college, your chances are slim—not impossible now, but slim. Despite the increased numbers of flying positions available, the search for a flying job is still extremely competitive. Don't expect instant success. When will you pick up your commercial license? How will you pay for it? What methods will you use to increase your flying time? What jobs are you willing to perform to reach each step of your goal? What will you not do? Would you take a short-term job? When will you complete your instrument rating? What's your organizational plan for sending out resumes? How will you locate the addresses of companies that hire pilots? Would you consider giving up a few years of your freedom to join the U.S. Air Force or Navy to learn to fly there? The list will grow as you spend more time on your plan, but very much like a business plan, a career plan is most likely never finished. Things will change, but the real trick is to know when to stick resolutely to your plan and when to take that detour. If there's any advice I can provide on this front, it's not to jump too quickly at anything, no matter how good it looks. Think how this decision will affect the rest of your plan. If you're not sure, ask a close friend to sit down and talk with you about an opportunity when it presents itself. Brainstorming, simply thinking about the possibilities without restrictions and networking, talking to everyone you know and meet about your job search will reap some incredible benefits in terms of information and, perhaps, even a job.

Where do you see yourself in one year, five years, ten years? Don't hold back here. Let your mind run free. Do you see yourself in the left seat of a corporate G-V (Fig. 1-10) flying Trans-Atlantic? Maybe it's the left seat of a 757 for American or Delta Airlines. Perhaps you enjoy flying but really enjoy being home more often. A regional airline or local charter job or possibly even some kind of utility flying is what might be best suited for you.

If you're a woman or a member of a minority group, there's good news for you, too. Besides the federal legislation that mandates that a company not discriminate against you because of your sex or race, most of the major airlines and corporations actively seek candidates from these groups. Unfortunately, as one Vice President of Flight Operations at a major airlines noted, "We don't even receive enough applications from either of these two groups to hire very many as pilots." If you're a woman, a place you can certainly begin your search for information about a career in flying could be *Women in Aviation International*, mentioned later in this chapter. For the members of other minority groups, all you need to know is that once you've completed your flight training, many of these companies will be waiting to interview you. Keep in mind that no one is promis-

Figure 1-10. Corporate G-V.

ing anyone a job. Personally, I don't usually believe promises of work "someday when you have a little more experience," as one chief pilot told me once. Most of these promises never seem to pan out.

The Airline Industry at a Glance

Pilots hired during the hiring surges of the 1960s are now beginning to retire in ever-increasing droves. Here's a look at what is happening within the industry. But to set the stage, a few definitions are necessary to understand what follows:

- A major airline is defined as one with $1 billion or more in sales.
- A national carrier is one that has anywhere between $100 million and $1 billion in sales.
- Jet operators show less than $100 million in annual sales.
- Many of the smaller, turboprop regionals also sell less than $100 million each year.

Kit Darby, president of Air Inc., the Atlanta, Georgia based aviation information service said, "In 1996, the previous record year for new airline hiring of pilots, 10,625 pilots found work. Of this number, nearly 3100 were hired by the major airlines alone. During this same time period, the national airlines and other jet operators brought nearly 5100 pilots on board. Nonjet companies hired 1825, while start-up carriers hired another 275."

Now for 1997, the best record year to date. Just short of 12,000 pilots found employment in 1997 thanks in part to a robust economy. This economic expansion fueled additional growth at the major carriers, as well as significant increases among regional airlines, many of whom are now converting a portion of their fleets to pure jet aircraft—like the 50-passenger Canadair RJ and the 50-seat Embraer 145. The hiring figures for 1997 break down like this: Major airlines hired 3,854 pilots, national and jet operators hired 5140, nonjet operators hired 1762, while start-ups hired 92.

Through the fall of 1998, airline hiring remained brisk. Estimates call for over 12,500 to find employment during 1998. By November 1998, 12,245 had been hired in the above categories, according to Air Inc. A number of things are driving this spree. A solid economy, which is fueling increased aircraft orders among airlines, which will call for additional crews to fly them. But, as most pilots know, the fluidity of the economy can become a significant factor in airline expansion and pilot hiring. Right now, however, there is another factor affecting airline hiring that is not dependent upon the current state of any particular airline's health—retirements.

According to Darby, the major airlines will retire pilots at an ever-expanding number, at least through the year 2010 as shown in Table 1-1.

Projections for pilot hiring (Table 1-2) are encouraging as well according to Air, Inc. By 2006, Darby believes "there will be some 35,000 new jobs flying large jet aircraft. Nonjet operator growth will be sizable as well, with some 40,000 new regional pilot positions being created thanks to retirements and carrier growth. Total numbers of aircraft will also increase significantly. By 2020, Airline Monitor, says U.S. passenger fleets should increase by nearly 60 percent to 6646 from 4204 today. In the world freight industry, the growth will be even more dramatic, to 3148 aircraft from 1464 today."

But despite these rosy growth forecasts, an aspiring pilot should remember the risks that are inherent in the airline industry, as well as how those risks impact other areas of the aviation industry as well, such as charter or corporate flying, as well as flight instruction.

A recession is an economic downturn that we either seem to have just gotten over or may be headed back into. Few believe, however, according to Darby, that a new recession could be as devastating as the one of the early 1990s that saw the end of a number of major carriers like PanAm, Eastern, and Midway Airlines, because that particular event was made all the more difficult due to the effects the Gulf War had on fuel prices and falling traffic.

Table 1-1 Major Airline Pilot Retirements by Year

Airline	Total Pilots	1998	1999	2000	2001	2002	2003	2004	2005	2006	2007	2008	2009	2010
ABX	766	3	2	4	4	13	12	19	24	29	29	30	25	32
AMR	9,488	216	226	164	121	153	160	208	207	262	315	403	459	370
AWA	1,407	15	18	17	15	39	48	48	39	34	48	50	52	49
CAL	4,245	97	102	116	113	115	144	115	160	258	317	246	219	209
DAL	9,667	360	374	364	368	391	263	247	231	219	290	326	280	270
FedEx	3,481	32	38	47	46	64	85	75	120	147	140	127	115	129
NWA	6,246	108	168	202	204	179	174	158	149	161	239	210	210	205
SWA	2,379	10	17	17	18	35	63	83	107	147	142	129	97	119
TWA	2,986	182	204	212	188	150	124	75	72	62	70	59	53	41
UAL	9,742	256	330	368	323	340	267	221	260	278	266	231	208	184
UPS	2,178	18	15	10	30	31	40	48	84	93	63	55	38	44
USA	4,979	65	97	94	113	127	149	146	161	229	273	197	232	195
	57,584	1,382	1,591	1,615	1,545	1,637	1,529	1,443	1,614	1,919	2,192	2,063	1,988	1,847

Copyright 1998 Air,Inc. - Atlanta, GA - 1-800-AIR-APPS

Table 1-2 Pilot Hiring Projections 1993–2007

	1993	1994	1995	1996	1997	1998	1999	2000	2001	2002	2003	2004	2005	2006	2007
MAJORS	483	1,266	2,377	3,080	3,854	4,300	3,760	3,323	2,891	2,660	2,819	3,327	3,593	3,780	4,082
NATIONALS	814	1,837	2,508	2,773	3,194	3,900	3,450	3,049	2,653	2,441	2,588	3,054	3,296	3,469	3,747
JET REG'LS	1,725	2,015	1,735	2,305	1,946	1,750	1,623	1,434	1,248	1,148	1,217	1,436	1,551	1,631	1,782
REGIONALS	1,913	2,566	1,745	1,825	1,762	1,700	1,602	1,416	1,232	1,133	1,201	1,417	1,531	1,610	1,739
UPSTARTS		288	76	276	98	135	100	88	77	71	75	88	98	101	109
TNG/FOR/CL	272	72	373	366	732	545	360	318	277	255	270	319	344	362	391
TOTAL	5,187	8,044	8,814	10,625	11,586	12,330	10,895	9,628	8,377	7,708	8,170	9,641	10,412	10,953	11,830

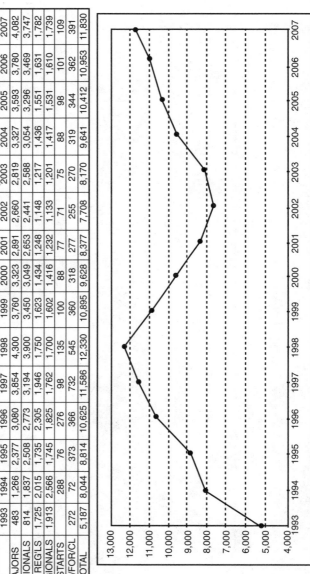

PILOT HIRING PROJECTIONS
1993-2007

There will be numerous labor agreements to be negotiated in the next few years. When they are successful, as was the eventual outcome of the American Airlines contract in 1997, there is barely a ripple that pilots will feel. If however, it becomes as turbulent as the Northwest agreement—or lack of it—eventually turned out to be, a long strike and a disruption to operations and hiring can be expected. Readers are urged to do their homework before an interview to be certain they understand the impact of a possible labor upset on their plans.

The Bottom Line

According to Darby, "The trial-by-fire of the last recession and the ensuing agonizingly slow recovery have left airline management teams far better equipped to cope with the risks of the industry than ever before. The transition from controlled market to the free enterprise system is almost complete. The world has watched our transformation and now cringes at the thought of competing with this lean, mean flying machine. More domestic consolidations are likely during the next downturn, while international consolidation is just beginning. The international market has grown 15%–25% per year and represents the growth opportunity for the future.

Only five to seven large airlines will survive. These will include the "Big Four" (American, Delta, Northwest and United) plus FedEx and UPS and one or two more passenger and freight carriers. The survivors will be huge and in general very stable companies. These megacarriers will be profitable and will offer their employees good wages and benefits and a stable career."

Decisions, Decisions
What Direction Are You Headed?

If you walk up to any average group of people at a party or event and tell them you either are a pilot or are thinking about becoming one, the most likely response—after they've told you just how cool they think flying really is—would be to ask what airline you fly for. That's because to the general public, being a pilot translates into being an airline pilot. There just simply are not any other options.

Nothing, however, could be further from the truth. There are many, many flying options for those of us who see our futures aloft.

In addition to the airlines, there is corporate aviation where you might find yourself in the cockpit of a Gulfstream 5 or Canadair Global

Express winging your way across the Pacific transporting your CEO from a dinner meeting in San Francisco to a morning meeting in Tokyo. You could be a charter pilot, on call, ready to take any number of people or parts or both, just about anywhere in the country in a wide variety of aircraft on a moment's notice. You might find yourself driving a Boeing 727 full of auto parts from Detroit to Memphis in the middle of the night. Then, of course, there are the unsung heroes of the industry, the flight instructors—the people without whom none of us would be able to sit back and talk about what we'd really like to fly. But the list does not end here.

There are pilots flying for the U.S. Customs Service and the FAA. Other pilots are out flying old cargo aircraft ready to drop a few dozen forest firefighters right in the thick of things. Then there are the pilots who drive the aircraft that drop the chemicals or water on those same fires with other kinds of aircraft. How about float plane pilots who carry passengers into and out of remote spots for business and pleasure trips. Or think about agriculture pilots spraying crops night and day, making sure the food we eat is safe. The list goes on.

But before you make the decision about what kind of flying you're interested in, you should think about your ultimate flying goal in terms of your interests and the type of flying that will keep you happy and productive. I discuss throughout this book that good, solid information is the key to an informed decision. This need to research should be an ongoing part of not only your job search, but also your flying career. But for now, you need to locate the information that's critical for the job-hunt and analyze it for the facts you need that support your goal.

Once this analysis is complete, you can begin the process of looking at your future from a dollars and cents standpoint. But don't forget that you'll need to consider the flying job you're interested in from an emotional point of view, as well. Only then will you be able to decide where to spend your flying career. Be prepared that your career perspective—the insights into the next direction your career may take—will change as you become older and more experienced within the industry, especially if you keep reading and learning what is happening outside the cockpit you're flying in.

Nancy Molitor, Ph.D., a clinical psychologist in Chicago believes, "For a pilot to choose the right direction for their career, it is important they ask themselves quality of life kinds of questions. Ask how much of a sense of control they want over their work environment, such as their schedules. An airline job is going to offer them fairly insignificant involvement with the passengers in the back of the airplane, while a corporate or charter flying position will bring the cockpit crew into direct interaction with passengers. They should consider the capacity for change a particular position offers. A unionized airline job is fairly regimented, for example. If you have an advanced degree and like to make decisions, other than basic flying ones, you will probably find airline flying somewhat of a disappointment, no matter how large the aircraft you fly or grand the paycheck."

In order to make the right decision, you should delve even deeper into what makes you tick. Ask yourself seriously, "What kind of environment do you feel most comfortable in?" says Molitor. "You'll go further and be a lot happier. Spend the time to learn what kind of working environment other pilots have. What kind of schedules do they work, or do they even have schedules? What are the opportunities for upgrade at the company? How will your job responsibilities change after you do upgrade? When you investigate all the aspects of a particular type of flying, you might find yourself changing your mind about where you're headed."

Other issues you'll need to wrestle with include how much you value money and why? Try to answer as honestly as you can, do you want a job that pays a ton of money just so you can tell the world "I've arrived?" Are you willing to do any kind of flying to get that cash? You may say yes now, but what about a year after you've been flying long, monotonous international routes with 12-hour duty days and only one landing. Would you be happier at a smaller carrier like Southwest, or in a corporate position, both of which pay substantial salaries, but offer you more actual flying time, more landings and takeoffs?

Many pilots tend to be rather tactical people, dealing with life and career problems much the same way they deal with the problems flying tosses at them—as they occur. This may work for some people, but in the end, you may find yourself only flying an airplane for a living. Why settle for this when with a little planning and forethought you can design a life that will offer you as many of the things you value in life while it avoids as many of the pitfalls as possible? Imagine planning the fuel load for a flight, filing the flight plan, updating weather information, however, but still being uncertain of your destination? You'd plan a flight to Denver much differently than you'd plan one to JFK. Why? Because the airports are different, the terrain and weather, as well as the traffic are all different.

To arrive at the right career decision, you need to know where you're headed. You might say your goal is that "I want a flying job." I challenge you that an objective like this is too simplistic. Dig a little deeper for the insight. What kind of flying job? "Big airplanes!" Still too simple. Keep digging. "Fly big airplanes on international routes." That's better. At least now, your potential flying job is beginning to take on a shape. Will it be airlines, corporate, freight charter, or what exactly? Think about those people involvement issues again.

Will you agree then that any job short of flying large aircraft on international routes will be only a stopping point? If the answer is yes, you should realize that you'll keep a job with Reno Air only for as long as it takes to get hired by American, United, Delta, etc. Flying a King Air from Janesville, Wisconsin is also not going to be your final career move then is it? What I'm after here is to make you to think through your objectives. Ask yourself why you want to fly large aircraft? Sounds crazy, but why do you want this? Does it also make you feel you've arrived? What do you want to prove and to whom? This kind of analysis will help you decide how you may need to—or want to—move from one job to another in that quest for Utopia.

Let's take a look at the characteristics of some of the flying jobs we spoke of earlier, because each has positive attributes, as well as a few drawbacks. And yes, even the major airlines have some disadvantages. I know that if you get on with the right airline—one of the big ones, and there are only about six or eight of them—and you successfully climb the internal ladder to larger and larger aircraft and eventually move from the first officer to the captain's seat, you could very well find yourself commanding a salary in excess of $200,000 per year while flying as little as ten or twelve days a month. So who wouldn't want this, right? Keep in mind that those little dollar signs have a way of swaying just about anyone, so don't think I'm crazy when I say that some pilots will look elsewhere for a career.

Let's look at the realities of flying for the following.

The Airlines

We might as well begin here since everyone thinks they want to be an airline pilot anyway.

An airline job often pays the highest salaries around for driving some of the largest airplanes in the world. They also offer some of the best training around. After the new Northwest Airlines contract becomes effective, some of their senior pilots will be pulling in nearly $250,000 per year for flying 747s over the longest routes. Not bad!

But what's the real cost of an airline job like this? First of all, only a very small percentage of Northwest pilots—or any pilots for that matter—will be making that kind of money. In late 1998, the average airline pilot salary was about $130,000 per year. Not bad, but it's not $250,000. Another factor you should consider is how much time you'll be spending away from home. If you're single, you might not care. If you're a newlywed, or have small children, being gone for long periods of time may cause some heartache for both you and your family, especially if you're going to be a two-income family. Who wants to hear about their daughter's school play while they're sitting in a hotel room in Tokyo? One Northwest pilot I know—a Boeing 747 captain—called me recently—just before she left on a 12-day trip. Now I'll grant you that she probably only flies one of these trips per month, but you'll still be removed from your family for quite awhile at a stretch when you do fly. Perhaps you should consider jobs that offer only a few nights away at a time? Airline flying also puts you on a schedule where you'll know which cities you'll see in a given month. While it is not uncommon for some pilots to bid Flying Line 121 at their company, for example, because they'll know their overnight locations well in advance, there is seldom any variety to the destinations from what is scheduled.

Will you live where you're based? If you and your wife and all of your families are from Columbus, Ohio, taking a job with United that bases you in Los Angeles, will also cause some sorrow. Your family may be upset if they have to move, leaving the rest of their family behind. You could choose to commute to and from LA, but be ready for that commuting time to eat into your personal time, just as it does to any other person trying to get to work these days. If you have a trip that begins out of LA at 9 A.M. on a Monday morning, you'll need to be in place in LA by Sunday afternoon at least. Where will you stay when you're operating out of your base, like just before a trip? A crash pad—a place where four or five pilots often live together—may be a cheap answer, but your privacy will be almost nonexistent.

While we're discussing commuting, consider the current state of airline travel; most airplanes are full—a lot! As a nonrevenue passenger trying to get to work, you'll get a seat only if there is an empty one. You might be able to ride in the cockpit jump seat, but if a more senior pilot comes by, he or she will get priority over you. You may need to be at the Columbus Airport early Sunday morning in order to allow for a safety margin of missed flights to be in position for work in LA. On the other end, if you finish up a trip at 11 P.M., will you be able to get home to CMH that night? Perhaps. But realize that with the flying time and the time zone changes, it will be tomorrow morning by the time you walk through the door, most likely exhausted. Your thoughts begin almost immediately too about when to leave for work again too. Many pilots do it, but most dislike commuting.

All Airlines Are Not Equal

There are more than a hundred airlines in the United States. But just 13 of them make up about 90 percent of the new flying jobs. Why not consider a smaller airline with a base in your hometown? America West, for example, still has a crew base in Columbus, Ohio. When you finish work, it would be a short drive to sleep in your own bed at night. There are other carriers flying large aircraft that are a much closer commute for you than Los Angeles. If you worked for ATA, you might be based in Indianapolis. In bad weather, or if all the connecting flights were full, you could drive to your base if you had to. Airlines tend not to be very patient with commuting pilots who don't make it to work because they got bumped (airline lingo here) off a flight. But, of course, you'll never make the kind of money at ATA that you'll make at United or Northwest. An L-1011 captain at ATA is making about $100,000 per year.

Remember, too, that time can cure many things. If you're 29 when you hire on with a major airline, you can put up with the problems for quite a few years before the base you want opens up and still have plenty of productive flying years ahead of you. An older pilot, however, may have fewer years until the magic Age 60 bell rings, limiting some of their options.

The Nationals

A national carrier—based upon the strictest definition—is one that has total revenues of more than $100 million and less than $1 billion. These are airlines that you might at first have considered to actually be regionals, such as Continental Express, Comair, and Horizon. It also includes companies like Kitty Hawk Air Cargo, ATA, Sun Country, and Reno Air.

Essentially, the differences you'll find are vast. The pay is less and very often the schedules are not as good. On many, there are no duty or trip rigs either that offer pilots any kind of pay guarantees for non-flying time spent at work. So why consider them, you might be asking? Because some offer a wide variety of flying experiences that you might not see at larger airlines, such as the opportunity to fly more approaches in a day, if you're someone concerned about sitting in cruise for too long. The pilot groups are smaller—United has about 10,000 pilots right now—so the camaraderie may be much better too. Why not find pilots who fly for some of these companies before you make any decisions.

A Regional

Flying for the regionals—formerly called commuters—is an adventure unto itself. While most pilots would point to money and benefits as THE reason not to fly for a regional, the schedules often tend to be somewhat more grueling than major airline flying as well. But regional pay has actually been on the rise in the past few years, although it will never reach the level of the majors. There simply are not enough seats in a 30-seat airplane to generate the revenue necessary to pay hefty salaries. But at Atlantic Coast Airlines, a United Express carrier based at Dulles Airport in Dulles, VA, for example, a captain on a 50-seat Canadair RJ earns about $65,000 per year, after 10 years of service, for flying an average of about 85 hours per month. But regional pilots are often home a little more too. Many live in the city they're based in and find themselves home every night. Is that important to you?

A note here too about guaranteed flight hours. An 85-hour month may seem fairly easy, but you need to understand how much duty time pilots are putting in to fly those 85 hours. This actually applies to ANY airline you talk to. From my experience, duty time is normally double the flight hours and in some cases nearly triple that—making for some pretty long workweeks. It can also make commuting to and from work rather difficult and sometimes fruitless when you have only two or sometimes three days off in a row. Do your research. One regional I looked at recently was in desperate need of both captains and first officers to fly some rather large turboprop aircraft. But the major airline they feed had only agreed to renew their code-sharing marketing agreement—and allow them to continue flying as an express carrier—for two years. After that, many jobs might be up for grabs.

But the upside for many regionals, is just that—their location. Some pilots with families are willing to forgo the big airplane, hefty salary magnet because they've come to realize there are other things more important to them.

Corporate

Corporate flying is as different from flying for the airlines as night and day. The relationship between passengers on corporate aircraft and their pilots is also much different from the airlines. On board a Part 121 airliner, the cockpit crew will normally have the door open until ready to start the engines, when the door is closed. The passengers never see the crew again until after landing. Depending upon the captain, the people in the back may hear an announcement from time to time about what is happening, but that could be it.

On a corporate flight, the crew does everything. They check the weather and file the flight plans, something that an airliner crew has a dispatcher to do. The corporate crew also makes sure the aircraft has all the supplies, such as food and drink, that passengers may want. They are also responsible for making certain the aircraft is clean and that the lavatory has been serviced. When the passengers arrive, the crew normally loads the baggage as well.

When a corporate crew starts engines, they will often leave the door open—that is if they even have a door, which many corporate aircraft do not. It is not uncommon during the flight for the chairman or the CEO of a Fortune 100 company to walk up to the cockpit to chat about the flight, the state of the company, or to ask the crew about their family. In general, the crew of a corporate aircraft is much more involved with the people in the back. When the weather ahead is bad or a delay is ensuing, a corporate pilot would probably not pick up the PA. They'd more than likely walk back and talk to the boss directly. To make this work, you need to like being involved with people and also be good at talking with them.

Corporate flying is not like airline flying in another sense—the regulations. The airlines, even the regionals, fly under Part 121. Most corporate crews fly under Part 91, yes the same regulations you flew under when you rented a Cessna 150 to go tooling around the skies in. The benefit—and some view this as a drawback—is that corporate crews have no duty and rest restrictions as do airline pilots. Airline pilots can fly only a specified number of hours per day, per week, and per month, depending upon which segment of the regulations their company operates under. A corporate crew could fly 12 hours a day if the company wanted them to.

But now for reality. Most corporations—at least the good ones—are governed by Part 91, but fly their crews to the same Part 121 standards the airlines use. No CEO of a billion-dollar company wants to put his life—or the lives of his senior officers—at risk by putting them on a $30 million aircraft with a crew who is likely to fall asleep just after takeoff. But the variations on corporate schedules can be vast. Some crews fly 50 hours a month, while others see only 15 or 20. There are often trips that leave early in the morning, fly an hour and a half away, and then force the crew to sit around for eight or nine hours before retracing their steps. But a corporate crew seldom visits the same city twice in a week either.

Corporate flying often demands some collateral effort from pilots as well, such as helping with the scheduling. Each day and each destination, however, is new to a corporate pilot. Another good part of that kind of schedule is that the time you have at the destination is usually your own to read, write, snooze, or call home. Salaries on the corporate side can also be substantial with captains of large jets making well in excess of $100,000 annually with all the benefits any other corporate executive might enjoy. A good corporation today also is more interested in your education than an airline will be, because they view you as a business asset to them, not simple as a limo driver, which is essentially what you can become as an airline pilot. In a good corporate position, there will be a future with the possibility of advancement into a managerial position within the flight department, if you hold the requisite education and skills.

Many corporate pilots report, too, that in times of financial difficulty, it can be quite beneficial to be flying around with the man or woman who makes all the decisions about their future. Any information you'd hear comes right from the boss's mouth. If you enjoy interaction with people and if you also want a voice in your future on a regular basis, a corporate job might be one worth looking into.

Charter Flying

Part 135 flying (see the story later in this book), uses a vast array of aircraft from Piper Navajos to a Gulfstream 5 or Canadair Challenger and everything in between. Charter companies, based all over the United States, act as a sort of hybrid corporate operation.

But the major difference between a charter department and a corporate operator is—again—the regulations under which they operate. A Part 135 charter pilot operates with flight time and duty time restrictions, as well as numerous safety and training considerations—much like Part 121. But the regulations are not quite as tough in a Part 135 situation; pilots can fly longer days and fly with much less training than a Part 121 carrier would be allowed to. It doesn't mean they are less safe, but just a little less regulated than the scheduled airlines, which brings us to the next major defining point in a charter company.

Part 135 regulations are mainly focused on unscheduled operations. That means that not only will flight crews often not know for certain where they are headed, but they often won't know when they are leaving or when they are coming back. Some of this kind of thing certainly goes on in a corporate flight department, but senior executives normally will plan out long trips ahead of schedule, simply because their time is so valuable.

A Part 135 pilot will usually carry a pager and sometimes rises at 1 A.M.to a phone call to get to the airplane ASAP. Most charter pilots will keep a bag in their car at all times. It can create somewhat of a chaotic life, but many love the variety and the fast-paced level of action. Schedules can be difficult for charter pilots. They are often on call a great deal of every month, but seldom fly all the days they're on call. The regs actually say a company only needs to give a charter pilot 13 days off in a quarter or about 1 day per week. Some run it close, while others do not. Depending upon the type of aircraft you fly, you might find yourself leaving home at odd hours of the day and night, but also find yourself sleeping in your own bed quite a bit too.

Having flown for a Part 135 charter company, I can tell you that you'll
quickly get to a point where almost no destination—nor time of the
day—will surprise you. Salaries on the charter side can vary widely too.
On the corporate side a Citation 2 captain might make $65,000 per year.
On the charter side—where there is a slim profit margin to contend
with—a captain might only bring in $45,000 to $50,000 per year. Again,
though, charter flying offers lots of hands on action with just about every
facet of flying and often into and out of short, remote places the airlines
don't serve.

The Fractionals

Sandwiched somewhere between the corporate and the charter sectors is
a quickly growing segment of business flying called the fractionals.
These include companies with names such as NetJets and Business Jet
Solutions. Essentially, fractionals sprang up out of nowhere a few years
ago to serve an emerging marketplace—companies that did not own a
business aircraft and could not justify the entire price of their own
machine.

Fractionals picked up their names because they sell parts of an airplane. A company can purchase a ¼th share in a Citation, a Gulfstream,
or a Hawker, even down to as little as ⅛th of the aircraft. What fractionals have done for business aviation—besides offer aircraft to new
companies—is to create an incredible demand for aircraft and pilots to
fill the needs of their customers.

The largest of the fractionals—NetJets—owned by Executive Jet
Aviation, based in Columbus, Ohio, was recently purchased by billion-
aire financier Warren Buffett. Buffet is so bullish on business aviation—
he also owns Flight Safety International—that he placed orders for
more than $3 billion worth of new corporate aircraft at the 1998
National Business Aircraft Show (NBAA) in Las Vegas. His order
included 10 Gulfstream 5s, with options for 12 more and 14 G4-SPs for
Executive Jet itself and 50 new Cessna Sovereigns—an aircraft
designed to compete with the Learjet 60 and the Hawker 800XP—and
options on 50 more. Net Jets employs nearly 400 pilots to crew some
130 aircraft. The good part about flying with Net Jets is that you'll be
flying—a lot. Net Jet crews normally fly a schedule of between four and
six days on and three to five days off. A Net Jet pilot will expect to log
somewhere between 60 and 80 hours per month. Pay is pretty good,
starting at about $27,000 per year. Senior captains can make close to
$100,000 here.

All sound too good to be true? For some it just might be. At
Executive Jet's Net Jets, pilots pay for their own training through

Flight Safety to the tune of about $10,000, but are type-rated in the aircraft after training is complete. We'll discuss the merits of pay-for-training later, but a pilot looking for a job here must be ready to finance that training. Net Jets remains out of the picture until the pilot has successfully completed their training. Commuting options are almost nil, so if you want to work for this fractional, you'll need to live near Columbus, Ohio. Another fractional—Flex Jets—is based in the Dallas, Texas area.

Freight Companies

If you're not a people person and want to make certain the captain never asks you to go back in the cabin to deal with some passenger problem, flying freight might be your cup of tea. Freight companies vary widely, from a local company flying newspapers in a twin-Beech 18 or MU-2 to United Parcel Service (UPS) or FedEx, flying cargo all over the world in large air-carrier aircraft. Traditionally, freight is flown at night, although some of the large carriers are increasing their presence during daylight hours. If you like flying at night when the air is smooth and the ATC system is not experiencing a coronary, freight could be for you. Then too, there is the rather closely knit brotherhood of night flyers that exists that many pilots say they will never give up. Very often, among the smaller carriers there is considerable on-demand cargo flying that occurs at all times of the day and night. Cargo companies operate under both Part 135 and Part 121 regulations.

The pay and days off at the top end of the freight spectrum where FedEx, UPS, and Airborne reside will be top notch. At last report, a second year captain will start out at about $135,000 per year at UPS, and about $113,000 at Airborne. At the other end of the spectrum small freight carriers will pay varying rates. One, Kitty Hawk Charters, pays $250 per week for a captain on a Twin-Beech 18. At AirNet—the largest carrier of canceled bank checks in the United States—a beginning Navajo captain will earn $18,000 per year. But AirNet (see story later in this book) pilots also fly only four nights per week and return to their base each and every night.

Flight Instruction

If there ever was a group of unsung heroes, it is flight instructors. Flight instruction—as you'll read about in Scott Spangler's editorial later—has always been a low-paid position that almost everyone (yes, I used to think this way too) thought was simply a way to build time until we could line up a real job. Nothing could be further from the truth, however, because when I look back on my instructing days, I realize that much of the pilot I am, as well as the person I am today, is based upon the experiences I had while teaching people how to fly.

The best pilots—in this writer's opinion—are those who rose from the flight instructor ranks. No matter what kind of a job you eventually land, you'll find that you'll be a flight instructor to some degree again. Pilots who never flew a PA-28 from the right seat trying to teach someone how to land, will be operating at a distinct disadvantage. Many of the worst corporate and airline pilots I've flown with avoided the flight instructing route, often because they believed the work beneath them. The true advantage of flight instructing—besides the time a CFI can log—is the training it offers them in relating to people, much the same kind of training they'll need later to relate to the people in the back seats as corporate pilots or perhaps as they train a new first officer on the airlines.

There is no doubt that flight instructors are not yet looked upon as the professionals they should be, but as the pool of well-qualified pilots continues to shrink, more and more flight schools may find themselves running ads like the one that Western Michigan University recently did (see box below). Unless we find enough instructors to train new pilots, that pipeline to the airlines, corporate and charter is going to begin to dry up.

Flight instructors often work five to six-day weeks and are required—at some schools—to be available at all hours during those days. Instructors need to be blessed with patience. If you have none, find another way to build time and save everyone else a great deal of grief. But if you truly enjoy flying—a lot—and you have a talent for teaching other people about flying and a desire to share the joy you feel about aviation, then an instructing position is something you should consider.

Part of the problem with flight instructing is that too often, the people who run flight schools are naive about what a good flight instructor job should be and often have little motivation to improve these jobs. Why? Because some flight schools are run by some pretty poor businesspeople. I hate to say it, but pilots do not traditionally make very good managers. A new CFI looking for work would do well to consider some of the same questions that student pilots are thinking about as they choose a flight school. Why would you want to work in a school that you could never imagine yourself training in? Avoiding these schools is the only way to help clean up this part of the industry and make instructing a more satisfying job. Remember, changing the role of perception of flight instructors does begin with you!

POSITIONS AVAILABLE

Western Michigan University—Career Flight Instructors

WMU's School of Aviation Sciences currently provides 4-year degree courses for 500 undergraduates and conducts 12-month airline pilot training courses through its International Pilot Training Center (IPTC), for 96 cadets from Aer Lingus, British Airways, and Emirates. State-of-the-art teaching facilities at Battle Creek's fully equipped W. K. Kellogg Airport and 60 top-of-the-range aircraft contribute to their training environment.

The School also invites applications from Career Flight Instructors to instruct in its IPTC. Applicants should possess a FAA CFII and ME Ratings with 800 hours of previous instructing experience. Candidates who successfully complete the WMU instructor standardization course will qualify for the salary of $45,000.

Information on WMU maybe found on our Web site at http://avs.wmich.edu. To apply for these positions, submit a letter of recommendation including resume with a breakdown of your previous flying experience to: Bonnie Sleeman, WMU School of Aviation Sciences, 237 Helmer Road, Battle Creek, MI 49015, USA. Fax 616/964-4676, e-mail: bonnie.sleeman@wmich.edu.

Summing It Up

Where do you see yourself in two years? In five or ten years? You should have enough variations on flying bouncing around in your head to make you think carefully before you make a decision about which direction to head. But remember, no decision is cut in stone. The real crime would be to remain in some segment of flying that turns out not to be a good fit simply because you've already put in a few years.

The decision is yours.

Profile: Corporate Pilot

Nick Verdea, Captain, Cessna Citation 3

"Never underestimate the power of networking," said Nick Verdea. His opportunities in aviation began at an early age when an uncle—then a Convair pilot for Mohawk Airlines—picked the young Verdea up in his V-tail Bonanza. The young aviation aficionado was hooked.

"I got my private out of high school and went to Spartan in Tulsa, Oklahoma where I picked up my commercial, instrument, multiengine, and CFI in one year. My uncle heard about a CFI job in Plattsburg, New

York, and I got it—until the company went out of business. I returned to my home in Cleveland where I broke my hand in an accident." While the accident may have been unfortunate, it helped Verdea make a decision that would prove to be critical to his future as a corporate pilot. "I decided to return to college at Ohio State in Columbus and finish my degree. I guess it was really a lucky break."

"One of my fraternity brothers at school had a brother who flew for a local bank. He introduced us. That's where I heard about the local pilot's association." Verdea joined and met a woman who needed a part-time co-pilot in an F-90 King Air. "I was paid $25 an hour for two trips a week between Columbus and Pittsburgh."After college, Verdea interviewed with Financial Air Express (now AirNet). "The guy that interviewed me said stay around and watch the sorting operation," Verdea recalled. "I planned to stay there until they hired me. They gave me a shot a week later. I began in a Baron and an Aerostar."

A year later, Verdea heard about a busy charter company from someone at the FBO in Louisville where he was based for Financial Air Express. "I interviewed with Executive Jet and they said they'd keep my name on file. But when I got back to Louisville that night, there was a message from EJ. I started two weeks later in a Citation. I was there two years and made captain after one." Verdea left EJ two years later with 4000 total time and nothing but good things to say about the company.

Nick heard about another job with a bank shortly after he got married. "It paid much better even though it was in a King Air 300 and I was home almost every night. Then we picked up a Hawker 400 and soon that evolved into a Hawker 800 and a Falcon 50. It was during my training in the 50 that I met a guy from ExecJet in Europe who has been my mentor ever since."

A big believer in the value of a solid education for a corporate pilot, Verdea took advantage of the bank's offer to pay for an advanced degree. "I got a masters in aviation management and operations from Embry-Riddle that I picked up by working over the Internet and watching videos of classes while I was on the road. In corporate aviation, the aviation department is becoming more of a business unit, not just a cost center." His belief in looking at a corporate aviation department as a business helped Verdea also win a scholarship to the Darton School of Business at the University of Virginia as well.

Verdea remained with the bank for nearly ten years when Jan Barden (highlighted elsewhere in this book) called to tell him about a Falcon 100 chief pilot position. "I interviewed, got the job, but was out of work a few months later when the company disbanded their flight department," Verdea said. "Even though I kept calling my network of contacts telling them I was looking, I was still unemployed for four months. Then the CAAP people called from Dallas and offered me the job I have now in

Chicago. The CAAP people were great when I lost my job. They got me prepared to go back out into the marketplace. They helped me keep my confidence level high."

Now that he's a chief pilot himself, Verdea spoke about the things that any new pilot should be thinking about too. "There is no such thing as luck. I believe a person creates their own. You never know who you'll meet in your networking efforts, either in-person or online through places like AEPS, corporatepilot.com or avcrew (all highlighted in the Web site section of this book). But networking also means that you must put something back. I try to help people out when I can."

"Quality of life issues are also important," Verdea added. "It's tough to separate your personal life from your professional one. It's important, for instance, to know how your family feels about you're being gone."

"There is a certain truth to being in the right place at the right time," he concluded. "Keep all the doors open. I contacted everyone I knew when I was looking for a new job. It wasn't luck that got it for me though. It was a lot of hard work."

The First Steps Toward a Career as a Professional Pilot

Before you even have a prayer of trying to become a professional pilot, you'll need the ratings. While there are numerous ratings available to you, depending on the type of aircraft you fly, there are only a few ratings that you really need on your certificate in order to get rolling. You certainly need a private pilot's license to begin. Follow that up with a commercial license. Your instrument rating will normally be a part of the commercial license. A multiengine rating will also be necessary in most cases. The flight instructor rating, while not required, certainly opens up a wide range of extra employment opportunities (Fig. 1-11).

The Medical Certificate

Before you can qualify for any of the certificates, you must also possess the appropriate class of FAA medical certificate for the license you're working on. In general, the medicals are categorized this way. The least stringent medical is the third class, valid for 24 months after it's issued and necessary to qualify for a private pilot certificate. Next comes the second-class medical, valid for 12 months and necessary for a commercial certificate, which

Figure 1-11. A pilot position on a Falcon 2000 is possible with the proper training.

is normally needed for the flight instructor rating. The granddaddy of medicals is the first class. Necessary for the issuance of an ATP certificate, the first-class medical is only valid for six months. Because an ATP certificate is necessary in most cases to be captain on just about anything these days, I'd find out very early on in my training where I stood in meeting the first-class medical requirements. One of the nice things about the FAA medicals is that they often serve more than one function. If, for instance, you were to obtain a first-class medical certificate and you were not exercising the privileges of your ATP certificate, that medical could be good for 12 months from the date it was issued and also be perfectly acceptable as a second-class medical. Hold on to your first-class medical for two years, and you'll still be able to enjoy the privileges of a private pilot. (Fig. 1-12.)

The Actual License Requirements

Private Pilot Eligibility Requirements

To be eligible for a private pilot certificate, a person must:

- Be at least 17 years of age.

Figure 1-12. Commercial helicopters are at home in the city.

- Be able to read, speak, write, and understand the English language.
- Receive a logbook endorsement from an authorized instructor who:
    1. Conducted the training or reviewed the person's home study on the aeronautical knowledge areas listed in Sec. 61.105(b) that apply to the aircraft rating sought; and
    2. Certified that the person is prepared for the required knowledge test.
- Pass the required knowledge test on the aeronautical knowledge areas listed in Sec. 61.105(b).

- Receive flight training and a logbook endorsement from an authorized instructor who:
  - Conducted the training in the areas of operation listed in Sec. 61.107(b) that apply to the aircraft rating sought; and
  - Certified that the person is prepared for the required practical test.
- Meet the aeronautical experience requirements of this part that apply to the aircraft rating sought before applying for the practical test.
- Pass a practical test on the areas of operation listed in Sec. 61.107(b) that apply to the aircraft rating sought.

Private Pilot, Airplane— Aeronautical Knowledge

General. A person who is applying for a private pilot certificate must receive and log ground training from an authorized instructor or complete a home-study course on the aeronautical knowledge areas of paragraph (b) that apply to the aircraft category and class rating sought.

Aeronautical Knowledge Areas

Applicable Federal Aviation Regulations of this chapter that relate to private pilot privileges, limitations, and flight operations:

- Accident reporting requirements of the National Transportation Safety Board
- Use of the applicable portions of the "Aeronautical Information Manual" and FAA advisory circulars
- Use of aeronautical charts for VFR navigation using pillage, dead reckoning, and navigation systems;
- Radio communication procedures;
- Recognition of critical weather situations from the ground and in flight, wind shear avoidance, and the procurement and use of aeronautical weather reports and forecasts.
- Safe and efficient operation of aircraft, including collision avoidance, and recognition and avoidance of wake turbulence.
- Effects of density altitude on takeoff and climb performance;
- Weight and balance computations;
- Principles of aerodynamics, power plants, and aircraft systems;
- Stall awareness, spin entry, spins, and spin recovery techniques for the airplane and glider category ratings;
- Aeronautical decision making and judgment; and

- Preflight action that includes—How to obtain information on runway lengths at airports of intended use, data on takeoff and landing distances, weather reports and forecasts, and fuel requirements; and how to plan for alternatives if the planned flight cannot be completed or delays are encountered

Aeronautical Experience

For an airplane single-engine rating. Except as provided in paragraph (i) of this section, a person who applies for a private pilot certificate with an airplane category and single-engine class rating must log at least 40 hours of flight time that includes at least 20 hours of flight training from an authorized instructor and 10 hours of solo flight training in the areas of operation listed in Sec. 61.107(b)(1) of this part, and the training must include at least—

- 3 hours of cross-country flight training in a single-engine airplane;
- Except as provided in Sec. 61.110 of this part, 3 hours of night flight training in a single-engine airplane that includes—One cross-country flight of over 100 nautical miles total distance; and 10 takeoffs and 10 landings to a full stop (with each landing involving a flight in the traffic pattern) at an airport.
- 3 hours of flight training in a single-engine airplane on the control and maneuvering of an airplane solely by reference to instruments, including straight and level flight, constant airspeed climbs and descents, turns to a heading, recovery from unusual flight attitudes, radio communications, and the use of navigation systems/facilities and radar services appropriate to instrument flight;
- 3 hours of flight training in preparation for the practical test in a single-engine airplane, which must have been performed within 60 days preceding the date of the test; and
- 10 hours of solo flight time in a single-engine airplane, consisting of at least—5 hours of solo cross-country time; One solo cross-country flight of at least 150 nautical miles total distance, with full-stop landings at a minimum of three points, and one segment of the flight consisting of a straight-line distance of at least 50 nautical miles between the takeoff and landing locations; and
- Three takeoffs and three landings to a full stop (with each landing involving a flight in the traffic pattern) at an airport with an operating control tower.

Commercial Pilot Eligibility Requirements (Abbreviated)

- Be at least 18 years old.
- Read, speak, and understand English.

- Receive the proper logbook endorsement.
- Pass the required knowledge.
- Pass the practical exam.
- Hold a private pilot certificate.

Commercial Pilot, Airplane—Aeronautical Experience

For an airplane single-engine rating. Except as provided in paragraph (i) of this section, a person who applies for a commercial pilot certificate with an airplane category and single-engine class rating must log at least:

- 250 hours of flight time as a pilot that consists of at least:
- 100 hours in powered aircraft, of which 50 hours must be in airplanes.
- 100 hours of pilot-in-command flight time, which includes at least— 50 hours in airplanes; and 50 hours in cross-country flight of which at least 10 hours must be in airplanes.
- 20 hours of training on the areas of operation listed in Sec. 61.127(b)(1) of this part that includes at least—
- 10 hours of instrument training of which at least 5 hours must be in a single-engine airplane;
- 10 hours of training in an airplane that has a retractable landing gear, flaps, and a controllable pitch propeller, or is turbine-powered, or for an applicant seeking a single-engine seaplane rating, 10 hours of training in a seaplane that has flaps and a controllable pitch propeller;
- One cross-country flight of at least 2 hours in a single-engine airplane in day VFR conditions, consisting of a total straight-line distance of more than 100 nautical miles from the original point of departure;
- One cross-country flight of at least 2 hours in a single-engine airplane in night VFR conditions, consisting of a total straight-line distance of more than 100 nautical miles from the original point of departure; and
- 3 hours in a single-engine airplane in preparation for the practical test within the 60-day period preceding the date of the test.
- 10 hours of solo flight in a single-engine airplane on the areas of operation listed in Sec. 61.127(b)(1) of this part, which includes at least—
- One cross-country flight of not less than 300 nautical miles total distance, with landings at a minimum of three points, one of which is a straight-line distance of at least 250 nautical miles from the original departure point.
- 5 hours in night VFR conditions with 10 takeoffs and 10 landings (with each landing involving a flight in the traffic pattern) at an airport with an operating control tower.

Airline Transport Pilot (ATP)

To be eligible for an airline transport pilot certificate, an applicant must

- Be at least 23 years of age.
- Be able to read, speak, write, and understand the English language.
- Be of good moral character.
- Hold at least a commercial pilot certificate and an instrument rating.

Airline Transport Pilot, Airplane—Flight Time. An applicant must have at least 1,500 hours of total time as a pilot that includes at least:

- 500 hours of cross-country flight time.
- 100 hours of night flight time.
- 75 hours of instrument flight time, in actual or simulated instrument conditions, subject to the following:
- Except as provided in paragraph (a)(3)(ii) of this section, an applicant may not receive credit for more than a total of 25 hours of simulated instrument time in a flight simulator or flight training device.
- A maximum of 50 hours of training in a flight simulator
- 250 hours of flight time in an airplane as a pilot in command, or as second in command performing the duties and functions of a pilot in command while under the supervision of a pilot in command or any combination thereof, which includes at least—
- 100 hours of cross-country flight time; and
- 25 hours of night flight time.
- Not more than 100 hours of the total aeronautical experience requirements of paragraph of this section may be obtained in a flight simulator or flight training device that represents an airplane.
- A person who has performed at least 20 night takeoffs and landings to a full stop may substitute each additional night takeoff and landing to a full stop for 1 hour of night flight however, not more than 25 hours of night flight time may be credited in this manner.

The Flight Instructor Certificate

To be eligible for a flight instructor certificate, you must:

- Be at least 18 years of age.
- Read, write, and converse fluently in English.
- Hold a commercial or ATP certificate, with the appropriate aircraft rating.
- Hold an instrument rating, if applying for an instrument instructor rating.

There's no specific number of hours required for the issuance of a flight instructor certificate beyond what's required for the issuance of the commercial, ATP, or particular aircraft rating. This means your instructor might not be terribly experienced, so be certain you discuss the instructor's experience before you begin.

Aircraft Type Ratings

To have an aircraft type rating added to a pilot certificate, an applicant must meet the following requirements:

- Hold, or concurrently obtain, an instrument rating appropriate to the aircraft for which the type rating is sought.

- Pass a flight test demonstrating competence in pilot operations appropriate to his or her pilot certificate and the type rating sought.

- Pass a flight test showing competence in pilot operations under instrument flight conditions.

The Instrument Rating.   A person who applies for an instrument rating must:

- Hold at least a current private pilot certificate with an airplane, helicopter, or powered-lift rating appropriate to the instrument rating sought;

- Be able to read, speak, write, and understand the English language.

- Receive and log ground training from an authorized instructor or accomplish a home-study course of training on the aeronautical knowledge areas.

- Pass the required knowledge test

- Pass the required practical test

Aeronautical Experience.   A person who applies for an instrument rating must have logged the following:

- At least 50 hours of cross-country flight time as pilot in command, of which at least 10 hours must be in airplanes for an instrument—airplane rating; and

- A total of 40 hours of actual or simulated instrument time on the areas of operation of this section, to include—

- At least 15 hours of instrument flight training from an authorized instructor in the aircraft category for which the instrument rating is sought;

- At least 3 hours of instrument training that is appropriate to the instrument rating sought from an authorized instructor in preparation for the practical test within the 60 days preceding the date of the test;

- For an instrument—airplane rating, instrument training on cross-country flight procedures specific to airplanes that includes at least one cross-country flight in an airplane that is performed under IFR, and consists of—

- A distance of at least 250 nautical miles along airways or ATC-directed routing;
- An instrument approach at each airport; and
- Three different kinds of approaches with the use of navigation systems.

Note: The preceding pilot certificate and rating requirements are only excerpts from the Federal Aviation Regulations (FARs).

Logbooks

In chapter 8, we'll be delving deep into the technology available to help you not only find a job, but to better manage the information you'll need to track during your search. You'll learn during your career, that one of the most important records you'll ever maintain is your personal pilot logbook. Guard your logbooks preciously. They are the only true records of the flight time and hard work you've put in toward reaching your goal of becoming a professional pilot. Here are some common definitions you'll run across in your flying career when it comes to logging flight time.

Solo—The only time you'll most likely log solo flight is when you're a student pilot, but the regulation says you may log solo flight time any time you're the sole occupant of the aircraft.

Pilot in command (PIC)—You can log PIC when you're the sole manipulator of the controls for an aircraft for which you're rated or when acting as a pilot on an aircraft for which more than one pilot is required under the type certification of the aircraft, or the regulations under which the flight is being conducted. An example could be when acting as a safety pilot for someone wearing a hood during instrument practice.

Instrument time—You may only log as actual instrument time that flight time during which you control the aircraft by total reference to the flight instruments in actual weather conditions. Simulated instrument flight, or hooded time, under simulated weather conditions must be logged separately.

Flight instructor—A certified flight instructor may log all time during which they act as a flight instructor as PIC. If you fly with an instrument student in actual instrument weather, the CFI may log the time as actual instrument time and PIC.

Second in command (SIC)—Pilots may log time as second in command when they act as SIC of an aircraft for which more than one pilot is required under the type certification of the aircraft or the regulations under which the flight is conducted.

Your Logbook: More Than Just Numbers

Reprinted Courtesy *Flying Careers* Magazine, Bessard Publications, Atlanta, GA 800 492 1881

One of the great advantages of aviation—or one of its more serious drawbacks, depending on your point of view—is that much of this industry is cut and dry. Since aviation is also one of the most highly unionized of industries, there is typically a procedure to follow for just about any situation. How many people can we carry with this fuel load? Check the aircraft's POH. What happens when the weather goes below minimums at your destination after you're airborne? Check the operations manual. There just is not a great deal of room for interpretation, until you come to a subject like pilot logbooks.

As Denis Caravella, safety programs manager, FAA Chicago DuPage FSDO said, "There is not much in the way of guidelines for logging time other than FAR 61.51 and the Flight Inspectors Handbook." The only advice WestAir pilot Dan Dornseif received was "be honest." Southwest Airlines pilot Alan Peirson said, "There is really no guidance about logging time. A great deal of it is just hearsay." Part 61.51 states that the only flight time you must log is the "aeronautical training and experience used to meet the requirements for a certificate or rating, or the recent flight experience requirements of this part must be shown in a reliable record. The logging of other flight time is not required." One unwritten rule of thumb is that your logbook is used to verify that the flight time you have documented is what you have actually flown. Any questions an interviewer might have about your experience should be easy to substantiate by reference to the remarks or the endorsement section of your book.

Some pilots believe logbooks are only logbooks. Actually, however, your logbook could be your competitive edge during an interview, especially since different companies look for different things in your logbook, such as exactly how your time should be totaled for their review. Bob Fiedler, supervisor for flight operations at United Airlines, said, "We have a different definition of dual time, for instance. We consider any dual received as student time." But, also, contrary to most pilots' opinions, most companies do not care whether the logbook you bring them was printed with a pen each night or with one swift keystroke of your computer. Rod Jones, assistant chief pilot in Phoenix for Southwest Airlines said, "While the computerized logbook shows much more attention to detail and much more effort, either version is fine for us."

Most commercially produced pilot logbooks are divided between the small plastic covered ones you might use for private flying and the larger, more expensive versions you will use to chart your career as a professional. Fiedler recalled a time when an applicant "brought in a box of small logbooks and loose pages tied together with paper clips and rubber bands." Jones remembered "an applicant who brought in a box of fuel receipts three inches high and said that was how he tracked his time." If you intend on chasing a career as a professional pilot, spend the big bucks for a professional flight logbook.

How important is neatness? That depends. United's Fiedler said, "To us, neatness really isn't that critical." However, when many people refer to neatness, they are actually referring to a book that is easy to read. Put yourself in the shoes of the person who must read your logbook—the interviewer for the airline or corporation that has a position you want. Your job is to make that interviewer's work as easy as possible. You'll accomplish this by making certain that there are a minimum of corrections and whiteouts. Jones added that "a correction on almost every page makes me wonder about that applicant's attention to detail."

Be certain the columns total up correctly. "We ask for a fairly detailed matrix of flight time on our application. That application information should be verifiable with the logbook," said Chuck Hanesbuth, director of Pilot Standards at Northwest Airlines. If you have a computerized logbook, offer a summary sheet that reflects the format of the application. Jones added that "some pilots bring in their logbooks up-to-date for the interview, while others have had no entries for months before. The latter tells me they didn't do much preparation for the interview and are probably a procrastinator."

What to Log?

What should you log? That's a question that has many answers depending on who is looking at your record. Most airlines and companies want to know at least the basics, including the date of the flight, point of origin and destination, the type of aircraft and tail number, the duration of the flight, the conditions—IFR or VFR—and whatever remarks you believe would be of interest in recalling something about the flight later for the FAA, the company, or a potential employer. Remarks could mention many items, like unusual conditions during the flight, CAT, severe thunderstorms, missed approaches, and perhaps the altitude you broke out of the clouds at on the approach. Five approaches to 200 and 1/2 display a different skill level than five approaches when the ceiling is 900 feet. But don't turn the remarks in to a soap opera of your flying life either. "Thank heavens they made the remarks section rather small," said pilot Randy Ottinger. "The comments in your logbook really should mature as you do as a pilot."

Another logbook issue that causes considerable confusion is logging pilot-in-command time. "We see many errors in logging pilot-in-command time," FAA's Caravella said. The problem often appears over the difference between when you are acting as pilot-in-command according to FAR 91.3 and when you can log flying time as pilot-in-command under Part 61. PIC as defined in FAR Part 91.3 refers to the person who holds the ultimate responsibility for the flight. There can only be one. Period. But to log PIC time, according to many sources, the pilot need only be acting as PIC-rated and the sole manipulator of the controls. A pilot who is typed in a Citation, for example, could be flying from the right seat. During the time he is actually moving the controls, he could log that time as PIC, even though he is not the PIC under the 91.3 definition. Essentially, the difference is who is moving the control wheel at any given time and who has the overall responsibility for the flight at all times. They might be the same person, but then again they might not.

And what exactly do the recruiters look for when they open up your logbook? Sometimes, you'll never know. During some airline interviews, like those at Southwest and United, for example, the logbooks are often collected from all the applicants early on during the interview process. The applicants never see them again until after the interviews are complete. This can have some serious drawbacks and could well force you to pay close attention to the details of your logbook before you hand it over, since you will not have an opportunity to explain any inconsistencies or errors in your work.

The best advice again is to make certain your work is easily decipherable by most anyone. Try offering your logbook up to a few other pilots for their input before the interview. They may well catch errors you have passed over a dozen times. This effort can pay large dividends, since a pattern of errors says a great deal about your attention to detail here and potentially in the cockpit later. Also, if you plan to buy interview preparation from a company specific to this industry, ask if they have a logbook evaluation as part of the service.

One saving grace for pilots who are mathematically challenged are computer logbooks. Southwest's Jones said, "Five years ago I probably saw one computer logbook in a hundred applicants. Today, that number is about one in 10." The reason is simple—flexibility. The electronic pilot logbook is essentially an electronic spreadsheet that can easily tabulate columns of numbers accurately and save you big dollars on correction fluid. If you've ever had to fill out an FAA Form 3710 for a checkride or insurance paperwork, an electronic logbook's ability to easily order the seemingly random sets of criteria the FAA and some airlines seem to want to see can make it worth the price. With an electronic logbook you can actually ask for how much IFR PIC time you flew in a Beech 1900 during the November 1993 to January 1994 period that included stops at MDW, DTW, MEM, and STL. The combinations are all selectable by the pilot before they run a report that takes just seconds to complete. Another potentially helpful aspect of the computerized logbook is that you can define any of the additional fields you want, such as PIC time just for the Boeing 737.

Few Drawbacks:

But there are a few drawbacks to a computerized operation. First, computers can only accomplish what you tell them to. Before they can begin any computations, they need the raw data that is your flight time up to the moment of the conversion. There is only one way to do that—type it in. While that effort may seem colossal at first, balance the work against the speed and flexibility you'll gain over your flight history for the remainder of the 20 or 30 years you might continue to fly.

But that convenience may not be enough for some pilots. WestAir's Dornseif said, "I thought about a computer logbook for a while, but I put it out of my mind. I've seen computers crash and do weird things. I don't want that to happen to my logbook." David Lewis, a Mesa pilot, however, said, "I like the flexibility and reliability of my electronic logbook. I use Aerolog. I know the math is always correct, and I can compile information for any application in just a few seconds."

To gain the flexibility of an electronic logbook, some pilots reported having tried a number of different routes to quickly enter the data into the system. One is a highly questionable single-entry-per-thousand-hours version. This technique puts one entry per year in your logbook for all the flying you've accomplished until this moment—e.g., "flew 652 hours in 1998." Think about the eyebrows that sort of entry would raise however if you read Part 61.51 (b) literally—"Each pilot shall enter...information for each flight or lesson logged." Another version goes the single entry one better. It takes the annual total and breaks down the amount by aircraft at least, "of the 652 hours in 1998, 386 were flown in the Boeing 727." That's better, but it won't suffice in the minds of airline recruiters. Imagine using these techniques with those airlines that don't offer you the opportunity to talk about your logbook. Would you want this to be the only impression of you a company walks away with?

The best—and the most accurate—method of making the switch from paper to computerized logbook is to type each entry in, one by one, verbatim from the paper logbook. You might consider paying someone else to type in your data, with you running through it later to verify the entries for accuracy. When it is all complete, simply run a summary report. If it does not agree with your paper logbook, you'll then have an option of believing the computer and moving on from that point or of trying to track the error. If the time difference is not great, you might want to consider simply taking the new total as gospel.

Finally, no discussion of logbooks would be complete without some mention of backups—whether you use a computer or a paper logbook. Dornseif remembers the time his regular logbook was stolen out of the flight bag in his car. Think about the implications of losing the only record of your flight time available anywhere. For people who pride themselves on planning ahead for all the possibilities of weather or fuel or mechanical problems, this strategy certainly has an odd ring to it.

If you have a paper logbook, consider what Ottinger does. "I make a fresh copy of each full page of my logbook when I've completed it." He then stores the photocopy version of his logbook at the home of his parents for safekeeping. If he ever did need to reconstruct a new logbook, the task would be relatively simple. If you are using a computer logbook, the backup process is normally as simple as inserting a blank disk into the A drive on your machine and hitting enter. Making two backup copies of your data is a good idea as well. Make certain you store one of them at a location away from where your computer normally stands, like a safety deposit box at a bank.

Jones summed up the logbook issue best, "Pilots should have some pride of ownership in their logbooks. Some treat them as just another document, while others I've interviewed show their log as a list of the accomplishments that got them to that interview. Their logbook says 'this is who I am.'"

Now you've seen the worst of it, the full set of requirements that will take you from rank beginner to airline transport pilot. This quest of yours will require time, money, and effort on your part if you're to be successful. In upcoming chapters, I'll be discussing all of these items to make certain you reach the rank of professional aviator with minimal fuss and maximum fun, because if you aren't having some fun along the way toward becoming a professional pilot, you probably should consider another line of work.

In every group there will always be naysayers, people who believe that you'll never be able to achieve your goal. "There are too many pilots now," they say. "The economy is too tough." "You're a woman" or "No one will hire a minority," they say. I happened across a story that I wish I had written. It's the story of a pilot who had a dream she never lost.

The following was printed with permission from UND Aerospace, Grand Forks, N.D., and Christy DeJoy, J. Patrick Moore, LaMaster, Farmer, Minneapolis, Minn.

Nine-year-old Jean Haley was to write an essay explaining what she wanted to be when she grew up. The assignment was simple; Jean knew she would be an airline pilot. But those dreams were nearly shattered when she saw the "F" at the top of her essay. Her third-grade teacher reasoned, "This is a fairy tale, not an essay."

The year was 1959, and the reality was that there was no such thing as a female airline pilot.

Fast forward to the year 1993, and Jean (Haley) Harper is a United Airlines captain.

The 34 years in-between have been filled with hard work, bravery, and commitment. Harper credits much of her success to her father. When young Jean brought her F-graded essay home, Frank Haley, a crop duster, said, "You can do anything you want."

She wanted to fly, so seven years later she began flying lessons. "I wanted to prove to myself I could do it. I figured I could prove the nay-sayers wrong by doing the things they said I couldn't do."

Harper determined that she needed to meet three criteria to become an airline pilot. First, she needed the appropriate licenses. So, at 20 years old, Harper took out a bank loan and spent six months getting her private, commercial, and instrument ratings. Second, she needed flying time—at least 1,500 hours. She started logging hours hauling skydivers and studied to become a flight instructor. And third, she needed a four-year degree. She enrolled in the aviation program at the University of North Dakota in Grand Forks on the advice of Mike Sacrey, a UND Aerospace graduate and the FAA pilot who checked her out as a flight instructor.

When Harper enrolled at UND in 1971, there were still no women airline pilots. "It was a scary feeling thinking I could spend the years and the money trying to attain it and still not be taken seriously. But not taking the chance of becoming an airline pilot was more scary," she said.

While female pilots were still a novelty at UND, she nonetheless became the school's first female flight instructor. "It was a first, but John Odegard was enthusiastic that I was there and was happy to offer me the position," she said.

During the next four years, Harper juggled her studies and a variety of jobs, including flight instructor, charter pilot, cloud seeder, crop duster, glider tower, and night airmail pilot. "I was spread thin," she admits, but was doing well enough to earn several scholarships and awards.

The best gift she received was from a UND student pilot named Tracy Van Den Berg, who told her in 1973 that Frontier Airlines had hired pilot Emily Howell. "I thought, 'Oh, my God. It happened. The last obstacle has fallen.' I finally had an honest-to-goodness role model." A few months later American Airlines hired Bonnie Tiburzi.

After graduation in 1975, Harper began writing to airlines asking for pilot application forms. She was getting little response until she started signing her name "J.E. Haley" instead of "Jean Haley."

Due to an industry down cycle, the airlines were hiring few pilots during this time. For nearly three years, Harper flew however she could—moonlighting as a flight instructor and flying with small cargo and commuter carriers.

At Meridian Air Cargo, Harper met a handsome pilot named Vic Harper. She used to switch shifts with his copilot so she could fly with him. Three years later they were to become matrimonial copilots.

In the meantime, airline pilot hiring increased. Harper interviewed with Delta, Allegheny, and United Airlines, and landed a job in Denver with United as a 727 flight engineer. She became United's third woman pilot, hired just weeks after the first two women. During that same time, Vic accepted a job with Frontier, also in Denver.

The person who most strongly supported Harper's dream of becoming an airline pilot never saw her in her United uniform. Harper's father died in an airplane accident not long after she graduated from UND.

In her 15 years with United, Harper has flown Boeing 727s, 737s, 757s (Fig. 1-13), and 767s. In November 1992, Harper, 43, flew her first line trip as a 737 captain—with her favorite first officer, Vic Harper, who has since joined United.

"Becoming captain was such an accomplishment for me. I'm so proud." She said her pride in her qualifications rubs off on other people, easing the minds of the few who might not be completely comfortable with a woman captain.

"The vast majority of comments I've received about being a captain have been very positive. Among the most enthusiastic congratulations I have received have come from the 'old timers'—the senior captains at United," she said. "I think my attitude has a lot to do with it."

In addition to her career achievements, Harper said she's also met her goals of a rewarding personal life. The Harpers juggle two airline careers and two children—Annie, 7, and Sam, 3$^1/_2$. In 1985, Harper became United's first pregnant pilot. United didn't even have a maternity leave policy for several years after that.

"I remember having serious concerns whether I could have this career and have a family, too. It bothered me that I might have to choose. No one would ever tell a man he couldn't be married, have children, and have a career," she said.

Harper has been able to have both career and family. "This is it. I have no more unfulfilled wants."

Perhaps Harper's third-grade teacher was right. She is living a fairy-tale life.

A Few Tips for Aspiring Female Pilots

BY SANDRA L. ANDERSON, VICE PRESIDENT, WOMEN IN AVIATION INTERNATIONAL

Women are making a difference! There has been a dramatic increase in the number of women earning ATP and Commercial certificates in the past 10–12 years. Women are making a positive impact in flight instructing, charter flying, corporate flying, and airline flying. If you are interested in any of these pilot occupations, it is very important to gather as much information concerning pilot careers and opportunities with people in the industry. Aviation organizations, groups, conferences, air shows, and libraries are excellent sources of information concerning various pilot occupations. Some of the tips listed below may be utilized in all areas of aviation.

No matter what your occupational objectives are in the aviation industry such as flight instruction, charter flying, corporate flying or airline flying, it takes passion, assertiveness, and perseverance to achieve those goals. Each individual female brings her own unique perspective, skills and capabilities to her particular occupation.

In order to achieve your goals as a flight instructor, charter pilot, corporate pilot, or airline pilot, it will take passion in your heart, assertiveness of your soul, and determination.

Flight Instruction

To secure a job in flight instruction, find a reputable FBO (fixed base operator), aviation school, or aviation university with good student enrollment figures. If you have graduated from an aviation university, consider the possibility of remaining at the university as a certified flight instructor (CFI). Consistent demand for aviation instruction will assist in securing your continuous employment as a flight instructor. Also, what is the opportunity for you to achieve additional flight instructor certificates and ratings such as a CFI-Instrument and CFI-multiengine land or sea while employed with a particular school or university. The more certificates and ratings you have as an instructor, the more valuable you are to your employer. Is your employer willing to help you achieve additional certification and ratings? If you enjoy teaching or working towards the next step in your pilot career, flight instruction is a very rewarding profession and an invaluable experience.

Figure 1-13. New-generation United Boeing 757.

Charter Flying

Jobs in the charter flying business vary greatly among companies and FBOs. The varieties include duty hours, benefits, maintenance quality, type of aircraft and avionics equipment, travel destinations, quality of training, and reputation of the company. Charter pilots enjoy an informal interaction between pilot and passenger. Passengers are greeted by the pilots and all aspects of the flight (itinerary, catering, and special requests) are reviewed with the customer. A charter business generally means day trips with some overnight stays.

Most charter flying companies do not require a four-year degree. However, the more formal training the pilot has enhances the chance of getting hired. Some issues to consider when applying for a charter job are duty hours, "on-call" requirements, benefits, pilot attrition rate, day trips, overnight trips, and destinations. Make sure you have researched the charter company you are considering. Are the company's aircraft logbooks and records in the proper order and has the company had any FAA violations or fines?

Airline Flying

Airline employment is highly competitive. One must ascertain the hiring requirements of the airline you are interested in prior to sending in an application and resume. It is vitally important to talk with other pilots who have gone through the hiring interview process. Before going for an interview, learn as much as possible about the airline company. Consider a membership in an industry organization that provides continuous information on what the commuters, regionals, and major airlines are doing.

Some areas to research are: what is the airline's route structure? What are the "pros and cons" in working for this particular commuter, regional, or major airline? Does the carrier fly domestic routes only or domestic and international routes? Will I have to relocate? Will I have to commute to my domicile? What is the pilot attrition rate?

Corporate Flying

Most corporate pilots seek employment in an area of the country where they like to live. The objective of a corporate pilot is to move people from an address to an address.

Depending upon the size and structure of the company and its flight department, corporate pilots may handle all aspects of the trip, from flight planning and weather to the destination transportation and all catering supplies. Pilots fly the same aircraft with usually the same passengers and occasionally new customers. Corporate pilots also have the opportunity to fly in some of the most advanced technological aircraft available.

Corporate and airline flying have similarities and differences that need to be considered when deciding which avenue to pursue. For both, you'll need an excellent background, preferably college education with a similar number of flight hours required for both—depending on the equipment the company uses. Some of the differences include scheduling—the airline schedule will be much more structured, the corporate aviation schedule much more varied. In addition, with corporate aviation, you become a part of the company's structure and culture with opportunities for advancement within the flight department as well as elsewhere in the company. Management and good communications skills can be very important even in entry-level jobs.

The International Women in Aviation Conference is a unique opportunity for women to network with other female professionals who provide a wealth of aviation experience. In discussing issues and opportunities with female professionals, prospective aviation women can discuss employment opportunities and issues, interviewing techniques, professional appearance, and make valuable contacts. Many of our WAI members are exceptional mentors not only in their chosen profession, but also as human resource contacts. Additionally, the WAI conference networking provides an opportunity for women peers to associate and network if they are looking to transition to another aviation opportunity, and to learn of employment opportunities through the various aviation companies who exhibit each year. Visit the WIA Web site at www.wiai.org. (Fig. 1-14.)

Figure 1-14. UPS loading operation. *Courtesy UPS.*

2

Flight Training

Before you can even think about flying for the airlines, a corporation, or the U.S. military, there's a more simple goal to accomplish . . . learning to fly. Now, simple is a relative term here, and I'm really not trying to treat learning to fly in a matter-of-fact way. How you learn to fly is not only important to your goal of flying professionally, but it is also crucial to the long-term goal of keeping whatever flying job you eventually land. Without the right flying education and the right attitudes, skills, and habits, your career could take a very early nosedive.

Learning to fly is a much more involved process than just running out to the airport for a few lessons once a week, taking a test, and picking up your license. During the months or years that it might take you to gain the experience necessary to leap to the next stage of training for that commercial, instrument, or multiengine rating, you'll be working on a regular basis with a number of different instructors. You might believe that the best way to progress through flight training is to find a "good instructor" and stay with that person all the way through to your first flying job. While that certainly is an option, it's one I would avoid.

That first instructor will seem like quite an authority to you in those early months of learning to fly. It's this first instructor who will, hopefully, sit patiently with you as you learn how the various parts are put together to form a real flying airplane. It's this instructor who will teach you first about lift and weather and federal regulations. But, along with those first bits and pieces of practical knowledge will come bits and pieces of an attitude. Student pilots tend to emulate what they see. If their first instructor is extremely careful about preflighting the aircraft before they fly or is constantly picking up the microphone while in flight to either use the air traffic control system or to check on a possibly threatening weather situation, the student will learn, too, that this is the way a well-managed cockpit is operated (Fig. 2-1). If, on the other hand, that first instructor is someone who flies carelessly, a person to whom just getting the job done is sufficient, an instructor who explains once and says, "Go back and read the rest on your own," and never spends the time to really find out what the student understands, the instructor is doing you a great disservice. If that same instructor chooses not to use a checklist, if that instructor has little regard for precise airspeed control in his or her flying, you'll come to believe this is the correct method for piloting an aircraft. Whatever the instructor does, you'll eventually find yourself mimicking.

And too, it's not just what the instructor teaches you that's important (Fig. 2-2); what that instructor leaves out of your training could be vital to your career and safety. One of the strengths of flying with more than a single instructor is the exposure that a student receives to another pilot's scrutiny. When the major aircraft manufacturers were designing their own learn-to-fly curriculums in the 1960s, they included a number of phase checks or minireviews. As a flight instructor in a Part 141 school, I was required to fly many of these phase checks. They were designed to give me, the assistant chief flight instructor, a look at not just how well the students were learning, but how well the instructors were teaching.

I remember a young female commercial student who arrived for a phase check one sunny Saturday afternoon. After a satisfactory chat about flying and regulations before the flight, I felt the woman seemed well

Figure 2-1. A professional pilot's career begins with a good instructor. *Comair Aviation Academy*

Figure 2-2. Today's cockpits are more automated than ever before. *Rockwell International, Collins Division*

versed in the subjects she needed to understand for the commercial flight test. After we departed the airport for the practice area, I quickly noticed her inability to hold an airspeed or a specified rate of climb. Initially, I put some of this off to nerves and flying with a new instructor, but within 20 minutes or so, I realized this was a real problem. I asked her to fly some more maneuvers for me, and I eventually realized she was not using a very basic tool, the trim wheel. The trim wheel in an airplane is designed to relieve pressure from the control wheel so an aircraft can be flown basically hands-off for a good portion of the time, if this operation is performed correctly. Performed incorrectly, a lack of proper trim in an airplane forces the pilot to work much harder than necessary. The result of this distraction in the cockpit forces even more work on the pilot when there are other operations to cope with.

I was amazed that somehow this student had almost reached her commercial flight test and had somehow managed to skip this very basic concept. After I spent a few minutes showing the student the trim wheel and what it could do for her, she was surprised. I saw her a month or so later after she had successfully passed her commercial checkride, and she told me that learning about trim was one of the things that really made life easier for her during the checkride. Without the opportunity for another instructor to check her progress, this woman would have missed some very important training. If you were a history major in college, you'd never think of taking every single course with the same teacher, so why give your flying career such a handicap? Good or bad, the best, well-rounded flying education is with more than one teacher.

Choosing a Flight Instructor and Flight School

If receiving a well-rounded, professional flying education is truly your goal, another important concern is "who" will teach you. Good pilots always use a checklist, a list that prompts them to be certain that they've not forgotten some important item like the fuel pumps or the landing gear. Good consumers use checklists of sorts, too, as they search for the best deal on a washer and dryer, an automobile, or a flight school. Let's take a look at some of the items you'll want to consider as you make the decision about which flight school will receive your training dollars.

Flight Schools are Not all Created Equal

JoAnn Watzke was stuck in a rut. After six years as an art therapist, she knew she wanted out. She was hooked on flying after 30 hours of lessons in a Grumman trainer. While she knew her future was in aviation, she also knew she needed a pocketful of licenses and ratings. But deciding where she should train for those items is the same problem hundreds of potential aviation professionals face. There are thousands of companies capable of putting a private pilot certificate or better in your wallet, from the local fixed-base operator to large schools like American Flyers and Comair Aviation Career Academy. What questions should a new pilot ask in the early stages of an aviation career? What answers should they listen for?

Considerations may include quality and quantity of training, accessibility, cost, etc. Lynetta Sowder, director of admissions at Southwind Aviation in Brownsville, Texas, said, "Go and look at the school. Tell them you want to look at the housing. Tell them you want to go out and fly with an instructor for an hour. Look at the people who run the school, too. Are they pilots who really know something about aviation?" Everything Flyable's director of operations, David Parsons, said, "People shouldn't be afraid to ask questions about their personal financial responsibility when it comes to insurance. Be sure and ask to see the aircraft's logbooks, too, so you can determine if the aircraft is legal to be flown."

But, before you make any solid decisions about where to train, you need a career plan. Comair Aviation Academy's former president, Michael Sullivan, said, "If your objective is to become a professional pilot, you'll want to search for a professional atmosphere to learn in." Exactly where are you headed professionally? Is your flight training aimed at placing you eventually into the cockpit of a B-737 for a major airline or do you see yourself flying a helicopter for a corporation? A school like Phoenix East Aviation offers a course to take a student from zero time through the ATP and a type rating in a B-737 if the student wishes because, as the school's vice president, Fred DeWitt, said, "We don't expect a guy with 200 hours to get a type rating, but they should start getting familiar with advanced turboprops and jets as soon as they can. After all, a doctor doesn't wait until he gets a job to start learning about advanced medicine."

If your goal is fixed-wing oriented, you might think the easiest method of paring down the list is skipping over schools that offer rotorcraft ratings, but that might not be the best plan of action. Some schools, like Everything Flyable, offer both. If your goal is to fly for an airline as soon as possible, you'll find that some flight training schools, such as Comair Aviation Academy or FlightSafety International, possibly can put you into those cockpits faster than others. Visit Flight Training magazine's flight school list at: www.ftmag.com.

The final advice about goals is: Write them down. When you put goals on paper, they take on a life of their own. They'll form the nucleus of your plan for becoming a professional pilot. Goals can be as simple as the choice of a climate to train in or as complex as trying to find financing for that eventual B-737 type rating or a good scholarship program.

You need to keep accurate records about each school you consider to make an efficient search. Don't trust your memory! Other questions to ask yourself include: Will you train near home or spend the extra money for room and board to get "better" training elsewhere? Is a big school in the North worth the time you may waste on canceled flights when the weather is bad? But, don't choose a Southern California training facility just because you like to surf in your free time. A school somewhere else may take you to your goal much faster.

Will your training be a full-time effort or a part-time struggle for ratings sandwiched between your current job? Parsons said, "Some of the local students make a pretty poor showing for lessons. If we had 25 local students all scheduled, only 15 to 18 would show up. Local residents (who train here) just have too many distractions."

Most pilots hired today hold a college degree. Recent Air Inc. figures show 87 percent of all surveyed pilots hired at global or major airlines held a four-year degree. That number is 62 percent at the regional airlines. Several large universities, such as the University of North Dakota, University of Illinois, Embry-Riddle Aeronautical University, Purdue University, and Parks College, run large flight school operations besides their regular degree courses. Before you invest in a school that provides only flight training, consider a college curriculum that could grant your degree as you earn your wings. A good place to begin is with the University Aviation Association in Opelika, Ala., an organization that represents all facets of collegiate aviation education.

How much time are you willing to spend to earn your ratings? Some schools believe a zero-time student can make the jump from first flight to flight instructor in about six months while other, smaller schools think the time is closer to 12 months. Phoenix East's DeWitt said, "All the way to the ATP is running about 24 months here." It may seem cheaper at first to stay home with Mom and Dad and train when you have the money, but that method could turn your quest for a flying job into a Herculean effort, both in time and money. Can you afford to learn that way? Are you willing to put your career on hold for the extra years it may take for you to earn your ratings on a part-time basis?

How will you finance this venture? Most schools can direct you to a financial institution to apply for a loan, but few offer financing directly. "I have a deal with the bank," said Ed Pavelka, owner of Ed's Flying Service in Alamogordo, N.M. "They don't offer flight training and I don't offer financing." How much money will you need? The prices range from a low of $17,000 to a high of nearly $30,000 to go from zero time through the multiengine flight instructor certificate. Phoenix East quoted an approximate price of $42,000, but this took a zero-time pilot through the ATP rating.

Students need to be aware not just of price, but of value for their training dollar. An inexpensive school could turn out to be less than what you need, while the most costly could offer too many choices. Your career plan needs a budget to keep your eyes from making the decisions your checkbook should.

Will you train at an FAR Part 141 school or under Part 61 regulations? Part 141 is very structured, with all paperwork constantly scrutinized by the FAA, while pilots can train in a less organized, though not necessarily less professional, atmosphere under Part 61. There are many fine, well-structured flight programs operating under Part 61 merely because a school doesn't want to deal with the hassle of paperwork and government "meddling" that comes with FAA approval under Part 141. Part 141 dictates how a school may advertise its services, the requirements for the school's chief pilot's job and how the school must keep records. The curriculum offers little flexibility over how a school handles training. Instead, Part 141 provides a set plan for what subjects the school covers and when.

A benefit of the Part 141 school over training under Part 61 is the total time needed to earn a certificate. An FAA-approved school (Part 141) can set you up for a private pilot license flight test after only 35 total hours. A school under Part 61 requires you to have 40 hours. A Part 141 student at the commercial-instrument rating level could complete the course with only 190 hours compared to 250 under Part 61. But check with the school for realistic statistics. The FAA says most private pilots, for example, need about 72 hours before they see their license, no matter where they train.

A Part 61 operation could involve a single flight instructor with a single airplane or, as mentioned earlier, a large school that doesn't want to hassle with the paperwork for Part 141. Technically, a school under Part 61 needs no chief pilot, nor does it need a standardized, orderly curriculum to operate. Sullivan said, "It's sometimes advantageous for a customer to be under Part 61, particularly if they are bringing in previously logged time." Part 141 generally forces you to start at the beginning.

One of the main reasons some students choose Part 61 training, however, is price. Generally, a Part 141 school costs more, possibly because its facilities and fleet are larger and therefore carry higher overhead costs.

When you pin down your flight school possibilities to just two or three, there is no better way to pass final judgment than to visit them in person, since only then can you decide if a school is right for you. You often can complete on-site inspections in a day if you will train locally. But, if one school is in Florida and another in California, you'll need cost-effective transportation to these visits. Don't think of skipping site visits. There simply is too much important, although subjective, information to gain from them.

When Watzke visited her first California flight school, she found the place "too sloppy for my taste, with books and papers scattered everywhere. I decided the rest of the school must be like that, too." She eventually chose Comair Aviation Academy after reading some magazine reports about it. Although she did not visit the Sanford, Fla., school personally before choosing it, and even though she is pleased with her career now, Watzke recommended, "A new student needs to interview the school. Talk to other students, current and graduated. Try to meet more than just those students the school sets you up with, too."

It's good business to make an appointment. When you arrive at the school, find the chief flight instructor for an indoctrination chat to gain a good, gut level feeling for the company. Did the chief keep you waiting a long time? Did he or she seem interested when you spoke, or were you quickly pushed off on an instructor to answer your questions? What did the school's lobby look like? Were the personnel friendly? Were you provided with added information quickly or did people pause when you talked about items like refunds? Take a look at the aircraft, too. Are they recent models? How many are IFR equipped? What condition are the paint and interior in? A poorly kept outside could mean the school skipped something under the cowling, too.

Check out the classrooms. Are they well lit and roomy? Can you imagine yourself spending several hours a day there? If you can't, find another school. Is there a place to eat nearby? Is there transportation to and from the school if you don't have a car? Does the school have simulators? How old are the simulators? Does the school teach ground school only one-on-one or does it use a computer or video-based system to supplement it?

Look at the flight schedule. Are airplanes and instructors kept busy or do you see large gaps in the day? Too many large gaps could indicate a problem picking up or keeping customers. Find out why. Ask to meet some of the instructors. Sullivan said, "Look at the interaction between the instructor and the student." How old are the instructors? What types of experience do they have? Do they seem friendly? Ask what the school's policy is on changing instructors if you don't click with the one they assign to you. Remember, your relationship with your instructor can be one of the most important of your career. You need to be nosy. Is the school worth the prices it charges?

You need only a few more items to round out your checklist before you make your final decision. Ask about each school's pass/fail ratio and how the school determined it. Some schools base the ratio on the percentage of students who pass the check rides the first time while others exclude from the figures people who drop out of the program halfway. Ask about references. Will the school allow you to speak to some current students and some graduates? Sullivan said, "One of our best advertisements is our former students." Any school should be happy to provide references of former students.

Check the schools remaining on your list with the local Better Business Bureau. While the BBB won't offer specifics about any incident, it will tell you if any complaints were lodged against a school recently. A check of a state's attorney general's office and the local FAA Flight Standards District Office also may provide information about a school, and FAA designated examiners can provide insight on the success of students from certain schools, and even certain instructors, when taking the flight exams.

Discuss money. How much will it cost you before training begins? Does the school want all the money up front, or just part of it? Will it allow you to pay as you go each month or after each lesson? Could a few more dollars on account deliver a lower hourly rate on its aircraft? What is the school's refund policy? Ask to see these items in writing before you sign anything. "Students can pay in full, up front, but I don't recommend it. I put a large sum of money up for a school once and four months later they went bankrupt," Parsons said.

Many schools will want you to sign a contract or an agreement before the training begins. A well-written agreement protects not only both parties if one of them does not live up to their word, but also outlines exactly how each party will handle certain situations. If the fine print seems too confusing, consult an attorney before you sign. Parsons recommended students "pick up a copy of the school's rules and regulations so they'll know what to expect."

Your checklist should be nearly complete by now. But, how do you make the final decision? As you look at each school's name on your list, think back to your personal visits. Which of the schools left you with only a so-so feeling? Pull those schools out of the competition first. You might call the presidents of the remaining schools to tell them you are deciding between their school and one or two others. Ask if there are any final items they want to mention about their school to help finalize the decision. You never know what an owner may offer to entice you to train with that school, but it always is worth asking. Usually, though, the better facility will stand out in some way.

One item Watzke was not ready for in her selection of a professional flight school—"They work you very hard and some students don't expect that. I thought this was harder than graduate school—but the training was very good." After following through on her decisions, Watzke now flies as a Comair EMB-120 Brasilia first officer. Those decisions, for her, resulted in the job that got her out of her rut.

Questions to Ask before You Choose a Flight School

- Did the school allow you to fly with an instructor for an hour?
- Did it have any housing? Did they show it to you?
- What about insurance, what are the terms?
- Did you see the aircraft's logbooks?
- Will the training curriculum direct you to the job you want?
- If affiliated with an airline, does that mean a guaranteed job with that airline, or just an interview?
- Is there an affordable place to live at the school or near it?

- Does it offer a four-year degree in conjunction with the flight training?
- Is it a FAR Part 141 school or does it operate under Part 61?
- Does it offer or recommend someone for financing?
- What are its prices?
- How many hours will a person need before getting the licenses and ratings they seek?
- Is everything neat and clean in the main offices?
- Is there a decent facility to house the school?
- Do the instructors seem interested when spoken to?
- Do the personnel answer questions freely?
- Are the aircraft recent models?
- Are the classrooms well lit and roomy?
- What is the pass/fail ratio and how did the school come to those figures?
- Were you allowed to speak to current students as well as graduates?
- Are there any complaints lodged against the school with the Better Business Bureau?
- What condition are the aircraft in?
- Does the school have simulators?
- How old are the simulators?
- How long has the school been in business at its present location?
- How much will the training cost be up front?
- Do you pay all the money up front or just half?
- Can you pay each month or at the end of each lesson?
- What is the school's refund policy?
- Does the school keep its instructors busy?
- What is its policy shout changing instructors if you don't click with the one assigned to you?
- What is in the school's training contract or agreement?

Training Standardization

One of the toughest parts of making it to the big time for many pilots is the transition from flying small, single-engine aircraft the way their flight instructors taught them, to piloting the fast, sophisticated aircraft that most corporations or airlines use. It's not that some are better pilots than others, but more often than not, it is a factor of training that separates the successful pilots from the unsuccessful. Why?

It's simple really. When you start flying a large turboprop or a jet, the qualification process involves performing the same kinds of airwork you did in a C-172—stalls, steep turns, NDB approaches, and even emergencies. Many pilots fail simply because they have no idea of how to perform the maneuvers. And many companies will not take the time to explain it all to you. They expect you to know what to do.

What follows, then, is an example of an airline training manual that explains precisely how this company expected its pilots to fly the company's aircraft—both in revenue service and during training. Learning the maneuvers procedure by procedure made learning the actual handling characteristics much easier and considerably less stressful. As you'll read about from one pilot, a lack of preparation in flying large, complex aircraft could cost you your job.

EMB-120 Brasilia, Flight Crew Standardization Manual

This manual and training curriculum has been implemented to develop high standards of training and coordination amongst all flight crew members.

Flight crews are required to observe these procedures for the continued safety and comfort of our passengers in daily line operations.

Each crewmember will be trained exactly to the same performance standards to ensure total conformity between all crewmembers.

The following material is a guide to the procedures, maneuvers and methods involved in line flying, flight training and flight checks. Specific maneuvers and standards are indicated for particular types of flights, as dictated either by company policy or by Federal Aviation Regulations.

The description of procedures, maneuvers and recommended techniques in this section complement the material contained in the Normal, Abnormal and Emergency section of the Aircraft Flight Manual (AFM). Pilots should be completely familiar with and operate in accordance with material in this section. Flying pilot is abbreviated as FP and Non-Flying pilot as NFP.

Flight maneuvers are to be flown to the following criteria for minimum proficiency:

Ability to maintain altitude (+ or −) 100 feet.

Ability to maintain airspeed within: (+ or −) 10 KIAS

0 to +5 KIAS engine failure climb

0 to +5 KIAS on approaches

Smoothness and coordination of turns

Constant degree bank

Recovery of turns within 5 degrees of predetermined heading

Precision Approach

Localizer and glide slope within one quarter scale

Nonprecision Approach

Localizer or VOR within half scale

RMI or bearing pointer + or −5 degrees

Tolerance on MDA +50 to −0

The Captain is always responsible for the aircraft, completion of checklists and cockpit procedures, regardless of who is flying the aircraft. Duties pertaining to aircraft operation may be delegated to the First Officer, but final authority is vested in the Captain.

Checklists

The checklist will be consulted at all times during ground and flight operation. On the ground, the First Officer will read the checklist using either the Challenge and Response (C&R) method or a participatory method. In flight, the nonflying pilot (NFP) will accomplish the checklist.

It is vitally important that we maintain good crew coordination by means of continuous communication, so at all times the Captain and the First Officer are fully aware of what is going on in the cockpit. This will greatly enhance the primary objective of our daily operation—SAFETY.

Aircraft Preflight

The First Officer normally will assume the responsibility of accomplishing the aircraft preflight. Both the cockpit checklist and external checklist will be utilized to full extent. The First Officer will check that the maintenance release has been signed on the Aircraft Flight Log (AEL). The logbook will then be signed by the First Officer upon completion of his preflight. Any problems found will be reported to both maintenance and the Captain. If the aircraft is deemed airworthy by the Captain, he will then sign the AFL and the logbook will be placed in the aircraft. All other paperwork should be checked including: Airworthiness certificate, Registration, Radio license, P.O.H., M.E.L., 135 Manual, Weight and Balance Info., and all aeronautical charts and plates.

Visually check the exterior of the airplane for its condition, security of access panels and for signs of damage. Check for any liquids on the ground that may indicate engine, hydraulic, or water system failures. Check all areas visible from the ground. Emphasis should be placed on go/no-go items such as NWS (nosewheel steering) oversteering pin, oxygen and fire bottle discharge discs, defuel switches safety-wired, etc. Ensure removal of landing gear pins.

Cockpit Preparation

The first flight of the day, the Captain and First Officer will complete the turnaround prestart checklist by the Challenge and Response method. Thereafter, the Captain or First Officer (at the Captain's discretion) will perform the checklist alone, it is still acceptable, however to complete the checklist by Challenge and Response if desired.

Engine Start Procedures

Each pilot will start his/her respective engine. Under no circumstances (online) will a start be initiated without ground marshaller approval. Normally, the #2 (right) engine will be started first, followed by the #1 (left) engine. Always visually clear the propeller area prior to start initiation, and ensure that all doors are closed prior to starting #1.

Taxi

Captains should practice taxi with and without nosewheel steering. NWS failure can be simulated by the IP holding the NWS disengage button depressed.

Ensure that the aircraft is not moved until removal of attitude flags and presentation of attitude is displayed on EADI's (electronic attitude indicator).

Set power levers to GND IDLE, release the parking brake then smoothly advance power to allow the aircraft to roll forward, then reduce power to GND IDLE.

Control taxi speed with the power levers in the beta range.

Rudder pedal steering is used while taxiing where small directional changes are required. After completing a turn and before stopping, center the nosewheel to relieve the tire loading.

Scan instrument panels and observe instruments for normal indications, including the magnetic compass.

Both pilots should verbally clear the propeller and wing area, at which time the captain will advance the power sufficiently to start the aircraft moving, then retard the power to ground idle for the initial turn away from the ramp area. Condition levers are set to minimum rpm. Limit Np to 65%. Brake smoothly whenever necessary, then release brakes to allow cooling.

Turns should be limited to the total steering limit of 57 degrees. Rudder pedal steering is available to 7 degrees nose wheel deflection. Locked wheel turns should be avoided as they cause excessive loads on the gear.

Before Takeoff

After leaving the gate, the Captain will call for the pretakeoff checklist. The checklist will be read by the First Officer, and the Captain will respond to all items pertaining to him. i.e.: altimeter setting, flight instruments, V-Speed review, etc. Preferably the crew will complete the checklist prior to reaching the end of the runway.

Notes:

A. Taxi speed will be slow (Jog). This not only enhances safety on the ground, but helps save brakes and tires and enhances passenger comfort.

B. During periods of lengthy delays, single engine taxi is permitted in order to conserve fuel. However, both the "starting engines" and "after start" checklists will be utilized any time an engine is shut down and restarted.

C. Crew briefing: The following items will be covered by the flying pilot:

1. Type of takeoff (normal VFR or IFR, crosswind, etc.)

2. Procedure to be followed in event of engine failure prior to V1, after V1 and after takeoff.

3. Procedures to be followed for any emergency with respect to weather, gross weight, mechanical failure, airport facilities, runway length and condition, etc.

Note: The Captain will always review the departure clearance regardless of who is flying.

V1. Takeoff: General

Takeoffs may be broken down into normal, engine out, crosswind, day or night, rejected, instrument (lower than standard) or any combination thereof.

Normal Takeoff

The Captain will align the aircraft with the centerline of the runway. The NFP will hold the ailerons to compensate for the crosswind conditions. The flying pilot will smoothly apply takeoff power to both engines and maintain directional control with the rudders. Before starting the takeoff roll both pilots will verify the takeoff clearance.

The flying pilot will:

Advance the power levers to 75% torque

Command set power at 75% torque

The nonflying pilot will:

Set the final takeoff power by 60 KIAS and call power set

Monitor the flight & engine instruments

Call out any malfunctions or abnormalities

Call 80 knots (FP cross checks his airspeed indicator)

Call V1

Call "Rotate"

Call Positive Rate of Climb (with positive indication on VSI and altimeter

Additional callouts as briefed

The use of nose wheel tiller steering after the aircraft is aligned is not recommended unless rudder pedal steering is inadequate.

Approaching V1, forward control pressure can be reduced to just maintaining nose wheel contact. At Vr smoothly rotate (approximately 2 degrees per second) to a positive angle of attack (10 degrees) then make pitch adjustments to attain V2 +10 KIAS or 10 degrees nose up maximum.

As soon as the aircraft is definitely airborne and climbing, the gear is retracted. Runway or noise abatement heading should be complied with initially. Climb should be made at a speed of V2 +10 KIAS or higher as necessary, limiting deck angle to 10 degrees or less. Accelerating through V2 +20 KIAS and out of 400' AGL call "Flaps Up" Establish a climb at 150-180 KIAS. If a close-in turn is required flaps will remain at 15 degrees until the aircraft is established on the assigned heading then proceed as 400'. At

1000' AGL the NFP will call "1000 feet." The FP will call "Set Climb Power, After Takeoff Checklist," at which time the NFP will comply.

Note: The after takeoff checklist will be completed in its entirety with the exception of calling company operations. This should be accomplished only after climbing through a safe altitude and away from any terminal area. During operations in a terminal area, both pilots should be monitoring ATC and devoting full attention to safe operation of the aircraft.

Use of Aircraft Lights

The anticollision lights are to be on at all times when engines are running. The strobe lights are to be used at all times in flight unless they cause distraction (i.e., in clouds). Taxi and/or landing lights will be used at the Captain's discretion, and below 10,000 unless cruise alt. is lower, especially operating in terminal areas. Use of wing ice inspection light(s) are at Captain's discretion. Logo lights are also a valuable anticollision tool and should be displayed at all times in terminal areas at night. Light bulbs are easily and readily replaceable. In effect, if you feel that you need lights, by all means use them! Aircraft lights are the most effective anticollision tools outside of your own eyes. Flight crews are expected to practice "See and Avoid" at all times.

Power Settings

Takeoff power should be maintained to 1,000 feet AGL, at which time the flying pilot will reduce torque to 84% (or T6 not above 720 degrees) and the nonflying pilot will reduce prop rpm (NP) as required.

Instrument Takeoff

The following procedures will be followed when executing an instrument or lower than standard takeoff:

No rolling takeoff will be approved. Align aircraft with centerline and position heading bug on lubber line heading (EHSI).

After positioning aircraft on runway, also check EADI for level reference mark, make initial power runup and stabilize.

Continue takeoff using normal procedures: Initiate T/O roll, NFP call out 60 Kts.—airspeed alive, monitor power settings and call V1, Vr.

At Vr rotate to +10 degrees pitchup EADI (reference mark +10 degrees). At 400 feet AGL minimum, lower nose to 7 degrees EADI and accelerate to 150 kts, while ensuring a positive rate of climb and constant heading.

NFP will call out deviations from target pitch, airspeed and attitudes.

Through first and second segment climb, maintain maximum takeoff power. At 1000 feet power reduction may be made, and normal procedures followed.

Emergencies During Takeoff. Takeoff is one of the most critical phases of flight. Therefore, crews will abide by the following procedures.

Any abnormal indication observed by either pilot will be immediately called and identified. i.e., Oil pressure right engine," etc.

Either pilot has authority to call for a rejected takeoff up to V1 by calling "ABORT! ABORT!"

The actions are to retard power levers to ground idle and apply positive forward pressure to control column. Using brake and reverse power (Beta), the FP will bring the aircraft to a stop. The Captain may, at any time, command control of the aircraft by calling "I have it." If the First Officer is flying, he will immediately relinquish control upon command.

If the takeoff is rejected because of a fire warning or other such critical circumstances, the aircraft will be brought to a complete stop, parking brake set, and condition levers placed at minimum rpm prior to further action.

If the takeoff is rejected due to a power failure of either engine, extreme caution must be exercised in use of reverse on the operating engine due to the yawing tendency which will be created. Maximum braking will be necessary, and again caution must be exercised to avoid skidding, especially when runways are wet or snow covered.

Crosswind Takeoff

Maximum crosswind component including gusts is 25 knots. (If braking is less than good, reference appropriate crosswind limits). The techniques required for this kind of takeoff are not much different than those used during the normal takeoff. The upwind wing will nave a tendency to rise and aileron deflection should be applied in the direction of the crosswind to keep the wings level. This deflection will not materially affect the takeoff performance.

Some forward yoke pressure should be applied and maintained to ensure positive nose wheel contact with the runway. (Avoid excessive pinning of nose gear.) Rudder deflection will maintain directional control with nose steering initially, and as speed increases, aerodynamically. Aileron input should be decreased as speed increases. The primary objective of aileron input during crosswind takeoffs is to KEEP THE WINGS LEVEL.

Engine Failure at or Above V1

If an engine failure occurs after V1 the takeoff will be continued. Use rudder to maintain directional control and aileron to counteract the roll. At Vr, rotate the aircraft initially to approximately 10 degrees pitch, and adjust aircraft pitch to maintain V2 speed. Retract gear when definitely airborne and climbing. Check Autofeather complete. If the engine has not feathered or is on fire, the memory items of the Engine Fire Checklist will be performed at this time. Continue to climb to 400 feet at V2. Level off and accelerate to V2 +20 retract the flaps, set maximum continuous power, and establish a climb. The Engine Fire Checklist should then be accomplished if previously started or the Precautionary Engine Shutdown checklist will be accomplished.

Note: Retracting the flaps causes the aircraft to sink. Compensate for this tendency in order to prevent altitude loss.

Note: The flying pilot will devote his full attention to the flying of the aircraft. The nonflying pilot will monitor the progress of the flight, retract the gear, check for fire and feather and make the appropriate callouts.

Caution: If an engine failure occurs, proper control inputs must be applied and maintained. All performance parameters are met even with the use of reduced power takeoff; however, if ground contact appears imminent, power on the operating engine may be increased even to the physical limit of power lever travel.

Consideration may be given to returning for landing on a single engine, proceeding to an alternate airport or attempting an engine relight.

Area Departure

The flight crew shall adhere to ATC clearances and use available navigation facilities and equipment according to established procedures.

Climb

Climb power is normally set after the aircraft accelerates through V2 +20 KIAS and flaps are retracted. Initial climb power setting of 720 degrees T6 may be used, reference the climb power setting charts as soon as practical. Maintain climb speed of 150-180 KIAS unless conditions dictate otherwise.

Single Engine Climb

Single engine climbs above acceleration height are made at maximum continuous power, until such time that the aircraft arrives at a safe altitude. At that time power may be reduced to max inflight torque until the aircraft accelerates to desired airspeed.

A safe altitude is determined by considering the airport traffic pattern, altitude assigned for radar vectoring, minimum sector altitude, MOCA, MEA, etc.

Turns

The pilot should be familiar with flight control inputs for coordinated turns in different configurations. This includes climbing and descending turns. The necessity for smooth and minimum control pressure use must be emphasized, especially at higher speeds and in turbulence.

Cruise

Care must be exercised in cruise not to exceed Vmo. Engine operation should be adjusted to maintain maximum or optimal cruise speed. Should a power split occur with power levers evenly matched, the power should be set according to equal torque values.

(A slight power split of up to $\frac{1}{2}$ knob width is not uncommon in engine installations, and from the company standpoint acceptable.

At all times, vigilance must be kept by both pilots to "See and Avoid." Remember, ultimately in or out of radar, this responsibility is yours! Per FAR 135.100, any operation "except normal cruise" below 10,000 feet MSL is considered the Critical Phase of Flight

Engine Failure En route

The single most important aspect of an engine failure en route is to maintain a safe airspeed. Power should be increased on the operating engine, and engine fire memory items should be accomplished until the propeller on the failed engine has feathered. After the propeller has feathered "Engine Fire or Precautionary Engine Shutdown Checklist" should be completed as appropriate.

Caution: At higher altitudes aircraft may be above single engine absolute ceiling. In this case altitude must be sacrificed to maintain a safe airspeed.

Descents

En route descents do not require the review of a checklist. However, the following procedures apply.

FP—Initiates descent, not to exceed Vmo.

NFP—Calls 1,000 ft. above assigned altitude, 500 ft. above assigned altitude, and target altitude. Will call deviation of plus or minus 100 ft. from target altitude.

Descent in Preparation for Approach—(Approach Descent) requires use of the Approach and Descent Checklist. The checklist will be called for by the FP within 30 nautical miles of the airport of landing and/or 15 minutes. In all cases, it must be completed prior to the final approach segment.

FP—Initiates descent and calls for approach and descent checklist.

NFP—Performs the checklist in command and response format.

During ATIS copying, or company calls, the NFP will always verify that he/she is "OFF FREQUENCY," meaning not monitoring or talking with ATC. Comm. 1 will be used for ATIS and company calls, Comm. 2 will be used for ATC as a company standard.

Crew Brief.

Approach Procedure—Will review appropriate approach plate

Any abnormal procedures

Altitude callouts

V-speeds (covered in Landing Data)

Missed approach procedures

Note: Items to be covered in an approach briefing (specifically)

Type of approach (ILS, VOR, NDB, etc.)

Calls for intercepting course ("Localizer Alive")

Calls for course deviation (+ or – dot or outside of 5 degree for NDB)

Call intercepting glide slope ("One dot to intercept")

Calls at FAF/Clocks started

Altitude calls of 1,000 feet to DH or MDA, 500 feet to DH or MDA; each successive 100 feet to DH or MDA

Runway in sight or missed approach

FP—Acknowledges checklist complete

NFP—Completes checklist and advises FP "Approach checks complete"

Landings

Maintain an approximate 3-degree glide slope on all approaches (VFR and IFR). (Always transition and cross-check from the electronic glide slope to VASI when available.)

Use correct approach profile and avoid low and slow/high and fast. Remember—a good approach leads to a good landing.

Land in the touchdown zone. Do not attempt the "First Turnoff."

Never position power levers aft of flight idle until the aircraft is firmly on the ground (all three gear!) When you are too high and fast—a go-around is the best choice. Never force a landing.

Retard power levers to ground idle smoothly prior to applying brakes.

Airspeed may vary in line operations for traffic requirements. Under no circumstances will airspeed limitations per P.O.H. (pilot's operating handbook) be exceeded.

Always tune navaids (localizer) to appropriate frequency for landing runway. Electronic glideslope will be followed to touchdown zone both IFR and VFR. If no localizer or glide slope is available, VASI will be followed (red over white).

If a flap setting other than Flaps 25 is to be used, it must be covered in the crew briefing and a new Vref calculated.

Radar Altitude is a good reminder of landing clearance, when cleared to land, set radar alt. to zero feet.

Parking and Engine Shutdown

The Captain will maneuver the aircraft into the gate following the ground marshaller's hand signals. (There must be a marshaller in order to park the aircraft.)

Turning off the taxi lights upon arriving in the gate area is a signal to the marshaller that he has assumed control to direct parking.

Upon coming to a stop at the gate, the Captain will set the parking brake and call for the "Shutdown Checklist." This checklist will be performed in the C&R method with the First Officer reading the action and the Captain responding.

Ensure that the flight controls are locked in order to prevent gust or jet blast damage.

Where a GPU or APU is not available, it is necessary to conserve battery power as much as practical. Crews should exercise good judgment regarding use of electrical power when on batteries only.

If the aircraft is left unattended, all doors must be closed and locked, switches in the off position, and if applicable, covers, plugs and gustlocks installed (overnight).

Engine Failures or Precautionary Shutdowns

Each emergency is unique; therefore, the circumstances prevailing at the occurrence cannot be predicted. Thus, standard procedures cannot be precise and well defined because priorities will vary according to circumstances.

The following notes are intended as a guideline to priorities in handling an engine failure.

Fly the aircraft "alter power" and configuration as necessary to achieve required performance.

Identify and confirm the failed or affected engine. In the event of a failure, the Dead foot, Dead engine rule applies in identifying the engine. Verify the engine with engine parameters (by both pilots). Generally, torque is the best indicator of an engine loss. In any event, Both pilots are instrumental in determining a failure, regardless of who is flying.

Carry out shutdown/fire vital actions per AFM. Standard terminology will be for flying pilot to first state "right" or "left" regarding switch or lever positioning. Example: Engine on fire Right (#2) Engine.

Crew Briefings

The first flight of each day will include a full briefing between pilots for both departures and arrivals. The briefing will be concise and specific for the circumstances of the operation. Avoid unnecessary details where both pilots already know the procedure. If either pilot does not fully understand the briefing, he/she will discuss the problem. The Captain will always review specific ATC clearances, but in all cases, the flying pilot will deliver the crew briefing.

Between two crew members who have not previously flown together, a predeparture briefing on Standard Operating Procedures should be given by the Captain prior to boarding the aircraft to ensure standardization of procedures.

When both pilots are fully conversant with the intended procedures under normal conditions (i.e., takeoff, landings), the items pertinent to each procedure may be shortened to "Standard Briefing." A full briefing will still be covered the first flight of the day.

Use of Checklists

The appropriate checklist will be utilized at all times. During ground operations, the First Officer will read and complete the checklist. All items necessitating a response from the Captain will be C&R method. Line Ups and After Landings should be accomplished silently by the First Officer.

Airborne, the NFP will read and complete the checklist once called for. The checklist will not be started until the Captain calls for it during ground operations, or the FP while airborne.

The "After Starting," "Line Up, After Landing" and "Shutdown" checks may be accomplished by flow pattern method. However, upon completion of the pattern, the checklist will be consulted to ascertain items completed. The pilot performing the checklist will always acknowledge "Checklist Complete" when finished with a checklist.

Do not become "checklist complacent." A wrong response or action must be corrected immediately.

The remainder of the standardization manual can be found in Appendix A.

Part 61 or Part 141?

Two sets of regulations apply to flight school programs: FAR Part 61 and Part 141. The first, Part 61 (whose basic pilot experience requirements were discussed in chapter 1) outlines the actual certification process for you to become a pilot. Part 141 is the regulation that tells a formal flight school—called an approved school—how it must operate, such as the requirements necessary for a chief flight instructor, how the school may advertise, and in what form the student records should be kept. Another advantage of the Part 141 school (besides the fact that it will be more formally organized) is that total times required before the flight test are less than if you were to pick up your license from a Part 61 operator. For instance, under Part 61, a private pilot must have a total of at least 40 hours of flight time before the flight test. Under Part 141, the requirement is only 35. For the combined commercial instrument rating program, Part 61 requires a total of 250 hours of flight time. Under Part 141 (Fig. 2-3), that requirement drops to 190.

So, then, is a Part 141 flight school better than the freelance flight instructor whose ad you might see in the local paper on the weekend? In all fairness to the Part 61 freelance flight instructor, and the Part 141 instructor, it can be pretty tough to say one is definitely better than the other. While the curriculum of the Part 141 school is certainly organized, it doesn't mean that a Part 61 operator is disorganized. While being able to pick up a private pilot license in 35 hours instead of 40 is a benefit at the Part 141 school, the statistics say that very few pilots complete training in 35 hours. The average private pilot picks up his or her license in about 72 hours total time. At the commercial, instrument-rating end of things, it might make a bit more sense, depending on how immersed in the training you really become. Realize, too, that while total times will be less, the price per hour at the Part 141 school will usually be higher.

If you're searching for flight schools that have been in business for a long time, you'll find that, for the most part, they'll be Part 141 schools. At

Figure 2-3. Part 141 flight schools offer formal classroom training. *Comair Aviation Academy*

a Part 141 school, you'll find yourself engaged in regular phase checks with other school instructors as a system of checks and balances on your education. You'll also have some assurances that the chief flight instructor needed a set amount of flight time under his or her belt before being put in charge of the school. A Part 141 flight school must even meet certain requirements for its pilot briefing rooms, as well as the ground-training rooms in which you'll be spending a lot of time. At a Part 141 school, the FAA pops in on a regular basis to be certain that the quality of the students is up to par, too. If you were engaged in any training to be paid for by the new Veterans G.I. Bill, the Veterans Administration would require that your training be conducted at a Part 141 school.

Why would anyone not want to attend an approved school, then? One of the first reasons could be that, in your location of the country, a formal Part 141 school doesn't exist. Another consideration is price, and that's often the most compelling reason people will choose a free-lance instructor. The reason the freelance price is usually less is simple business economics. A freelance instructor might only own one aircraft and possibly even rent an office for ground training. Of course, while there might not be all the amenities of the Part 141 school, the benefit comes at the end of the lesson with the lower overall bill. Recently, I happened across an ad for a freelance instructor while I was traveling in a major East Coast city, so I called up the teacher. She used an older model Cessna 172 that she didn't own. Another friend owned the aircraft and leased it to the instructor with the insurance. The instructor charged $60 per hour for the airplane and $20 per hour for her time. At the same time, a nationally known flight school charged $78 for a C-172, plus $25 an hour for the instructor. The bottom line is that the facilities are nice and the Part 141 schools are more regulated, but you certainly will pay for them.

Freelance flight instructors might also be a little tougher to locate than a Part 141 school. Try searching the classified ads in your local newspaper, and you'll probably see, "Private flight instruction. Your airplane or mine, $20 per hour, call John . . ." Make the phone call, just like you did at the Part 141 school; make the appointment to visit the aircraft and make your decision. In both cases, I'd ask for the names and phone numbers of a few recent graduates to speak to as references. While the privacy laws in some cases could make the operator a bit wary of giving out these items, any legitimate school shouldn't have any trouble putting you in touch with people who have used their services. If they refuse, I'd walk out.

While prices could be considerably cheaper for the freelance instructor, there are some potential problems you should consider. If the instructor is only equipped with one aircraft, it might be pretty tough to book that airplane for some of the long cross-country flights you'll need to complete your license. What if the aircraft has a mechanical breakdown somewhere along the way? A large school will simply switch you to another airplane while the freelancer might need to switch you to another day entirely while the aircraft is repaired.

Consider the difference. I've taken instruction in both kinds of operations, and I've also worked in both as a flight instructor. I once found that a simple cup of coffee drunk standing near the wing of a Cessna 150, listening to a freelance instructor tell me some of his adventures and discussing what I wanted out of my private license was enough to make me work with him. Later on, for me at least, when I started working on my instructor ratings, I found working at a formal Part 141 school fulfilled my needs better because I thought it was better organized.

Finally, reserve the right before you even walk through the door of the flight school, that if things just don't either look, feel, or sound right, you'll leave. There's usually going to be more than one flight school near your home, so make sure you like the one where you'll possibly be spending thousands of dollars.

How Much Is This Going to Cost?

Asking how much it will cost to become a professional pilot is a bit like asking how much it's going to cost to purchase your first home; it depends on how many extras you'd like. If you were to work with a freelance flight instructor in a Cessna 152 from zero time to your commercial, the 1998 figures in a major city might look like this.

250 hours of total flight time is required, of which at least 100 must be pilot-in-command and at least 50 hours of dual. Keep in mind that this price is based on FAA minimums with all training performed in the aircraft, as many Part 61 operators might offer today. But, as always, do not simply compare bottom-line figures when choosing your flight training. It's the overall value that counts and that is measured differently by everyone.

Commercial Pilot

240 hours Cessna 152 (aircraft rental only) @ $53 per hour = $12,720

50 hours dual-flight instruction (instructor fees) @30 = $1,500

10 hours Cessna 172 RG @ $81 per hour = $810

50 hours ground instruction @ $30 = $1,500

FAA checkride: $200 (approx.)

Commercial Pilot Certificate cost from zero time = $16,730

No commercial pilot certificate will hold much value for you without an instrument rating to go along with it. If you picked that up from a freelance instructor, the totals might look something like this. I'm assuming that most of the rating is conducted in the aircraft. Although the price may be less than a Part 141 school, I would search hard for an instructor who at least had access to a simulator, even a PC based training device that you'll read about in chapter 8. There are simply too many distractions in an airplane - noise, bumps, etc., to make it worthwhile to do all the training while airborne. A good simulator offers you a place to learn and practice the procedures you'll refine in the airplane. My estimate includes some simulator time.

Instrument Rating

40 hours in Cessna 172 @ $70 per hour = $2,800

25 hours of ground instruction @ $30 per hour = $750

FAA checkride = $200 (approx.)

Instrument Rating Total = $3,750

While the commercial certificate under Part 61 requires 250 hours, that does not guarantee that you will be able to learn all the maneuvers proficiently enough to pass the exam in that time. Most pilots take a little longer. If you look at the total Part 61 price of $20,480, it seems like a great deal of money, and it is. Another benefit, though, in working with a Part 61 flight school is that the owner might be willing to negotiate the rate down for you from the $20,480 total, if they know you'll give them this much business. This would not normally be something you could plan on accomplishing with the larger Part 141 operators. However, some of the large Part 141 schools might offer advantages that a smaller Part 61 operator might be incapable of, such as better classroom or simulator facilities.

Note: The price of the written knowledge exams is not included in the above costs because those can vary significantly from locale to locale.

The Knowledge Exams

All the pilot certificates and ratings I've talked about also require you to pass an FAA written exam, now called a knowledge exam, prior to the day you take your final flight test for each rating. One rating—the multiengine—doesn't include a written. These writtens have always been a requirement, but I personally question their value in deciding what kind of a pilot you'll become. I know of some pretty spectacular pilots who freeze up in a written test situation. Personally, I've never been good at them.

The writtens can be passed in a number of ways. You can buy any of a number of excellent books on each rating and study them on your own. You could sit down with an instructor for enough ground instruction to pass the test. You could also take a weekend ground school and spend 10 hours on Saturday and Sunday having the needed material crammed into your brain so you'll be ready for a Monday morning written exam. Another method, becoming more and more common today, is to buy a set of ground-school video tapes from any of a number of companies that sell them, like King or Jepp. The ads for their tapes can be found in many of the aviation magazines. The major benefits to this system are that you not only work on the ground school at your own pace, but you can also spend as much time on a subject as you need to make it sink in.

Here's a sample of the video tapes available from the King Schools to help you prepare for the FAA Knowledge exams. I've also used the King Computerized Exam Reviews to help me study questions I once took on my commercial, instrument, and ATP writtens. It's amazing how many times questions from these exams reappear during the interview process.

Written Exam Video Courses

Private Pilot

Instrument

Commercial, Flight Instructor

ATP-135

ATP-121

Flight Engineer

Instrument Instructor

Computerized Exam Reviews

Private

Commercial

Instrument

Flight Instructor—Airplane

Flight Instructor—Instrument

ATP-135

ATP-121

Flight Engineer

Call King at 800-854-1001.

Commercial Flight Schools

Today, large commercial flight schools can provide many more different types of services than they could just a few years ago. Sanford, Florida-based Comair Aviation Academy is just one of a number of large schools that specializes in just about everything a new pilot could need to point them in the right direction toward that eventual cockpit job. Comair Aviation Academy, a Part 141 school, provides, first of all, an accelerated professional pilot training program from absolutely zero time through the certified flight instructors ratings. The training is airline standards-based, which is reflected in academics, simulator, and flight instruction programs.

Comair's Air Carrier Training programs provide initial first officer training to Comair Airlines for pilots who meet the minimum requirements of 1,200 hours of total time, of which 200 hours is multiengine experience. The cost is $10,995. The Academy's Flying Service division is a Part 61 school that works with international students and people who cannot commit to the flight training environment on a full time basis. Although the Flying Service is certified to operate under Part 61, all students are trained to Part 141 standards.

Headquartered in Sanford, Florida, where the weather is great most of the year (could save you money by shortening your total training time!), the academy operates 90 aircraft at their Central Florida Airport location. Comair Academy flies the C-152, C-172, C-172RGs, and Piper Seminoles with the help of their 115 flight instructors. The school also uses three Frasca 141 and two Frasca 142 (multiengine) simulators, as well as two of the new PCATDs you'll be reading about in chapter 8.

The standard of training at all schools might obviously vary some around the Part 141 regulations, but how those variances take their form is the crucial point. The Comair Aviation Academy, owned by Comair Airlines, a Delta Connection carrier, has chosen to market itself as an airline training school. Right from the start, academy students are taught with manuals written just like those provided at Comair, the airline. The Academy's Training Doctrine and Training Course Outline are based on the central concept of standardization, as the term is understood and applied in commercial airline operations. This concept is reflected in training aircraft standards manuals and checklists, in construction and content of all courseware, and in the selection, initial training, and transition training of flight instructors and the details of operations. Even the most basic aircraft in the Comair training fleet, the Cessna 152, comes equipped with a Cessna 152 Flight Standards Manual, just as a first officer in a Brasilia would receive.

What's the point of all this airline consistency? Why should a prospective professional even consider a place like Comair Academy for his or her training? Most people need only to look at the chance provided to some of Comair's graduates of the school's Airline Qualification Course for their answer. The course, built around the Embraer EMB-120 Brasilia, is primarily designed for pilots with some experience who are looking for that first step up into the right seat of an airliner. Currently, minimum qualifications are set at a commercial certificate with multiengine land and instrument ratings, 1,200 hours total time, and 200 hours multiengine time. The course will qualify pilots as first officers in the Brasilia. Here's an opportunity to qualify for and get hired into the right seat of an airliner. Don't expect this to be a Boeing 727, however. It will be a turboprop regional airliner or possibly a regional jet operated by Comair Airlines itself. Comair strategic plan calls for the carrier to go all jet in the near future, replacing the Brasilias with an ever-increasing number of Canadair RJ Regional Jets. New first officers may find themselves moving up very quickly into the right seat of a jet rather than a turboprop when they complete the course.

While Comair Aviation Academy is not the only flight school around that will train you and later assist you with the interview and hiring process at a regional, they're unique because they're owned by one of the hiring airlines themselves.

A number of regional carriers charge pilots for their training. In some cases a third-party training organization, like LaGuardia, NY-based Flight Safety International, has assumed responsibility for an entire training department.

The process begins with an evaluation that might take the better part of a day to accomplish. At their own cost, the applicant flies to the training center and pays a fee of between $250 and $300 to fly a simulator, take some exams, and have a personal interview. If the pilot is found competent, and if one of the training organization's contracting airlines indicates a need for new pilots, the applicant might be offered a conditional position with the airline, valid if the pilot completes the training program successfully. The applicant would then pay for training in a specific turboprop aircraft, currently costing somewhere between $8,000 to $10,000 for the complete course. If successful, the pilot will most likely be hired by the airline they interviewed with. The majority of the risk, however, is the pilot's. If the pilot doesn't complete the training, the airline has lost nothing. At major airlines, however, the company is still paying for the training of its pilots. Whether or not this system is good or fair is really for the individual to decide. What the self-funded training has accomplished is the weeding out of financially disadvantaged pilots. If you can't pay, you can't have the job. Certainly there are training organizations that will finance the training, but your credit must be sufficient to carry the debt, often at a time when the school loans from the previous round of aviation training are yet to be paid off. Comair Academy also performs the initial screening exams for client companies such as Mesaba and Express 1 Airlines.

Certainly a school such as Comair Aviation Academy will move you closer to your goal of becoming a professional pilot in a short amount of time, but the $10,995 price is significant. Here's how the costs break down:

Ground School

	Hours	Rate	Total
Crew Resource Management	16	$17.98	$ 287.74
Basic Indoctrination	40	$17.98	$ 719.35
Systems Training	120	$17.98	$2158.06
Cockpit Training	10	$17.98	$ 179.84

Aircraft/Simulator

Training		
28 hours—14 PIC, 14 SIC—in an EMB-120		
Brasilia simulator	$ 831.25	$5,818.75
Checkride		
2 hours PIC & 2 hours SIC	$ 831.25	$ 831.25
Aircraft Checkride	$1000.00	$ 1000.00
Total Cost		$10,995.00

This price does not include housing or travel expenses. Contact Comair Aviation Academy at 800-U-CAN-FLY.

Pay for Training

During every airline hiring boom, thousands and thousands of well qualified pilots move closer toward what they believe to be the perfect flying job—an airline cockpit. As these airman move ahead to replace senior pilots retiring from the majors and nationals, as well as to support airline expansion, they leave behind them thousands of pilots' positions that must now be filled by qualified pilots who have not yet had the opportunity to make it to the majors.

For example, when a pilot moves to majors from a regional, someone gets a chance to move up to captain—perhaps someone who is currently a first officer. This change then, spells opportunity for the regional pilot. And as this soon to be captain is trained as a captain and allowed to fly the line, some junior first officer will move up in seniority to fill the vacancy of the new captain. This will also mean that the airline will probably hire a new class of first officers as well.

But while all of this moving around spells success for pilots, it spells red ink for the company they leave behind. The red ink comes from the extra dollars that must be spent to train that new pilot to make them ready to assume command.

How much it costs to train a new pilot varies significantly by the type of aircraft and the size of the company in question. To retrain a pilot to fly a Piper Navajo, will probably cost no more than a few hours of flight training and some ground school. Let's assume that for a small charter company, bringing someone on payroll and training them to fly single pilot in that Navajo costs the company $2000.

To train a new captain on a Gulfstream 4 to replace a pilot who left for the airlines could set a company back $30,000–35,000 once travel time to Flight Safety, hotels, and payroll are figured in—a pretty substantial sum. At the regional level, training a new class of captains might fall somewhere in between these figures.

Some of the overall cost can be absorbed at various levels because most airlines have a training department of some sort that will normally include ground and flight instructors whose job it is to make these new captains ready for the line to replace those who have left or are expected to. No matter how you slice it though, training costs can be a significant part of any aviation company's budget. And remember that small company who was only going to spend a few thousand perhaps to ready a new captain? While that amount may seem small compared to the others we've looked at, that small sum might well represent a sizable percentage more of that small operator's budget than for the G-4 operator.

As small airlines and corporate and charter operators watched pilots leave their employ—some less than a year after they began—a few decided to take action. While most realized they could not stem the tide of pilots leaving the company for better pay and benefits, as well as the lure of larger aircraft, they did understand the need to control their costs and demand a better return on investment (ROI) than they had been willing to settle for in years past.

One solution for some companies came to be known as pay for training. Now, let's be certain not to confuse this with the method you'll use to pay for your flight training, because the two situations are vastly different. Pay for training has also been called "Buy Your Job," by some, because only pilots with enough money to pay for the training were offered work, leaving many well-qualified pilots on the sidelines.

Essentially, an airline takes a look at your paperwork and decides to hire you. But before they actually extend a regular job offer, they demand that you successfully complete training in the aircraft that company flies—but you pay the bill.

Some new generation pilots who have graduated from college in the not to recent past may see this as an opportunity if they have the money to pay for their training. Indeed, some new pilots have found employment with regional carriers around the United States with unbelievably low total hours. One recent new hire at Chicago Express Airlines, the commuter carrier for ATA, based at Chicago's Midway Airport, had just under 300 total hours when he began training in the Jetstream 31 aircraft the company flies. He saw it as a chance to bypass the dues paying that most pilots have done in the past. He spent very little time as a flight instructor when he was interviewed with Chicago Express.

The company made a tentative offer of employment to this pilot with the understanding that it would be withdrawn if the pilot did not complete training successfully. A week later, this pilot was on his way to Flight Safety's St. Louis Training Center to begin ground school. Total cost to the pilot was about $9000. This pilot was successful and began flying the line for Chicago Express just a few months later.

You may be reading this and scratching your head trying to figure out what the problem is then. Guy or girl wants to get ahead, gets the training they need, and moves ahead. They get a bill just like going on to grad school, right?

Well, not exactly. One issue with pay-for-training is that—and this may not seem like much of a factor to some newer pilots—is that traditionally in the airline industry, pilots have not paid for their own training. When you reached a point in your career when an airline gave you the thumbs up, they paid for the training.

Pilot qualification is also a factor in pay-for-training. During the recession of the early 1990s, thousands of well-qualified pilots were put out of work as carriers like Eastern, Midway, and PanAm ceased operations. While many of these pilots were qualified to fly aircraft that many of the regional airlines were flying, they could not be offered a position unless they were willing to write a check to pay for their training—often in an aircraft that they were already qualified in.

So why would an airline want to put a pilot in the right seat of an airplane when they have just four or five hours of time in it, when a pilot who has a few hundred or even a few thousand hours in that aircraft cannot get the same job. Quite simply, it's all economics. Paul Berliner, President of the Berliner-Schafer Aviation Consulting Group says, "Pay-for-Training is also an ethical and a moral issue." Berliner is a Boeing 727 captain for a national airline.

Many regional carriers began outsourcing their entire training departments to companies such as Flight Safety—this has not been much of an issue with the larger jet carriers normally, although a few do require a pilot to pay for training like Vanguard—they looked upon the surface as if they were simply contracting out for a service that was too expensive for them to provide. This was really not the case however. If you look through any of the magazines that offer flight training services—Flying

Careers, Airline Pilot Careers—you'll see advertisements for dozens of companies that provide type ratings in large aircraft, like the Boeing 737 for example, for a price.

The intrigue becomes more visible as you work some simple cost analysis figures. If you can purchase a type rating in a 737 for $6500, why should it cost nearly $10,000 to train you to right-seat standards in a Jetstream 31 (a 19-seat turboprop) where a newly trained Jetstream pilot leaves Flight Safety with no type rating to show for his or her efforts, although the company will type-rate some pilots for an extra fee? Berliner comments that "A doctor takes his or her diploma with him or her after he or she completes his or her training. In a pay-for-training environment, the pilot walks away with nothing." If the pilot is later furloughed from the job he or she has paid for, he or she may get some help from Flight Safety to find a new position, but there is no guarantee.

So where is all the extra money that pilots pay for their training going? Certainly Flight Safety is entitled to a fair profit for the work they are performing upon the part of the airlines that contract with them. But what is a fair profit when you compare what a pilot receives versus what he or she spends? There can be only two places that the solid profits of pay-for-training are going—either as a bonus to Flight Safety or back into the airline's pocket itself. Pay-for-training then has transformed what was a cost center at some airlines, to a profit center at others. That simply is not right when you consider that these airlines don't make other professionals pay for their training in order to work there.

Berliner adds a note of caution too, that pilots who see this as a quick way to climb the ladder should consider before they sign on the dotted line. "There is backlash from the pilot community at the majors (about pay-for-training). On every pilot hiring committee that I know of, there are pilots who are vehemently opposed to pay-for-training. They don't believe it is fair for a pilot to bypass the traditional dues-paying method of gaining flight experience because they have money." Each and every applicant who gains their job through a pay-for-training environment must ask themselves whether they want to go into an interview at a major airline knowing they are blackballed to some extent. "Even a one-percent chance is not worth it," according to Berliner.

But the economy is changing and with it—perhaps—the way some companies view pay-for-training. While many pilots would like to believe the change has evolved through some sudden moral insight at the airlines who use it, the law of supply and demand has simply begun to run its course. As more and more pilots are being brought into the major and national carriers, the supply of well-qualified pilots willing to pay-for-training has dropped beneath the level that some carriers can use to still keep their operations running—despite low-time pilots who are willing to pay-for-training.

Significant changes to pay-for-training have occurred at USAir Express carrier, Commutair (highlighted elsewhere in this book), and Atlantic Coast Airlines. Even Continental Express, the regional airline for Continental Airlines, a company that tried to make pay-for-training palatable for some by agreeing to cover training costs for pilots with more than 2000 hours total time, has now reduced—but not eliminated—the requirements to qualify for free training. Effective October 1, 1998, all Continental Express applicants with more than 1,200 hours total time and 250 hours multiengine, will no longer have to pay for their jobs. Berliner adds that "our staff believes that all pay for training schemes will be completely eliminated within 12 months." And may pay-for-training never rear its ugly head again.

So how can companies recoup their training costs from pilots who were not staying with a company long enough after training to make it all balance. What we are left with as pay-for-training begins to make its exit, are training contracts. These contracts—while not forcing a pilot to pay for the training out of their own pockets before they begin to work—do require a pilot to sign a legally binding document that asks for money in return should a pilot leave employment before a specific number of months after completing training.

A Citation 3 captain, might be asked to agree to remain employed with a company for at least 12 to 18 months after that company pays for the type rating. Some contracts are worded in such a way that the pilot is denied any right to legal recourse and normally contains no provision for getting out of the contract once it is signed, no matter what happens. Many would force a pilot to pay for his or her training in full, if he or she left one month before the agreement was set to expire, with no proration clause available. Legally speaking, most training contracts have been found to be unenforcable in a court of law and of more cost to prosecute than the company stands to gain if they win.

But the policy of asking—or forcing—pilots to buy their jobs still exists. Watch for it when a job opportunity looks a bit too good and avoid it at all costs. It can be dangerous to your career!

This Letter to the Editor appeared in the November 9, 1998 Aviation Week & Space Technology. It was called "Profits on the Pilot's Backs." It read, "I'm glad my employer—Continental Airlines—has had its 14th quarter of record profits. In the meantime, the 1,287 pilots at Continental Express—who will soon be flying the first all-jet regional—are in their 18th month of working without a contract. The majority were required to pay a $9,280 "training fee" to get a job that started at $13.25 per hour. Any wonder how this company can be so profitable? Name Withheld by Request.

Ab Initio Training

Another program you might consider is the ab initio system. Basically, the ab initio system takes a pilot from absolutely zero flight time up through the right-seat job in a regional airliner or, at the very least, all the professional pilot certificates, including the various flight instructor ratings. (See Figs. 2-4 and 2-5.) Although the expense for a flight training program such as this is considerably higher, the benefits are usually a flat price for all the ratings as well as a specified time period for completion. In Europe, this concept of training someone from the beginning in one location with a specific curriculum is used by major airlines such as Lufthansa. What the airline eventually ends up with when a student pilot completes the training (which often takes years) is a pilot who is totally immersed in this particular airline's methods of operation. The program also provides the airline with a certain amount of employee loyalty.

One young Italian pilot I recently met came to the United States as a part of his country's ab initio training because the price to train a pilot in the United States is considerably less than comparable training in Europe. At the completion of his training (with a total of about 700 hours, much of it logged here in the United States), this Italian pilot returned to his homeland, where, with a final few months of training, he picked up a type rating in a Hawker-800 and then began flying in the right seat of a DC-9.

Figure 2-4. The Beechcraft Bonanza and Baron are often used in some ab initio training. *Beach Aircraft Company*

Figure 2-5. See Fig 2-4. *Beach Aircraft Company*

While a great deal of controversy surrounds the ab initio training from an operational status, there's no doubt in my mind that being totally involved with your career, be it medicine or airplanes, is the way to go. From an operational standpoint, many pilots don't believe a relatively low-time pilot belongs in the right seat of a high-performance turbo prop or pure jet aircraft. They believe that only actual experience, total logged hours, can really indicate a pilot's ability to cope with a difficult situation aloft. We've only to look at the military operations of the United States for another view of this controversy. In the United States, pilots with only a few hundred hours of total logged time are out flying high-performance supersonic fighter aircraft by themselves. Obviously, then, low-time pilots can become productive members of a cockpit crew.

The only problem with ab initio training is the price, but even that is relative, because again, value is the most important aspect of training, not simply the cost. Let's take a look at Comair's version of this concept. This information should help you when you research other schools about their programs too. Keep in mind that a flying club may also offer a great compromise between the standard freelance instructor and a regular flight school. One club in Chicago offers a Cessna 152 for just $37.50 per hour, a C-172 for $47.50 per hour and a C-172RG for $54 per hour.

Table 2-1. Professional pilot course at Comair Aviation Academy

Private pilot single and multiengine land

Instrument airplane

Commercial pilot single and multiengine land

Flight instructor: Airplane single and multiengine land & instrument

Ground instructor: Basic, advanced, and instrument

Flight time	Hours	Rate per hour
Cessna 152	48	49
Cessna 172	42	72
Cessna 172RG	5	95
Piper Seminole	29	160
SE Simulator	20	57
ME Simulator	7	72

The complete package (as of December, 1998) comes to $22,156. After students complete their training at Comair Aviation Academy, 85 percent of the school's graduates are hired as flight instructors. Comair Aviation Academy guarantees each graduate an interview. In fact, the only people who can be flight instructors at Comair Aviation Academy are graduates of the school. As they progress in responsibility, the Comair instructors build their flight time, both multiengine and single engine; they also build their bank accounts because they're now employed in aviation. The goal is to reach the magic numbers of 1,200 total time and 200 multiengine. When each student arrives at that point, they're guaranteed an interview as a potential first officer with Comair Airlines, definitely one of the benefits of attending a school owned by an airlines. Comair Academy Vice President, Susan Burrell, reports, "Ninety-seven percent of the students we send for interviews at Comair are hired. We give the airline a proven product in our pilots. They know the standards we train them to." Additionally, The University of North Dakota and Western Michigan University (see their Web pages listed in chapter 8) offer ab initio pilot training.

There are definite differences in the curriculum of any school, as well as differences in price. You'll have to be a good shopper and spend the time to check out everything one school offers in comparison to another. But remember, too, that price shouldn't be your only deciding factor. Location can be important, as will the kind of housing you'll have while you're there. When you've pared your decision down to just two or possibly three schools, I'd take the time to visit the campus and speak to some of the people who run the school as well as to some of the students. Price and paperwork are certainly going to give you some direction, but again, only an on-site visit can make the final decision for you (Figs. 2-6 and Fig. 2-7).

But FAA ratings aren't the only items necessary to become a professional pilot today. While a pilot's position used to be reachable without

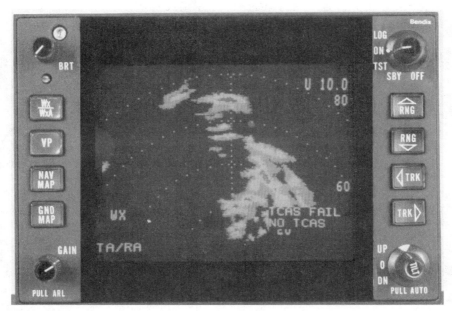

Figure 2-6.

Figures 2-6 and 2-7. Good training involves learning how everything in a modern cockpit works. *Bendix/King*

the benefit of a college degree, that sheepskin is effectively a requirement in most segments of flying. Today, many airlines still list a college degree as encouraged, while other airlines and many corporate operators list the degree as a requirement. Either way, if you intend to compete with others in the aviation game, you're going to need a degree. There are many fine state universities such as Purdue, University of North Dakota, University of Illinois, Embry-Riddle University, and others that combine a standard four-year degree with programs specifically designed to offer the student all the professional flight ratings they need to make their mark in the profession.

It's a tough call whether or not picking up your ratings on the side while you complete your college degree is less effective toward the goal of flying for a living than attending a four year university that also happens to run a flight program. Chief pilots I interviewed varied considerably from only caring whether or not the pilot held the ratings plus the degree to asking what the degree was in. Many corporations are currently asking about the degree itself, so a pilot can become a more useful member of the corporation during the time they aren't flying. Some chief pilots did seem to hold a soft spot for pilots who attended the same school as themselves, at least long enough for the new applicant to get his or her foot in the door. After that, the applicant needed to compete on the same footing as everyone else.

Financing

Most careers offer the student not only the chance at a lifelong job that they'll hopefully enjoy, but also a method of spending substantial amounts of cash in a relatively short period of time on the necessary training. Where do students look for the financing they need to cope with the big bills that accompany this kind of training? The best place to search for money initially is to talk to the school you intend to train with. If you're planning to work with a local freelance instructor, the chances of picking up the financing you need are pretty slim. If, however, you approach the people at some of the larger flight schools or universities, they should be able to steer you in the right direction. While Comair Aviation Academy is certainly dedicated to the high-quality products and services they sell, they must stay in business to be able to offer those services, something that might be pretty tough if they had no way to help their customers pay their way. I doubt Sears or Wards would try to sell you a washer and dryer without offering you a way to finance the purchase.

Most local banks offer student loans normally backed by federal financing. Whether or not the rates are competitive will be a matter for you to decide. Because of the variables in financing (like interest rates and repayment plans) shopping for a loan can become just as big a project as finding the school in the first place. Depending on the state of your finances, be ready for the fact that many financial institutions might not be willing to make you a loan without a cosigner to guarantee repayment of the money. Regular state universities will offer tuition financing through the financial aid office on campus, so the best advice on financing is to first locate the school you're interested in, and then talk to their financial aid counselor. Remember, too, that while the lowest monthly payments might at first look to be the best way to finance your education, a slightly higher payment or a slightly shorter repayment period could shave thousands of dollars off the total amount of money to be repaid. Be sure to ask about these possibilities.

If you head to your local library's reference desk, you'll find a number of good books available to locate financing for your educational venture, so don't just depend on your local financial or educational institution for help. One school we interviewed was willing to finance a pilot's flying, but at annual rates near 14.5 percent, considerably higher than a regular student loan. People actually took them up on a regular basis because of one reason . . . the financing process was relatively quick and easy to complete. Many students, the financial adviser admitted, never even asked about the interest rate, being more concerned about how long the repayment period would be and just when the payments would actually begin.

Another possibility could be scholarships. A nice bunch of people at the University Aviation Association put together a 52-page directory of all the information you'll need about scholarships for college. The price of the "Collegiate Aviation Scholarship Listing" is currently $12.00. Write the University Aviation Association, 3410 Skyway Drive, Auburn, Alabama or call 334-844-2434.

The New G.I. Bill

If you're a veteran, you might be eligible for government assistance that could pay as much as 60 percent of the cost of adding additional flight ratings. As with any government program, there are requirements and possible red-tape delays, so if you think you might qualify, contact your local Veteran's Administration office as soon as possible. The Montgomery G.I. Bill is considered the new G.I. Bill because it's only available to veterans who served on active duty in an armed force of the United States after June 30, 1985. You must also have made deposits to the G.I. Bill program at the rate of $100 per month for the first 12 months of that active duty. To begin G.I.-Bill-qualified training, you must hold a private pilot certificate and meet the physical requirements for a commercial certificate. In addition to the school you select being FAA approved, that school must also meet the VA school requirements. Just being FAA approved doesn't necessarily mean the school is VA approved, either.

Internships

Internships are a career path that's open to you, but there aren't a great many of them, nor are they easy to win. Internships in most careers have been around for years as a method to expose young people to a possible career track while at the same time giving a potential employer a chance to see how the young person performs in the real world.

Most internships are arranged during the third or fourth year of college, as is the one we'll take a look at here: the pilot internship jointly arranged with United Airlines (Fig. 2-8) and participating colleges (see the following list). At United, this program, officially called the Flight Officer College Relations Program, has been running since 1983. A student enrolled at one of the participating universities fills out an application. If, after interviews, the applicant is accepted, he or she takes a full semester off from school to work at one of United's facilities. In the fall of 1998, United Airlines had 22 interns on board around their system. United's Instructor Training Manager and the intern program's director, John Bauserman said, "We interviewed 60 students to come up with 22 interns. To apply, a student must be in his or her senior year of college, already hold a commercial pilot certificate with an instrument rating, and have a grade point average of about 3.0 on a four-point scale. In the past, about 250 interns of the 400 who have applied, have successfully joined United as pilots. One student I interviewed did his internship at Washington, Dulles International Airport (IAD).

Intern and now Boeing 737 first officer Arnie Quast began working in flight operations at IAD. "I had a chance to work closely with the pilots and get a good idea of what it was like to be a commercial airline pilot,"

Figure 2-8. The United internship program could lead you right to the cockpit.
Robert Mark

Quast said. But what made Quast's internship truly valuable was that, "I had the chance to jump-seat on all the United aircraft that flew from Dulles and observe firsthand what crew coordination was all about. I also had the opportunity to fly many of the 27 United simulators in Denver," he said.

After graduation, Quast continued to build his flight time as a flight instructor while he waited for the call from United. With just over 1,000 hours total time and just 25 hours multiengine experience, Quast received a United class date for the Boeing 727. He attended the United Training Center in Denver and found the training to be top-notch.

Before you sign up for a particular university program, check for information on these intern programs. A phone call to some of the other airlines will reveal whether or not they're involved in some type of internship program too.

United Airlines Flight Operations College Relations Program Participants

Arizona State University
Mesa, AZ 85206-0903

Central Missouri State University
Warrensburg, MO 64093

Daniel Webster College
Nashua, NH 03063-1300

Delaware State University
Dover, DE 19901

Embry-Riddle Aeronautical
University
Daytona Beach, FL 32114

Embry-Riddle Aeronautical
University
Prescott, AZ 86301

Florida Institute of Technology
Melbourne, FL 32901-6983

Kent State University
Kent, OH 44242-0001

Louisiana Tech University
Ruston, LA 72172

Metropolitan State College
Denver, CO 80217-3362

Middle Tennessee State
University
Murfreesboro, TN 37132

Ohio University
Athens, OH 45701-2979

Ohio State University
Columbus, OH 43210-0022

Parks College of St. Louis
University
St. Louis, MO 63156-0907

Purdue University
West Lafayette, IN 47906-3398

San Jose State University
San Jose, CA 95192-0081

Southern Illinois University
Carbondale, IL 62901-6621

St. Cloud State University
St. Cloud, MN 56301-4498

University of Illinois
Champaign-Urbana, IL 61874

Utah State University
Logan, UT 94322-6000

University of North Dakota
Grand Forks, ND 53202

Western Michigan University
Battle Creek, MI 49015-1682

Alpha Eta Rho, a Professional Aviation Society

Alpha Eta Rho is a collegiate fraternity founded to bring together those students having a common interest in the field of aviation. Started in 1929 at the University of Southern California, the fraternity has grown to more than 70 chapters nationwide, pledging nearly 1,000 new members each year.

The fraternity serves as a contact between the aviation industry and educational institutions. It bands outstanding students, interested faculty, and industrial leaders into one organization for the purpose of studying the problems of everyday life as influenced by this modern industry—aviation.

Alpha Eta Rho serves to actively associate interested students of aviation with leaders and executives in the industry. This close association, strengthened through the bonds of an international aviation fraternity, establishes opportunities for all members to inspire interest and cooperation among those in the profession who are also members of Alpha Eta Rho (Fig. 2-9). Alpha Eta Rho continues to grow, and it serves as a lasting tribute to the farsighted understanding and vision of its founder, Professor Earl W. Hill.

Philosophy and Goals of Alpha Eta Rho

- To further the cause of aviation in all its branches.

- To instill in the public mind a confidence in aviation.

- To promote contacts between the students of aviation and those engaged in the profession.

- To promote a closer affiliation between the students of aviation for the purpose of education and research.

The constitution of Alpha Eta Rho declares that the success of aeronautics depends on its unified development in all the countries of the world and on the cooperation of different phases of aviation with each other. Alpha Eta Rho affirms its character to be international and declares that

Figure 2-9. Alpha Eta Rho members can be found in many different kinds of aircraft. *Learjet Corp.*

eligibility for membership is not dependent on race, religion, nationality, or gender.

The address for the Alpha Eta Rho national office is:

Palomar College
1615 Gamble Lane
Escondido, CA 92029
attn. Kent Backart

Profile: Steve Mayer, First Officer, Boeing 757

Northwest Airlines

"Flying was always a part of my life," says Steve Mayer. "My dad is a former military pilot and now flies a 767 for Delta. Before that he was a pilot for National and PanAm."

Mayer was a keen observer of events around him as well as a pilot who listened when a more experienced voice offered advice. "On my 20th birthday, I looked at my dad, a 727 captain at the time. One of his friends, however, who had never gone through the military, was a 747 captain.

The 747 captain was making a hundred grand a year more than my dad."
Mayer decided to forgo the military route to an airline job.

"I really didn't start flying until I was 18 or 19 because I just didn't
have the money to do it," Mayer recalled. "A pilot told me to get my
commercial as soon as possible and I started when I was in college. I
worked a lot every week while going to school and got my degree in
political science."

It was after college that Mayer again listened to a more experienced
pilot who told him that the best way to an airline cockpit was with a ton
of multiengine PIC time under his belt. "I decided to hop an airplane to
the Virgin Islands because I heard they were always looking for pilots
there. I arrived one night with my backpack and nowhere to sleep and
just walked up to the first hanger I saw. I asked if they needed any pilots
and some guy said, 'Come back at 4 AM and you're hired.' I showed
up and got the job, delivering newspapers, bank checks, fish—anything
you could put in the Aztecs and Aero Commanders we had. I worked
there for two years, logging nearly 2800 hours, most of it PIC. I think I
worked seven days a week actually."

Although the work was excruciating at times, Mayer had a plan to
get himself through the long days. "I had a theory it was just a time
builder. After I left this company, I got hired into the right seat of a
Jetstream 31 for Express Airlines 1 and stayed there two years but never
upgraded."

But during his time as a regional pilot, Mayer says he "applied to
every airline imaginable, here in the United States and even internationally. I sent Northwest an application when they were not even
hiring. I filled out the application in pencil, believe it or not. I just
wanted to get into their database. Then out of the blue, Northwest
called me. Later, during the interview with Northwest, they asked me
about the penciled application. I just told them the truth. I also got
the job."

My advice to applicants is that the "airlines don't care what kind of a
four-year degree you have, just as long as you have one. Avoid the fancy
flight-school colleges. I have friends who are still paying off school loans
and will be for some years to come. I thought maybe I was making a mistake by just trying to build PIC time anyway I could, but I'll be a captain
here at Northwest when some of my friends are just being hired. But I
also paid the money to get some simulator time before the interview at
Northwest and it was worth it. I also bought a subscription to Air Inc. to
get a leg up on any information I could. Getting an interview at the airlines is like hitting the lottery, so you have to be prepared." (See Figs.
2-10a and b.)

Figure 2-10a. 727. *Courtesy Brian Schiff.*

Figure 2-10b. Thunderstorm. *Courtesy Brian Schiff.*

Is the Military a Possibility?

Since the winding down of the Cold War in Europe, the United States military has also entered what some experts call a slow-growth mode. This actually has become the same kind of downsizing that corporate America began experiencing in the early 1990s, with thousands less personnel being kept on active duty and many military bases around the world being shut down or reduced in size. What this change to the military means to you, if you're considering the military as a possible place to pick up your flight training, is best outlined in a story from *Aviation Daily*.

Military Pilot Pool Will Shrink under DOT Cuts, Industry Told

The nation's airlines, which traditionally recruited 65 percent or more of their pilots from the military services, will face a shrinking supply later in the decade due to Defense Department budget and manpower cuts and as the services entice more of their pilots to remain on active duty, Senator John McCain (R-AZ), ranking member of the aviation subcommittee, warned yesterday. In the near term, however, airlines will reap a bonanza as the military downsizes in the wake of the breakup of the Soviet Union, McCain yesterday told a committee looking into possible future shortages of pilots and maintenance technicians.

James Busey, DOT deputy secretary, told the panel, which is expected to issue a report in about a year on what measures can be taken to ameliorate possible shortages, that a shortfall in either category could have a profound impact not only on the industry, but on the nation as a whole as far as its ability to compete on a global scale. "We're about to lose an important source of trained pilots from the military. The balance has now shifted and the supply will not be there," Busey said.

Due to the slowdown in the air transportation industry, it's now getting only about 45 percent of its pilots from the military, a figure that could get lower as the military pilot supply dwindles, said John Sheehan of Phaneuf Associates, the consultants to the panel that's chaired by Kenneth Tallman, President Emeritus of Embry-Riddle. He said air carriers will need about 2,400 pilots per year for the next ten years and that regional airlines will need about 28,000 pilots over that period. . . . The panel was originally to be a joint military-civil effort, but the downsizing military services aren't as concerned as they previously were. This will prove ominous for the industry later in the decade as a prime source of pilots disappears, both Sheehan and McCain said.

If you plan to try to enter the U.S. military at some time in the future, you need to know that there will be fewer flying jobs and tougher competition for those remaining cockpit positions. Certainly you'll need to talk to your Armed Forces recruiter when you're seriously ready to make this decision about your career because the volatile situation of the American military community could change at any time.

Let's look at the United States Air Force, for example. Because all Air Force pilots are officers, you'll need to apply and be accepted at Air Force Officer Training School (OTS) before anything else can happen. The U.S. Air Force Officer Training School is small and highly specialized. Candidates selected to attend OTS are college graduates. OTS prepares them for positions of responsibility so they can lead the Air Force of tomorrow. The school's motto, "Always with Honor," reflects the ethical and professional standards expected of Air Force officers.

OTS is a fast-paced, three-month course. You complete it on-campus at Lackland Air Force Base in San Antonio, Texas, studying communication skills, leadership, management, military history, Air Force customs and courtesies, world affairs, and more. You take part in organized sports and physical conditioning to develop your confidence and teamwork abilities. The course is designed to aid your transition into the Air Force way of life.

To be eligible for OTS, you must be a United States citizen, 18–29 years of age, in good health, and able to pass a physical exam. You must be a graduate of an accredited college or university and have excellent moral character. You must also score well on the Air Force Officer Qualifying Test. To enter a technical or nontechnical career, you must be commissioned before age 30. Pilot or navigator training applicants must pass a flight physical and enter training before age 27 $1/2$.

Pilot training selectees without a private pilot's license attend the Flight Screening Program before going to OTS. This four-week course identifies your potential to complete undergraduate pilot training. Training time includes cockpit time in some single-engine aircraft like the T-3A Firefly that is used to evaluate a potential candidate's talent at coping with the traditional flight environment during the Air Force's Flight Screening Program. Although the T-3A is the aircraft preferred by the Air Force to conduct this evaluation, these tests are being conducted by civilian flight schools in some location such as Colorado Springs because of a temporary grounding of the T-3A (the Air Force hopes to return the aircraft to flight status soon). If you have already hold a private pilot certificate when you apply for undergraduate pilot training, the screening exam is waived. (Fig. 2-11.)

Training time includes cockpit time in a single-engine Cessna 172, called a T-41 (Fig. 2-12) by the Air Force. You'll need a solo flight to

Figure 2-11. B-2.

Figure 2-12. USAF flight-screening candidates fly a T-41. *USAF*

complete the Flight Screening Program. If you already have a private pilot license, Flight Screening Program is waived.

After OTS, you'll begin a one-year, intensive flight training program (Fig. 2-13). (See chapter 7 for more details on Air Force training.)

Once accepted for OTS, you'll enlist in the Air Force in the rank of Staff Sergeant. The Air Force pays your way to OTS as well as your flight training costs. While in training, you'll be paid nearly $500 every two weeks. You live on base and eat in the OTS dining hall. You pay only for personal items such as laundry, postage, and telephone calls. A clothing allowance helps you defray the cost of uniforms. Upon graduation, you're commissioned a second lieutenant. OTS graduates who attend undergraduate pilot training incur an additional Air Force commitment of 8 years after training is complete.

To learn whether you might be eligible to compete for a military flight training program, contact the information service of the branch of your choice at the numbers listed in Table 2-2.

Figure 2-13. Successful USAF pilot training progresses to the T-38. *USAF*

Table 2-2. Information service numbers

Air Force	800-423-8723
Air Force Reserve	800-423-8723
Navy	800-327-6289
Army	800-872-2769
Marines	800-627-4637
Coast Guard	800-424-8883
Air National Guard	800-424-8883

FLYING FOR THE MILITARY

I recently spoke to a young Air Force pilot, 1st Lt. Michael Fick, about some of his experiences during his Air Force flight training since he graduated from school. I think his answers about the training will tell you how viable this part of aviation is as a career field, although you certainly need to check into each different branch of the service for its particular requirements (see chapter 2). In 1983, the Air Force enlisted 1,590 new pilots. In 1993, that number had dropped to 700. The Air Force reports that of that current 700, 35 percent will be flying bombers and fighters while the other 65 percent will serve out their tour of duty in airlift/transport aircraft.

Q: Why did you choose the USAF, Mike? A: I had a great desire to continue my education after high school and very little money to do it with. I didn't want to be burdened for years with school loans, so I decided to apply to the U.S. Air Force Academy. My flight experience was limited prior to applying to the Academy, but my desire to fly was strong. I was turned down for the Academy my first year, but received a partial Falcon Foundation Scholarship to New Mexico Military Institute. This prep school provided me with the chance to prove my continued interest in the Academy. I graduated at the top of my class a year later and won an Academy slot.

Q: Do you plan on making the Air Force a career? A: Right now I do. But the final decision depends on how well the Air Force treats me in the future, much like any other job.

Q: Would you tell me a little about Air Force flight training? A: Sure. The training really began for me, though, at New Mexico Military Institute, where I learned about the attitude necessary to be a professional. That became even more firmly set at the Academy. My actual flight training began in the summer between my junior and senior year at the Academy, when I went through the required preliminary flight training in the T-41 (Cessna 172). Performance in the program, both my G.P.A. and M.P.A. (Military Performance Average) allowed 125 out of 600 students to apply for the Euro NATO Joint Jet Pilot Training at Sheppard AFB, TX. Forty were selected. The program ran 13 months, where standard Undergraduate Pilot Training ran 12 months. A good majority of the training revolved around fighter tactics.

Q: What aircraft did you fly in training? A: I flew the T-37 Tweety for the first six months and the T-38 Talon (Fig. 2-14) for the second six months.

Q: What effects have the military cutbacks had on your career? A: Due to the cutbacks there were very few fighter assignments when I graduated from training. Most were for tankers, transports and instructor pilots. Some of the pilots were taken out of flying for a few years first, before being given their assignments. I had the C-130s, C135s, C-12 or T-37 Instructor pilot jobs to choose from. I chose the C-12 (Beech King Air 200) assignment at Andrews AFB, MD, just across the river from Washington National Airport. This assignment should be followed by one in the C-141.

Figure 2-14. Air Force student pilot and instructor plan a T-38 flight. *USAF*

Q: What did you enjoy most about training? A: I think the two- and four-ship formation flying provided the most serious adrenalin rush, as well as two-ship low-level, where I flew at 420 knots about 500 AGL.

Q: Have you thought about your flying once you return to civilian life? A: Most definitely. One of the driving factors because I couldn't get into fighters was the flying experience that will transfer to civilian life. I'll have turboprop time, in the C-12, as well as multiengine jet time in the C-141. And the C-12 time is in some of the busiest airspace in the country to boot.

As I end this section, I'd like to give you a little food for thought in case your plans have wandered towards flying outside the United States. If you've not given that any consideration before, perhaps you should.

About to Leave Military Service?

Since so much of this book is devoted to pilots sharing their experiences with other pilots, this message sent by my friend Derek Martin, an ex-U.S. Navy pilot and now a first officer for Southwest Airlines should be of immense value. Derek was passing on some advice to his old boss about what to expect when he left the Navy and began searching for a job as an airline pilot.

Hi Skipper,

Got your message—yeah, life is good at the show (the airlines). Unfortunately, I have good news and bad news.

First, the bad news—if you're retiring in the fall, you're behind the power curve by about six months (minimum) if you want to have a job waiting for you. The time is now to flood the market with your resumes and/or applications, and I mean everybody. This includes carriers that you have no real intention of working for, for it's going to be the interview experience that you'll have under belt that's going to pay big dividends when United, Delta, or American come calling. They're probably not going to call for 6–12 months after you put your app in, so use that time by practicing the art in interviewing with others.

Currency—an issue that depends on the carrier. Most look for being recent in complex airplanes (i.e., jets or turboprops), but flying regularly in a Cessna or Piper is the next best thing. Most carriers look not only at your total time, but also at your time in the last six and twelve months. Allowances might be given if you last tour was flying the proverbial desk. My advice is to join one of the flying clubs and go out and start flying ILS approaches with an instructor to get you going on flying in the civilian world. There is a difference.

Now the good news—your timing is still good. Don't even think twice about the age factor. In my class at Southwest, our oldest guy was 52 and there were 4 military retirees plus two 15-yr retirees. American just started hiring and all the others are in full swing. It's not going to be the heyday of the mid to late 80s, but the numbers are still really good. It's all a matter of doing the work to position yourself for the opportunity when it comes calling.

From my limited perspective, the most difficult thing you're going to have to do is break yourself out of the cocoon of comfort you've spent your whole career building. You've still got a job to do in the Navy, but you're going to have to put yourself on a war footing that things are going to be tough and difficult, and that you're going to do whatever it takes to get you (and your family) to the promised land. The wife and kids are going to have to realize that some big sacrifices are going to have to be made as far as time on your part is concerned. The stress level is going to go up the closer you get to retirement and will not abate until you get hired. At this point it will be replaced by training, IOE (Initial Operating Experience), and your off-probation check-ride stresses. Include her in the process, have her quiz you with flash cards, let her run you through some mock interviews if she's good at that. This will go a long way in maintaining peace and harmony.

Where to start? Join Air Inc. as step one, and buy the full package ($160–200). The binder and the monthly newsletters are chock-full of good gouge, as well as resume boilerplates and a bunch of other odds and ends.

Next buy yourself a copy of the FAR/AIM and study, study, study. Parts 91 and 121, and the whole AIM is where to concentrate. Make flashcards and study in the car or wherever.

Study your aircraft flight manuals. They're not going to ask you to draw a schematic of the S-3's electrical system, but they will expect you to relate an aircraft system to them, usually in the form of an emergency that you experienced.

Learn how to read Jepps charts and plates. In the sim ride if they offer you a choice between Jepps and DOD plates, take the Jepps.

Get on CompuServe or AOL—it is one big gouge machine. In the airline forums you will find a ton of information on what's going on, what the latest interview was like, etc.

Get a subscription to *Aviation Week & Space Technology* and save every article or blurb on any company you've got an app/resume in with. Prior to any interview, you can catch up on current events. Nothing will impress an interviewer more than when you know more about his company than he does.

Read the book HARD LANDING, I believe by Thomas Petzinger, Jr. This book is required reading as far as I'm concerned, for anybody wanting to get to the show. It's a well-written history/business lesson on the workings of the airline industry. Once you know the environment, then the craziness of this industry will be a heck of a lot easier to understand.

Interview suit: dark navy, white shirt, black wingtip shoes, muted tie with maroon in it. It always amazed me how many guys showed up for interviews with their dream airline "out of uniform" and no haircut. The interview was somewhat predictable after that.

The art of the interview. All the gouge in the world doesn't mean anything if you're not ready for the interview. Practice and real experience will lower the stress level, but for most carriers (or companies for that matter), the delivery of the answer is just as important as the data itself. What most captains are evaluating when they sit across the table from you is "could I spend 8–12 hours a day, 3–4 days at a stretch, over the course of a month with this guy?" Frame your answers in a situation-action-result format. Lose the acronyms unless you know your audience. Be yourself. They understand you're under a lot of pressure and they want to see how you handle it. If you don't know the answer to a technical question—say so and where you would go to find the answer. Irv Jascinski has a somewhat dated book on airline interviewing, but the gouge and mindset that's needed are still valid.

A word on app fees. It's pricey, it's expensive, and it doesn't seem fair. Pay the fee and get on with your life. Last week I had a Captain jumpseat with us from the commuter I flew with between the Navy and Southwest. I asked him if he had an app in with American, and he replied no because he thought the $100 app fee was ridiculous. Well, I think the $100 app fee is stupid too, but I'm not going to forego a career with a pretty stable airline flying some pretty good equipment making a pretty good salary over a $100 principle issue. It's nothing more that an investment that could pay some pretty good dividends. Look at it that way and the check will be a lot easier to write.

Flying the commuters. Don't you let anybody tell you that a military-trained pilot is better than a civilian-trained pilot. I flew for WestAir/United Express, and some of the best and most professional (and most fun) pilots I ever saw were captains I flew with, most of whom were my age or younger. The pay stinks and the lifestyle might be bordering on horrible (at least for an FO), but the education and experience I gained were priceless and I wouldn't trade it for the world. How good or professional a pilot is is completely up to the individual, no matter what his training or background might be.

Remember, every dollar you spend on your career transition: buying flight time, Air Inc, subscriptions, AOL, your suit, dry-cleaning that suit, etc., is all tax deductible under professional expenses.

Finally, what to do if it's Day 1 of retirement and the phone's not ringing. It's easy to say, "Don't worry," because you will and that's OK. I'm a firm believer in making your own luck. Every effort you make, no matter how small or insignificant, will eventually bear fruit. Some rewards happen right away, others will take time to manifest themselves, but if you want it—it will happen.

Hope this helps and good luck Skipper!

No matter what route you use to gain your ratings, realize that everyone else started out pretty much the same way toward their goal of becoming a professional pilot: sweating it out on long cross-country trips building the flight time for a commercial certificate, or under the hood with a view-limiting device over their heads as they worked towards that instrument rating. There's no easy, quick way to a job in this industry. But all the work is worth it. You might not really believe that during all the tough months of training, but the first time you step into the cockpit of an airplane as a required crew member, you'll know that all the work was worth it.

3
Ratings
What's Really Involved?

Understanding what ratings you need, as well as how you'll pay for them, may seem like your only goal right now, but you can't forget that there is also a considerable amount of work necessary on your part to pass the exams necessary to win those ratings. But the training process can at times be overwhelming to people who are unaware of just what they're getting themselves into. Choosing the right flight instructor is crucial here, for without a strong one-on-one relationship that allows the student to question or ask for and receive regular guidance, all the hard work necessary for these ratings will become more difficult. Let's take a look at the ratings, then, piece by piece, and give you an insider's look at what really makes them tick.

The Commercial Rating:
When to Begin

In their quest toward a professional pilot career, most people begin with the commercial pilot certificate. Perhaps they begin here because this rating holds the first glimpse of things to come; it's usually the most time-consuming of the ratings, and some pilots want to tackle the toughest first. Personally, I enjoyed the commercial training more than just about any other. It seemed more like play to me than work, but then many professional pilots look at flying, any kind of flying, as anything but work. The commercial pilot certificate training takes you from the rank of novice, the private pilot, to the rank of someone who is really beginning to understand what makes an airplane fly in many different kinds of situations. The commercial pilot training could begin right after you receive your private license, but I encourage you not to go that route.

Training for the private pilot certificate takes a great deal of time and energy. By the time most pilots receive their certificate, they've logged an average of 65 to 70 hours total time, with considerably more on the ground. This is when you should take a training break. Go fly for 25 or 35 hours around the local area and on a few cross-country trips. Before you bog yourself down with more flight training, classroom instruction, and written exams, have some fun learning to use the skills you've spent many hard months learning. You need to experience the joy, the fun of flying, before you become embroiled in too much work. Total cost is another advantage of putting some time in after you've received your private license. If you're working in a nonapproved school again, you'll need 250 total flight hours before you can be recommended for the commercial exam. With a special exemption, an approved, Part 141 school will only require 190 hours. If you pick up your private at 65 hours and start right in on the commercial package, you'll end up spending dual instruction prices for time leading to the requirements that you could have logged solo. If you start the commercial a bit later and have logged 25 hours on your own, that's 25 × $23 per hour (the instructor's fee), or nearly $600 you could save. That's a pretty hefty savings in anyone's book. After you have about 90 to 100 hours or so, sign up for the commercial course with the school of your choice (Fig. 3-1).

Figure 3-1. Much of the flying for a commercial license can be in a Cessna 172. *Comair Aviation Academy*

The PTS

Before you're through with aviation, hopefully many years down the road, you'll have run up against more acronyms than you care to even think about. Here's one you're going to become very familiar with on the road to becoming a professional pilot. PTS, or Practical Test Standards, is the label given to a small book, one for each and every certificate and rating the FAA offers. These books describe exactly what subjects you'll need to know to pass the flight test for the particular certificate or rating you seek, in this case the commercial. You must be able to perform and will be asked to perform each and every task set forth in the PTS. That might sound pretty straightforward right now, but when people were picking up ratings a few years back, the booklet in use was the FAA's flight test guide.

The problem with that publication was that the examiners could pick and choose which items they wanted to test you on. They were required to choose some, but not all. Unfortunately, as happens with human nature, some examiners picked the same maneuvers each time they gave an exam, and the word eventually got out. If you were weak on NDB approaches, find Fred; he never asks you to fly one anyway. Herb never cared about Chandelles on a commercial, so if you had trouble with those, you'd do your darndest to locate Herb to give you the test. To say the least, this made the entire test procedure rather unfair and at times much too easy to predict, which meant that sometimes people of various degrees of skill were awarded the certificate.

With the establishment of the PTS, much of the ambiguity of the testing procedure was eliminated because each applicant was told, long before the test, that they would be tested on everything in the book. This actually made things a great deal easier, even if somewhat more complex, because if the students could perform all the maneuvers to the required standards, they were prepared for the flight test . . . period. The FAA says this about the PTS concept . . .

> FARs specify the areas in which knowledge and skill must be demonstrated by the applicant before the issuance of a rating. The FARs provide the flexibility to permit the FAA to publish practical test standards containing specific tasks in which pilot competency must be demonstrated. The FAA will add, delete, or revise tasks whenever it's determined that changes are needed in the interest of safety. Adherence to provisions of the regulations and the practical test standards is mandatory for the evaluation of pilot applicants.
>
> An appropriately rated flight instructor is responsible for training the student to the acceptable standards as outlined in the objective of each task within the appropriate practical test standard. The flight

instructor must certify that the applicant is able to perform safely as a pilot and is competent to pass the required practical test for the rating sought.

The examiner who conducts the practical test is responsible for determining that the applicant meets the acceptable standards as outlined in the objective of each task within the appropriate practical test standard. This determination requires evaluation of both knowledge and skill because there's no formal division between the "oral" and "skill" portions of the practical test. It's intended that oral questioning be used at any time during the practical test to determine that the applicant shows adequate knowledge of the tasks and their related safety factors.

Additionally, people complained a number of years ago about some of the methods examiners used to determine just how qualified an applicant was. The main complaint was the examiner's use of distractions during a flight test. Some applicants believed it was unfair to toss in little questions or problems during a time when the applicant needed to concentrate on more important things. In the new PTS, the FAA addressed this issue . . .

Numerous studies indicate that many accidents have occurred when the pilot's attention has been distracted during various phases of flight. Many accidents have resulted from engine failure during takeoffs and landings where safe flight was possible if the pilot had used correct control technique and divided attention properly.

Distractions that have been found to cause problems are: preoccupation with situations inside or outside the cockpit, maneuvering to avoid other traffic or maneuvering to clear obstacles during takeoffs, climbs, approaches, or landings.

To strengthen this area of pilot training and evaluation, the examiner will provide realistic distractions throughout the flight portion of the practical test. Many distractions may be used to evaluate the applicant's ability to divide attention while maintaining safe flight. Some examples of distractions are:

- Simulating engine failure.
- Simulating radio tuning and communications.
- Identifying a field suitable for emergency landings.
- Identifying features or objects on the ground.
- Reading the outside air temperature gauge.
- Removing objects from the glove compartment or map case.
- Questioning by the examiner.

This is what the examiner will be looking for when he or she gives you the flight test for the various ratings. Let's take a look at the commercial PTS and what's required to be successful at the test. As you can well imagine, a commercial pilot certificate allows you to carry passengers or cargo

Figure 3-2. There's a PTS for helicopter flight tests too. *Bell Helicopters*

for compensation or hire. Another way of looking at the commercial is a rating that allows you to make money for flying. But, besides the ability to make money at flying, the capture of a commercial certificate for your wallet means you've begun to look at aviation in a whole new way—professionally. The maneuvers you'll be asked to perform as well as the subjects you'll learn are designed to offer you a significantly higher level of knowledge of flying, which you're required to demonstrate on the flight test. You'll find, too, that even though some of the maneuvers might appear similar to those you learned for your private ticket, the tolerances you'll be expected to fly to are much tighter (Fig. 3-2).

Profile: Warren Cleveland, Regional Jet Captain, Mesa Airlines

"My dad used to fly remote-control airplanes when I was a kid," remembers Warren Cleveland. "That was my first exposure to flying. He always wanted to fly but never had the opportunity before he passed away."

The younger Cleveland took the money from his fourteenth birthday and bought a flying lesson from a flight instructor in Baton Rouge, Louisiana. "I started lessons on a pretty limited basis. You could see years pass on one page of my logbook," Cleveland added. He got a job at 15 washing airplanes and had much of his salary credited to a flying account at the FBO he worked for. He soloed at 17.

Looking back on it, he recalls that "I really didn't have a clear plan of how to get my ratings. I wished I had a mentor. I went to college because someone told me I'd need it to be a pilot. I even applied for Navy ROTC but was turned down. I really hated college." But he did meet someone who had enrolled in a flight program at San Juan Community College in Farmington, NM. It was Mesa's Airline Pilot Development program.

Cleveland spent some time talking to the people there and felt at ease early on. "I felt they had a more organized view of how to help me get where I wanted to go. They told me that 100 percent of their grads were getting jobs. Embry-Riddle was just too expensive for me."

He enjoyed the New Mexico program from the beginning. "We began flying brand new "A" 36 Bonanzas with HSIs and RMIs. It was sweet! I picked up all my ratings and even completed a turboprop transition course before I graduated. You got 10 hours in the right seat of a Beech 1900 and 10 more watching your partner fly. If you had a "B" average at graduation, you were guaranteed an interview with Mesa. I got hired and flew as a first officer in the Brasilia for 18 months but could not upgrade because I wasn't old enough (you must be 23 to hold an ATP). But eventually, I made captain on the Beech 1900 and then the Brasilia. Now I've just upgraded in the Regional Jet. As it turns out, I was the youngest RJ captain in the world when I upgraded."

And how does Cleveland like his new position flying a jet at FL410? "It's really hot stuff! Mesa's new management team, headed by Jonathan Ornstein, is really great too. He communicates with the pilots. I think

Mesa is on track to being one of the top regionals in the country. It's a great place to work."

Cleveland offered aspiring pilots a few insights. "In training and in the hiring process, attitude is everything. You'll get more opportunities with a good attitude than anything else. Show enthusiasm, make lots of contacts, and never and never, never burn your bridges behind you."

Commercial Subjects

I'm not going to cover every single subject on the Commercial Flight Test; I'll cover just those that I think I can offer some particular insight into. As in the private pilot test, the commercial pilot will be asked to prove that the aircraft is airworthy by displaying and being able to discuss the various sorts of paperwork involved in making the aircraft legal to fly. This includes a possible discussion of the Minimum Equipment List (or MEL) for your aircraft, not to be confused with the FAA-required minimum equipment for day and night VFR and IFR flight. You'll also be expected to be able to prove that you're legal to conduct the flight test, right down to being certain that the information on the FAA Form 8710-1, the Application for Airman's Certificate, is correct. One problem that almost caught one of my students recently was the location of the ELT in his new Bonanza. During basic questioning at the preflight, I asked the student where the ELT was located. When we looked in the small door in the rear, where the switch to arm or turn off the ELT was located, the student found nothing. He couldn't prove that the ELT was there. It took a mechanic with a flashlight to finally locate the device for us. Better for me to ask than to have been caught on a flight test with this question, so be certain you don't just listen to your instructor; take an active part and make certain you really understand what's being taught. Learning to fly is no place to be passive.

In every aircraft I've ever checked out in, I've always found the systems test to be the most difficult part of the new exam or, in this case, the commercial flight test. When it comes to a systems test, the key to passing this section is to not just understand that the landing gear, for instance, is hydraulically actuated, but also to understand the troubleshooting aspects of the system. What happens when the system falters? How do you lower the gear when the hydraulic pump fails? What's the backup if the alternate system fails? If you can't lock the left main, but the right and nose are down, should you land on two or bring them all up? If your landing gear is electric, take the time to learn more than simply the emergency gear extension. If the discussion turns to electrical, know what a voltage regulator is and where it's located on the alternator assembly. Realizing what kinds of electrical problems you can cope with is important for the oral, but so is being able to tell the examiner when your system has run out of options.

When it comes to weather, the examiner will be looking for an ability to read prognostic and radar charts as well as sequence reports and forecasts. The key here is being able to tell the examiner the trends from what you read. I don't mean forecast the weather for the next 12 hours, but be able to relate what you're reading with how that weather will affect the flight you're planning, in terms of routing, winds aloft, altitude selection, fuel required, and icing considerations. Sure, you might be able to tell the examiner that rotation about a low is counterclockwise, but on a cross-country trip, what does that mean regarding the weather and the changes to that weather that you'll encounter? When will you encounter a head-wind or tailwind, when will the pressure change, and which way will the altimeter move? (Fig. 3-3.)

Another area that often causes a great deal of stress is emergencies. This shouldn't be such an unusual reaction because emergencies, by their very nature, are anxiety-producing situations. I've found that, for the commercial, the most common emergency is still total or partial engine failure.

Figure 3-3. The use of automated weather reporting systems is becoming more common. *NOAA*

(Most instructors will talk about an engine failure right after rotation in a single engine aircraft, but they shouldn't be even simulating this event below 500 feet above the ground.) Know the memory items of your aircraft checklist and be able to recite them while you're being distracted with something else because that's most likely how the engine will die. It could quit in a steep spiral or during an eight around a pylon. Be able to tell the examiner what you'll do if the door pops right after rotation. I remember the time I learned how to show the examiner a simulated in-flight fire recovery in a BE-55. When he yelled fire, I chopped both throttles and almost rolled the airplane on its back as I headed down toward the ground. He quickly grasped my respect for the crisis.

When it comes to cross-country planning, you actually get a break. Although you can expect to spend a portion of the oral explaining how you determine fuel burn and proper altitude and whether or not there are NOTAMs current for your route, the big factor here is how to plan for the unusual. You'll spend half an hour looking up data for Fred's Airpark and the examiner will look it over and say, "Yes, but now I want to go to He Haw International."

Find all the airport information and explain what you find. Before your flight test, be sure your instructor has quizzed you on every single symbol, color change, information box, and navaid symbol. It's not good enough to only point and say that's a VORTAC; you must understand what it means. So, here's the break. You don't actually have to fly a cross-country trip like you did for your private. You just have to be able to explain it.

During ground operation questions, be ready to tell the difference in colors between the fuel. Know that no color could be jet fuel in your tank instead of avgas. Be ready to explain just how frost might actually form and how you should remove it, as well as what you shouldn't try if the contaminant were ice (like beating it off with the handle of an ice scraper, which could dent the leading edge). Do you taxi and pay attention only to taxiing, or do you try to write ground control instructions or ATIS while you're moving? Both of these would show some serious judgment problems. Before you take off, how well do you organize the charts you use? Do you put them somewhere where you can easily reach them but they won't blow around on takeoff? Do you have a pen close by as you fly?

On your pretakeoff check, you'll be expected to demonstrate a professional attitude toward the checklist. If there's any one thing I see students slip on, time and time again, it's checklists. Too often, students read the item and look, but they don't actually check to see if the proper operation was actually performed. For example, "altimeter. . .set." They look, but they don't actually change the setting in the Kohlsman window.

When it comes to radio and ATC communications, you'll be expected to handle the radio like a pro. That doesn't mean you never miss a call or always understand what air traffic is trying to communicate, but it does mean that when you only catch part of the call, you take the proper action to repair the situation, like asking ATC to, "Say again!" That doesn't sound tough, but you'd be surprised how many pilots just sit there when they missed something on the radio or look over at the examiner and say, "Did you catch what they said?" Not a good idea. If you live near class B, C, or D airspace, be absolutely ready to explain when and how you'll request and enter the airspace. One student was recently caught when the examiner asked if the ATC facility repeating his call sign for entry into a class C meant he could keep going toward the airport. The answer is yes. When asked if it worked the same way in a class B, the student said sure. Wrong! Without teaching you the airspace designations, know what happens if one type of airspace overlaps with another? (I already know. You look it up.)

When it comes to maneuvers, I think there are really only a few techniques to use to perform each one. When you're performing eights around pylons, you either understand how to compensate for the wind or you don't. You either know where the reference point is supposed to be on eights on pylons or you don't. What's probably the most important part of the commercial preparation is that there's simply no substitute for plain old practice. Too many students want to rush it as soon as they meet the requirements. Whether it's crosswind takeoffs and landings or eights on pylons, you must get out and practice if you intend to display competence on the exam. Personally, I'd be out practicing a few hours, three times per week before the test, to become proficient enough to pass the flight test. Finally, in this discussion of the PTS there's one more simple concept, at least it seems simple to me. Know the tolerances of the maneuvers. For a lazy eight, know that when you're selecting your reference point, as the maneuver progresses you should never drop below 1,500 AGL, or that the altitude tolerance is plus or minus 100 feet at the 180-degree point, or that airspeed really must be within 10 knots at both the 90 and 180-degree point. The commercial license is the level where you're expected to fly professionally. Finally, If you don't understand what's expected of you at any point during the exam . . . ask!

Sometimes, too, you may not know exactly what to ask, but you will know that something is confusing—most likely you at this point. Again, say something about it.

There is also no particular order to the sequence in which you earn your ratings, as long as you begin with your private. You might follow that with your commercial and then the instrument, or do it the other way around. You might earn your multiengine rating in between both. It makes no difference, as long as you complete all those that are necessary—the commercial, multiengine, and instrument rating being the minimum required for almost any flying position. (See Fig. 3-4.)

Figure 3-4. Cessna Citation II. *Courtesy Richard Neville*

Now That You Have the Ratings, What's Next?

This seems like the grandest question of them all, but once you've picked up these various ratings, what do you do next? The most important answer is to keep flying. That might sound a bit simplistic, but do anything you can to keep logging and building your total time as well as the variety of your experience. One way to accomplish this is to stick close to an airplane any way you can, whether it's by buying an airplane of your own or renting. It all counts toward your total time.

Let's talk about that coveted logbook for a moment. Everywhere you go in this industry, you'll hear people asking how much time you've logged. It seems, however, that the problem of logged hours is not quite as significant for people who have surpassed the point of 3,000 to 5,000 hours or so. Less than 1,000 hours pretty much puts you in the range of unproven more than anything else. When you're working with total hours in the hundreds, then, total time to a potential employer is very significant, depending on the job you're trying to win. A pilot with 400 hours total time might find that the kind of jobs they'll be limited to will be sight-seeing demo rides, or perhaps glider-towing missions.

But all is not lost for those of you with less than 500 hours logged. Don't forget that companies such as Comair and Flight Safety do run first-officer qualification programs to hire low-time pilots.

Realize that a potential employer is limited, usually by insurance, as to what you can and can't fly, so be ready when this problem appears.

Another significant hurdle you'll run into during your career will be the amount of multiengine time you have under your belt. Multiengine time is much tougher to come by because it's not only expensive, but also difficult to obtain because, again, most insurance policies written on twin-engine aircraft pretty much prohibit low-time pilots from flying them solo. What you might be left with is trying to find a charter operator who will allow you to fly second in command (SIC) on one of their airplanes or perhaps a small company that will allow the same kind of SIC time in their twin. I won't pull any punches with you here. This isn't going to be easy. I remember the struggle: begging, borrowing, and running around the airport at all hours of the day and night to try to pick up that multiengine time.

I started out flying right seat on a light twin for a small Chicago manufacturing firm and eventually ended up in the left seat. The airplane was an early Piper Seneca 1, but it might as well have been a 747 to me. After a hundred or so hours in that aircraft, l found myself in the right place at the right time to be able to ferry some aircraft for an aircraft distributor, which eventually led me to a Piper Navajo job. As a ferry pilot, I would fly around the country with the power pulled back to economy cruise and everyone was happy. The aircraft burned less fuel on the trip and I logged more time. Even today, I still ferry aircraft just for the chance to fly someplace totally unscheduled.

Another option you might consider could be flying club operation with a twin. You might have to fly with an instructor for a while before you can take the aircraft alone, but once you qualify, the rates will probably be relatively cheap. It will take some time and effort on your part to search out the bargains and opportunities, though.

Just when you find yourself with a few hundred hours of multiengine piston time under your belt and you start shopping for some of those flying jobs, you might run smack into another category of time someone will ask for: turbine time (Fig. 3-5). Turbine is basically a generic word for jet engine time. The time can be logged in two forms, however. The first is turbo prop time, where the aircraft is powered by a jet engine connected through gearing to a propeller. This could be a King Air or Cheyenne type of aircraft. The next is pure jet time. This would be Cessna Citation, Learjet, Boeing 737 time. After I had accumulated about 1,200 hours of turbine time in a turbo prop aircraft, I felt like pretty hot stuff. Then, when I went looking for a job, some of the ones I really wanted asked for jet time. It seems there's always going to be a carrot in front of you somewhere. As soon as you have jet time, someone will probably say they wished you had more time in that jet, or they'll tell you they wished you were type rated in some other kind of jet. That's just the way this industry works.

Figure 3-5. Flight time spent piloting a turboprop counts as turbine time. *Jetstream Aircraft Inc.*

An Inexpensive Method of Building Time

What about that pilot who has only a few hundred hours and is still seemingly too far away from twins and turbines and jets to even be concerned? What's the best way to build time, any kind of time, toward that first 1,000 or so? Looking back on what I know now, if I were a relatively low-time pilot, I would find a partner and purchase an airplane of my own and fly the pants off the thing. You might at first be aghast, considering how expensive you might believe airplanes to be, but I'm not talking about buying a twin Cessna or a Beech Bonanza for $100,000.

Recently, I saw an ad in a local Chicago newspaper for a Cessna 120. The aircraft, a taildragger and slightly slower version of the Cessna 152 (Fig. 3-6), was reportedly in good condition, and the asking price was

Figure 3-6. You could buy a single-engine aircraft to build time. *Comair Aviation Academy*

$13,500. Let's assume there was no dealing at all and you actually paid the asking price. You'd be required to put down 20 percent of the purchase price as a down payment, or about $2,700. Your financed amount would come to $10,800.

A recent chat with the folks at the AOPA sent me to the Maryland Bank, who quoted the following rates on this deal. Financing $10,800 at the current fixed rate of 10.9 percent would give you a payment of $184.98 per month. If you chose their variable rate of 8.75 percent, the payment dropped to $172.40 per month. The rate on the variable note changes quarterly with Maryland Bank and is always 2.75 percent above prime rate as seen in the *Wall Street Journal*. With a monthly payment of roughly $185 per month, I added another $75 per month for tiedown and about $40 per month for insurance. The aircraft burns about 4 gallons of fuel per hour, at about $2.00 per gallon. Oil is consumed at about a quart every 5 hours, so we'll add another 40 cents to the hourly cost. The Cessna 120 is a relatively simple aircraft to maintain, but we must put something on the side just in case, so we'll add another $4 per hour for possible maintenance costs. The grand total of how much it will cost you to fly this machine each hour depends greatly on how many hours each month you truly take to the air.

I'm going to assume, because you're trying to become a professional pilot, that you'll take advantage of having an airplane around to fly. For my calculations, I'll assume you fly the aircraft 30 hours per month, or about an hour a day.

Here's the cost breakdown:

$185 per-month mortgage
$ 75 per-month tiedown charges
$ 40 per-month insurance
$300 per-month fixed costs / 30 = $10 fixed costs per hour

$4.00 per-hour for fuel
$ .40 per-hour for oil
$4.00 per-hour for maintenance
$8.40 variable costs per hour

Total cost to you per hour would be $18.40 per hour.

One of the best parts about this is that you could take in a partner who also has the same goals and wants to fly as much as you do. With one partner, your down payment is half, about $1,350. All the fixed costs are halved too, although your hourly costs would remain constant. So with a partner, you pay just $150 per month in fixed costs, which works out this way.

$150 monthly / 30 = $5 per-hour fixed costs
$ 8.40 per-hour variable costs
$13.40 per-hour flying costs

Imagine, with a single partner in an airplane of this size, that you could be out building time at the rate of $13.40 per hour. I guarantee that you won't find a deal like that at any flight school anywhere. Keep in mind, too, that these aren't just pie-in-the-sky numbers. If you fly the airplane only one hour a day, you can certainly have time left over for your partner to fly one hour per day. Perhaps a better solution would be to alternate days. There are plenty of options, but remember, the pilot who succeeds is the one who's willing to take a chance on something a little bit different.

By the way, in case you think I skillfully skipped over the subject of down payment on that Cessna, I didn't. I just assumed that some of the money you saved on not spending $45 or $50 per hour to rent a time-building Cessna 152 could be used to pay back your parents or your brother or your savings account or whomever you used to help you get the plane in the first place. If you do buy a taildragger, have fun. Nothing makes a real pilot like a taildragger.

One word of caution about building time in a single-engine aircraft. It has its limits. If you and another pilot are both interviewing for the same position with 1200 total time and all of yours was in the Cessna 120 and the other pilot's is half single-engine and half multiengine, the other pilot is going to look much better qualified because of that larger aircraft experience. Ballpark figure? I wouldn't want much more than 500–700 hours of single-engine time in my log before I began trying to find some multiengine time. The exception to this, of course, would be if you are instructing in single-engine airplanes. Companies view instructing time differently, but only up to a point in a single-engine aircraft.

The Job Hunt

So, you've spent a ton of your hard-earned cash to pick up a commercial and instrument rating as well as a multiengine and perhaps even a flight instructor certificate. You've even managed to build some flight time, maybe even some multiengine time, and you think it's time to start looking for a way to bring in some money instead of just writing all the time. But where do you look, and furthermore, once you do find some job listings, what's the plan of attack to get the interview?

When it comes to finding a flying job, there seem to be two kinds of pilots; those who look for themselves and those who find someone to search for them. I've used both methods personally, and I think each has its benefits and drawbacks. If you search on your own, you'll be a busy person, indeed, at least if you work the way I do. I believe that a person who is not working really does have a full-time job. Their full-time job is looking for a job. Looking for a flying job (compared to talking about looking for a flying job) really requires not only hard work, but also a real plan of operation as to how you'll reach this new goal. There's that word goal again. Funny how it keeps creeping up, but I believe that, without clear-cut goals, a personal flight plan, if you will, you'll end up making emotional decisions that might not be in your best long-term interests.

Being involved with airplanes in general tends to be a rather emotional experience, so make sure you've some idea where you're headed before you go marching off toward a new job. What's your ultimate goal? If you're relatively low-time, almost any flying job will probably be beneficial, as long as you can afford the cost. And sometimes the cost can be dear, everything from asking you to move a few thousand miles away to starting wages that appear to rival poverty scale. Only you can decide.

If you're going to begin looking on your own, I suggest boring ahead with full steam because there are many places for you to look for that new flying job (Figs. 3-7 and 3-8). In this section, I'll tell you about some of the places you can locate information on companies who use pilots. Some will be publications, some will involve the use of a personal computer to access a database, while yet others will be full-fledged pilot recruitment services anxiously awaiting your signup so they can give you their

Figure 3-7. Is the corporate cockpit your goal? *Falcon Jet Corp.*

Figure 3-8. Corporate jet cockpit. *Falcon Jet Corp.*

knowledge . . . for a fee. You can quickly accumulate a wealth of information about who owns aircraft and which companies are hiring.

Notice too, I don't mention "not hiring." If a company owns aircraft, they all must hire, eventually. It's a matter of finding out who they are, first of all, and then planning your appearance on their doorstep at the proper moment in time. Some of the magazine classified sections contain job ads, while some are ads for training schools. But, sometimes too, just talking to people at the training schools can generate some ideas to help your plan take shape. Many of the training schools also use 800 phone numbers, so call and ask questions. But your selection process shouldn't be just limited to the sources mentioned in this section. What follows are some of the sources. These sources are not organized by their importance. They are all important. But most important of all is what you do with the information you're provided, for no agency or publication in the world will be of any help if you don't act on the information you locate.

Can You Really Find a Job With Your Computer?

OK. Here's the deal. A computer won't find you a job, but your chances of finding a job without one are slim. But don't confuse online methodology as a substitute for good-old in-person networking. It's not. Online job searching is simply an evolutionary improvement in how pilots look for work. Let's take a look at some of the major companies that offer a presence on the Internet—as well as a look at how their products and services can help you find the right cockpit position.

This is not a complete list of companies that offer help to aspiring pilots through traditional and online methods. These are simply the ones I've personally dealt with.

Air Inc. (print and online)—Air Inc. is probably one of the most well known of the pilot information services with products and services to be found in both a print version and online.

Air Inc.'s services focus on the airline industry with one major product—the Airline Pilot Career Development System—and a host of supplemental books and services. The Career Development System includes the Airline Pilot Career Decision guide that looks at past, present, and future hiring issues and statistics, as well as Interviewing Dos and Don'ts. Included too is a comprehensive Airline Information and Address Directory with all the information an applicant needs to apply. The system includes a pilot application handbook that offers sample resumes and cover letter tips. Another book—some applicants say the most important—is the encyclopedic US Airlines Salary Survey. Air Inc. offers a number of other publications that range from an Airline Simulator Training Manual to a personnel interview guide that offers 172 technical questions that any applicant might be asked.

One of the major benefits to a service like Air Inc.—besides access to a ton of useful information—rests with their counseling services, the ability for a pilot to speak one-on-one with an information specialist who can explain exactly what the interview process at TWA or Atlantic Coast Airlines will be like, from how many people they'll interview with, to what a typical simulator profile for that carrier will look like. Air Inc. also holds regular seminars around the country that offer in-person talks on subjects such as Crew Resource Management, Military/Airline Transition, or an Interview Survival Course. Best of all, these seminars are also packed with representatives from most of the major and regional airlines around the United States who are willing to begin the hiring process right on the spot.

The initial Airline Pilot Career Development package is currently priced at $199. All other books, as well as the one-on-one airline interview preparations are available at an extra cost. Purchase of the initial package also entitles a new member to a subscription to *Airline Pilot Careers* magazine, as well as to a subscription to *Airline Pilot Job* Monthly.

Online you'll find information broken down into two categories, the first is the public sector where you'll get a no-cost glimpse of what Air Inc. has to offer, such as the newsletter, magazine, pilot development system, and counseling services. The other section is a private area just for members and includes items such as a hot news section, a rumor control area where you'll learn whether the word on the street has any teeth to it and even online version of Air Inc.'s "Airline Information and Address Directory", monthly newsletter and the most recent issue of *Airline Pilot Careers* magazine. The system also offers you the opportunity to search the previous two year's worth of *Airline Pilot Careers* issues. Air Inc. also offers e-mail updates of hot news items right to your computer as they happen.

For more information call Air Inc at: 800-JET-JOBS, 800-AIR-APPS, or visit their Web site at: www.airapps.com

Flying Careers **magazine** (print and online)—*Flying Careers* magazine and pilot information system is a significant competitor to Air Inc. because it costs much less and because it also explores other piloting careers besides the airlines, such as corporate and charter flying.

Flying Careers magazine is the solution publication for aspiring professional pilots, providing airline features, training articles, and other vital information essential to an applicant's efforts at finding the right flying job.

Airline and Corporate Employment Survey (ACES)—This directory of employment information, published once a year, provides the necessary contact names and addresses for aspiring airline and corporate pilots. The directory is also regularly updated with supplements sent to all members.

Flying Careers Employment Newsletter—A monthly listing of job opportunities that detail who is hiring, how many, and how to apply. The newsletter also includes key airline application window dates, special news coverage, and industry statistics.

The cost for the premium package costs just $99.95. For more information, contact *Flying Careers* at (800) 492-1881 or visit their Web site at: www.flyingcareers.com.

David Jones, editor-in-chief at *Flying Careers* related his vision of on the flying-jobs market for the future. "If I were a pilot with a four-year degree and at least 2000 hours total time seeking a job in early 1999, I'd be very optimistic. A pilot who is out there building quality time and picking up the right ratings will be very employable. Some carriers are modifying their minimums, but overall, qualifications are not coming down significantly because there is still an ample supply of pilots."

Jones added that "For younger pilots who are tempted to build time any way they can, rather than finish their four-year degree, they should know that the statistics show that the number of pilots hired without a four-year degree is low. Occasionally, a few high-time pilots might get in without one, but they are few in number. The difference in flying and other jobs, is that everything you do—on and off the job—can count for something too. Don't do something stupid, like pick up a DUI. And as always, your attitude is critical. A bad one will come back to bite you."

The Berliner/Schafer Aviation Consulting Group (print and online)—The Berliner/Schafer Group is a network of over 25 airline pilots and aviation professionals combining four decades of experience to help their clients obtain airline and corporate pilot jobs. For the past six years The Berliner/Schafer Group has helped pilots by providing a personalized, one-on-one approach to the job search process. The entire Berliner/Schafer Group staff fly as line pilots and check airmen for some of the largest carriers in the country. The founder and president of the Berliner/Schafer Group is Paul M. Berliner, a captain for a national U.S. airline and holder of type ratings on the B-767, B-757, and B-727.

Berliner/Schafer does not subscribe to the theory that there is a pilot shortage or that there will be some miracle hiring boom in this decade. They are on the front lines of airline hiring and say there is still unbelievable competition that exists in today's marketplace.

They also believe that the current trend of having to pay for your airline training is "dangerous, disgraceful, and completely unethical!" Berliner/Schafer believes that buying your job is one of the quickest ways to ruin your airline career. They claim first-hand knowledge that five major airlines have disqualified applicants simply because they "bought" their jobs. "It's expensive enough to go through flight school these days let alone having to pay another 8 to 10 thousand dollars just to get an entry-level job," said Paul Berliner. "We are active lobbyists against this practice!"

Most importantly, Berliner/Schafer claims to have more career airline pilots on their staff than any other consulting company with nearly 120,000 flying hours of flying experience. Berliner/Schafer does not guarantee that you will get a job tomorrow. However, they believe that if you follow their techniques and advice closely, you will significantly reduce the time it takes you to get an airline job. "We all knew there was a tremendous amount of resources out there," said Paul Berliner, "but without someone to help you put it all together, you'll be lost. We wanted to help pilots get that dream job and give them a structure to help avoid the pitfalls along the way.

The interview process was often the most overlooked part of the entire process. An applicant needs to practice the presentation and delivery as much as possible. For instance, applicants often want to talk about how great they are. The airlines want to know what a great captain you'll make. How will you deal with passenger and system problems.

The biggest mistake we also find that applicants make is waiting too long to send out resumés, thinking they need the minimums before they begin. Most airlines save everything you ever send them. If you've been sending them a resumé every 45 days, that tells them you want the job.

Berliner offered two final thoughts to ponder as you develop your career plan. Young pilots think they don't have enough hours to apply and older pilots think they are too old. Both groups are wrong."

Online, you'll find back issues of the company's newsletters, as well as information about finding the right corporate pilot position. Also included are details of publications and services of value to pilot applicants. For more information, contact Berliner/Schafer at 888-745-6899 or visit their Web site at: www.pilotswanted.com

Aviation Employee Placement Service (AEPS—online only)—AEPS is only available online and claims approximately 2,000 member companies. In case you're thinking that seems like quite a few airlines to represent, you should know that AEPS does not work exclusively with pilots. The member companies also range from airlines, to charter and corporate operators as well who may be searching for mechanics and other aviation professionals. Company president Jim Dent, himself a former PanAm pilot, claims some 56,000 members, about 9,000 of whom are pilots in search of cockpit positions.

"AEPS is a change in the way pilots get jobs," said Dent. "In our system, you can restrict who views your information. The companies who search our database also don't get unsolicited resumes and only see your resume if you fit the qualifications they want." His advice to applicants is to "keep your information in our database as current as possible."

It does offer an easily customizable data record for a pilot's qualifications that a new member fills out online when they sign up. Then, when a member company has a position to be filled, they simply put the necessary criteria in their computer and perform their own search. The benefit to both the pilot and the member company is that they each get results that meet their requirements. If a regional carrier wants pilots with 2000 hours total time, no records of pilots with less than that amount of time will be sent, saving everyone a lot of time and energy.

AEPS also sends out a weekly aviation industry news update—via e-mail—published in conjunction with *Air Transport World* magazine. Whenever a member company performs a search and extracts the information on a member, AEPS sends that member an e-mail update. Members can also read profiles on the companies AEPS represents. AEPS charges a monthly membership fee—currently at $12.95—but does not charge for updates to the member database. Visit their Web site at: www.aeps.com

Fltops.com (online only)—(See Fig. 3-8a.) By its very name, fltops.com tells you that this is an aviation information company that has a presence only on the Internet. But don't let that sway you, because the man in charge of fltops.com—Louis Smith—is an experienced aviation-information provider, as well as a DC-9 pilot for Northwest Airlines. Smith is the former president of FAPA (Future Aviation Professionals of America).

One of the advantages of an all-Web-based information service is their lower overhead translates into a service affordable by most any applicant at $4.95 per month. But, one caveat to fltops.com, is they only track information on the top 13 airlines in the country. Those are Airborne Express, Alaska Airlines, American Airlines, America West, Continental Airlines, Delta Air Lines, Federal Express, Northwest Airlines, Southwest Airlines, Trans World Airlines, United Airlines, United Parcel Service, USAirways.

Fltops.com offers up-to-the-minute news on the Big 13, as well as a Pilot Pay Update, a Big 13 Fleet Update, and Pilot Screening "gouge" from pilot candidates, a comparison of the Big Six Airlines, and Monthly Pilot Hiring Snapshot.

Monthly service also includes access to the *Employment Directory*, a continuously updated profile of the "Big 13" U.S. airlines that details hiring trends and projections, contact, and background information.

The Job Report includes the following:

- Pilot Hiring: Monthly new-hire tracking, trends, and projections.

- Pilot Screening: Testing, Interview, and Simulator check information,

Figure 3-8. Fltops.com.

- Industry Comparisons, including:

  Crew pay: The most detailed and current airline pay rates available by airline, aircraft type, crew position, and tenure and sample pay rates.

  Cockpit career comparisons: Detailed pay and benefit, job security, profit sharing, and "airline-within-airline" comparisons at Big Six airlines.

Fltops.com also offers Flash Job Updates—electronic mail notification of major occurrences affecting pilot employment among Big 13 carriers.

Louis Smith said, "What makes us different is that fltops.com is not based upon the print model of an aviation information magazine. We only cover the Big 13 in an Internet format that pilots simply jump in to." But Smith was also vocal about what fltops.com is not. "You won't get individual evaluation and we won't give career advice. Pilots just want to know where the jobs are and we provide that information from the key people doing the hiring at the airlines. We've transferred a lot of the work to our customer. Pilots make a capital investment and in return, they get a really good product at a really good price. We update the Flight Line weekly and soon it will be daily."

Smith's advice to applicants, "Networking is still #1 in finding that job. You have to get to know someone at the companies you want to work for.

You can't beat someone telling a potential employer about you before you're hired." For more information, visit the Web site at: www.fltops.com

UPAS (online only)—UPAS stands for the Universal Pilot Application Service. UPAS is a subsidiary of ALPA, the Air Line Pilots Association and serves only the needs of pilots, claiming nearly 18,000 active members. Like AEPS, UPAS members fill out an extensive electronic form listing all ratings, experience, and past employers. Like AEPS, UPAS companies perform a search on members based upon their particular needs at any given time. When a member's records are presented to a company, the pilot applicant receives a message like the one below.

To: " 'rob@mark-comm.com' " <rob@mark-comm.com>

Subject: UPAS Company Search Notification

Dear Valued UPAS Subscriber:

We are pleased to inform you your file was downloaded by TWA on 10/1/98.

Their search criteria was Commercial Certificate with Multi-Engine and Instrument Ratings or ATP, F/E Written, 1,000 Total Time (No Heli), First Class Medical with no waivers, and Four Year Degree or TWA Intern.

Thank you for subscribing to UPAS.

UPAS charges an initial fee of $150 for new members to register that includes three free updates to member pilot records. UPAS charges a $100 annual renewal fee and a $25 fee for each update after the first three each year.

For more information, contact UPAS at 800-745-6827 or visit their Web site at: www.upas.com

Avcrew.com (online only)—Avcrew.com is a compact, easy to navigate Web site that offers corporate pilots and potential employers a place to find each other. And that's what this is all about. Jobs are posted as they come in and pilots can post their resumes here. Elsewhere in this text you'll see an example of the dialogue found on the interactive portion of AvCrews's site. It's a place to talk about the issues of the day as they relate to corporate flying. A recent question asked what toll airline flying was taking on corporate flight departments, while another asked for opinions on fractional operations.

Richard Harris, publisher of AvCrew.com and an NBAA member, said "AvCrew was the first site devoted only to corporate pilots. As employers learn more about the benefits of technology when searching for qualified pilots, this medium will become even more popular (to locate pilots).

For more Web site listings, see the Web site review section elsewhere in this book.

The Publications

Next I suggest you locate *The World Aviation Directory*. Published by McGraw-Hill, the WAD, as it's affectionately known, is about ten pounds of more information on the aviation industry than you ever thought about or possibly even wanted to know. The WAD is a world-renowned directory to all things aviation, from airlines, to fixed-base operators to flight schools to parts suppliers to government aviation agencies. And best of all, the WAD covers the aviation industry worldwide. Let's take a look at some of the facts you can locate inside the WAD.

From the Industry Trends and Statistics section:

- Worldwide Carriers Systemwide Traffic
- U.S. Majors and National Carriers
  - Domestic Financial and Traffic Summary
  - Financial Indicators
  - Expense Indicators
- Fleet Analysis
  - U.S. Major Carriers
  - U.S. National Carriers
  - Worldwide Carriers
- U.S. Major Carriers Aircraft Operating Expenses
  - Domestic
  - Systemwide
  - International

One section lists the name of every airline flying in the world, as well as their home base. These can then be easily cross-referenced to the air carrier list, which not only lists the correct address and phone and FAX number for an airline, but also the major executives of that carrier and the current fleet size, something that might come very much in handy during an interview. You'll be suitably impressed with the information that's also available on aircraft manufacturers and charter companies. Finally, if you know the name of a particular aviation person you're trying to track down, the alphabetical list could be just what you need. A current WAD (they're released twice a year) is not cheap, about $200 per year, but there probably is no better source of information in book form available. Most library reference sections will carry a *World Aviation Directory*, so you can go take a look before you shell out your hard-earned dollars.

Here's a glimpse of a few of the magazines in the aviation industry now. The best part about all of these is that they not only have classified sections that will steer you toward even more information for your job search, but they also contain stories about what's happening within the industry, such as a new airline start-up or a corporate flight department opening or closing. That may influence your decisions. These magazines and information sources are listed without regard to price, since that might change.

My advice on the use of any publication's information about flight schools is that I'd be very careful about schools promising work after training, unless you have some way to verify those claims. I can provide the information about sources, but you'll have to do the footwork to verify what you learn (Fig. 3-9).

Professional Pilot **Magazine**

Based in Washington, D.C., *PRO PILOT* calls itself the monthly journal of aviation professionals. The magazine contains spotlight articles on air-

Figure 3-9. Some of the aviation publications available. *Robert Mark*

line and corporate operators, as well as technical articles on state-of-the-art electronics and flying techniques. The slant is definitely toward professional pilots, although newly rated pilots could benefit from the educational aspects of the publication. A recent issue contained these stories: a piece on aviation in Ohio, a story on the Canadair RJ jet, an avionics piece on Flight Management Systems, a profile on a hurricane-hunting NOAA pilot, a story on alcohol abuse, as well as a piece on Martin State Airport, and a technique article on learning from experience. Regular features include an IFR chart refresher quiz and basic aviation news. The magazine's classified section recently listed various flight training schools that might or might not advertise in other publications, as well as a few listings for help under the jobs-available section. Contact *PROFESSIONAL PILOT* magazine at 3014 Colvin St., Alexandria, VA 22314 or 703-370-0606. This publication is not available on the newsstand.

E-mail Pro Pilot magazine at: editorial@propilotmag.com or visit their Web site at www.propilotmag.com.

Flying

FLYING claims to be the world's most widely read aviation magazine. The emphasis is towards general aviation, although many articles highlight the newest aircraft and equipment in aviation, be it corporate, airline, or—occasionally—military. The magazine uses a number of regular, highly respected columnists, each with an interesting perspective on what's happening in the industry. The magazine's classified section has grown, and it now covers seven full pages. In a recent issue, no less than 46 flight schools were advertised. The classified section also included 13 ads for employment information, everything from an ad that asked for low-time pilots to one specifically aimed at corporate flying jobs. Recent articles in *FLYING* included a story on GPS, as well as other stories on the Bell LongRanger helicopter, the Falcon 2000 corporate jet, the Jetstream 41 regional airliner, Hypoxia, and how to update the IFR equipment on your instrument panel on a budget. Contact *FLYING* at 500 West Putnam Ave., Greenwich, CT 06830; call 303-604-1464 for subscription information. *FLYING* is available on the newsstand and at most libraries.

Business & Commercial Aviation

Business & Commercial Aviation is written for the crews of corporate aircraft, although many of the articles cover subjects of interest to all pilots. Recent stories included a piece on flying in the CIS, a discussion of the V1 flying speed controversy, and a pilot report on the TBM 700 turboprop.

BCA has one of the better monthly aviation news sections too. The marketplace section is small, but many ads of various categories are scattered throughout the magazine. Contact BCA at 4 International Drive, Rye Brook, NY 10573. Subscription information is available at 800-257-9402. BCA is not available on the newsstand.

Aviation Week & Space Technology

Av Week, as this magazine is known, is the weekly bible of what's happening in all facets of the aviation industry worldwide. With editors scattered around all parts of the globe, you can expect the most comprehensive description of aviation events that could affect your career. Coverage of the general aviation end of flying is a bit skimpy at times, but the magazine more than makes up for it with their coverage of airlines, military, and corporate aviation. If you follow aviation, this publication is something you must have. Their classified ads often contain some rather major flying jobs, along with employment services in the industry. Contact *Av Week* at PO Box 503, Hightstown, NJ 08520-9899 or 800-525-5003.

Air Line Pilot Magazine

This publication is the voice of the Air Line Pilots Association (ALPA), so the stories tend to be oriented toward matters for airline pilots as well as the day-to-day politics of a union with more than 50,000 airline pilot members. The classified section is pretty short on anything of interest to aspiring pilots. What makes this magazine particularly interesting is the insiders' view—from a union perspective—of what's happening within the airline industry, such as how new FAA regulations could affect a pilot's job. Recent stories in *AIR LINE PILOT* included a look at the life of a regional pilot, a study on a recent EMB-120 Brasilia crash, an interview with the new Department of Transportation secretary, as well as regular comments by ALPA's President. *AIR LINE PILOT* is not sold on the newsstand. For subscription information, contact ALPA at 535 Herndon Pkwy, P.O. Box 1169, Herndon, VA 20172 or 703-481-4460. E-mail: 73714.41@compuserve.com.

Flying Careers

Flying Careers is the career information source for professional pilots because it covers many varied facets of the aviation job scene besides the airlines. Recent issues have delved deeply into how to find a job in the corporate sector, as well as interviews from pilots who are already there.

The magazine has regular columns on interviewing techniques and legal affairs. Recent articles have included a look at the new Midway Airlines, United Feeder Services' operation through ORD, and a report on what to expect during your probationary period with any company.

Flying Careers is not available on the newsstand. For subscription information, call them at 800-492-1881. You can e-mail *Flying Careers* editor-in-chief David Jones at fc4editor@aol.com.

Air Transport World

Air Transport World is the place to learn about issues relevant to the airline industry, but from a management perspective. The magazine's editors focus on U.S. and international carriers each month with a wide variety of topics ranging from a highlight piece on at least one individual airline per issue, discussions of how companies finance aircraft purchases, new engine performance, and the aircraft manufacturing industry itself.

I'd add a subscription to ATW simply to offer you a different point of view from the managers at some of these carriers you might be interviewing with—for instance, their stance on labor issues.

Contact *Air Transport World* subscription at 216-931-9188. ATW is located at 1350 Connecticut Avenue N.W., Suite 902, Washington, DC 20036.

AOPA Pilot

This is the publishing voice of the Aircraft Owners and Pilots Association. The magazine is focused on general aviation, but in the past year, it has added a very interesting and useful feature called "Turbine Pilot" to explain the intricacies of flying jets and turboprops to people used to piston engines. This feature alone could make the magazine worth the price to a relatively new, yet aspiring professional pilot. The magazine is offered as part of the annual membership dues from AOPA, currently $39. If there's anything to be said about AOPA, besides the fact that nearly 60 percent of all active U.S. pilots are members, it's that the $39 will bring so many extras that you'd be crazy to pass it up. Recent articles included flying with over-the-counter medicines, lost communications procedures, a report on Cessna's Caravan, and a look at what's doing at Lycoming and Continental Engines. The classifieds tend to be rather lean, but they would still be a good source for training school ideas. Contact AOPA at 800-USA-AOPA.

Airline Pilot Careers

Airline Pilot Careers is published each month by Air, Inc., in Atlanta. Air, Inc. (which we'll talk about later in this chapter), is a company that's 100 percent devoted to bringing the latest information on airline pilot jobs to their members. The Air, Inc. gurus sift through thousands of pieces of information and produce plenty of forecasts for the industry. Each month, *AIRLINE PILOT CAREERS* highlights at least one airline, usually two, and tells all: history of the company, hiring requirements, fleet size, pay ranges, pilot comments, and tips on making it to the first interview. Recent articles include a cover story on United Parcel Service, how to find international flying jobs, how pilots get conned in investment scams, and Trans World Express. Just to keep you up to date on who really is being hired each month, *AIRLINE PILOT CAREERS* offers a new hire feedback section. Here, recent pilot hires are interviewed and relate their experience levels at the time they were hired. Here's a listing example from a recent issue, in which I switched my name with the real new hire. Rob Mark, 35:2, 20/100 vision, United. B-727 FE, 7,150 hours, 1,000 jet, 3,000 turboprop, ATP, FEw, CFII, MEI. B.A. degree. Cargo. LR-36, DA-20, BE-99, PA-31, Burbank, CA.

Broken down, this means the pilot was 35 years, two-months old, with 20/100 vision. He was hired by United as a B-727 Flight Engineer with 7,150 total hours logged, of which 1,000 was in jets, 3,000 in turboprops. At the time of hire, this pilot held an ATP certificate and had passed the Flight Engineer written exam. He also held the instrument and multiengine instructor ratings as well as an associate degree. He flew mostly cargo in Learjets, Falcons, Beech 99s, and Navajos. He lived in Burbank, California. *Airline Pilot Careers* is not available on the newsstand. Contact Air, Inc. at 800-AIR-APPS.

Aviation International News

This is the big news magazine. It's not only big in the sense that's it's chock full of news about airline and general aviation, plus fixed-base operator topics, but it's also big physically. The *Aviation International News* format is glossy-magazine style, but twice the size of a regular magazine. While the content of the magazine is news, this publication allows for much longer articles than other magazines. Recent stories included a rather comprehensive job report about what the job market really looks like for pilots, as well as an in-depth look at Warren Buffet's purchase of Executive Jet. The news stories make this magazine important, especially because there's no classified section. Not available on the newsstand, contact *Aviation International News* at 214 Franklin Ave., Midland Park, NJ, 07432 or 201-444-5075. E-mail: ain@compuserve.com.

The Flyer

This tabloid is normally devoted pretty much to general aviation. Upon reading a recent issue, I looked through their "Pink Sheet," which is *The Flyer*'s classified pages, and I found some ads that could be of interest to a new pilot looking for methods to build time. Under the "Help Wanted" section, I saw, "Pilots fly 1,000+ hours per year with great pay. Alaska offers seasonal and full-time employment . . ." No promises here, but an opportunity in a place that a future professional pilot might not look. Also, *The Flyer* can be an excellent source of ads for the parts necessary to keep that airplane of yours in top shape. This tabloid is the sleeper of the publication list, so be sure to check into it. Contact *The Flyer* at 800-426-8538.

Flight Training Magazine

This monthly magazine provides constant coverage of the flight training regime because the motto here is "back to basics." The issues are designed not only for the student of various ratings, but also for the instructor. The magazine produces an excellent national list of flight schools. Contact *FLIGHT TRAINING* at 405 Main Street, Parkville, MO 64152-3737 or call 816-741-5151. The magazine was recently purchased by AOPA.

The Wall Street Journal

(See Fig. 3-10.) No publication subscription list would possibly be complete in your search for a new position without considering the *Wall Street Journal*. The journal is published five days a week and can be delivered right to your door each day or purchased at most newsstands and drug stores.

There is nowhere else where you can read the kind of up-to-the-minute reporting on issues that hit home to you if you're in the job hunting and hence the interviewing mode. This immediacy is the *Wall Street Journal*'s competitive edge over many of the other publications here. While the *Journal* is certainly a business newspaper, the quality of the reporting is excellent and normally portrays quite balanced views, even on labor issues.

Recent stories have included in-depth looks at the struggle at FedEx between management and its pilots' union, a major series on labor in America, great coverage of the strike at Northwest Airlines, and the involvement of the unions in United Airlines' recent dismissal of its company president.

Subscription information can be found by calling the *Wall Street Journal* at 800-JOURNAL. And don't forget that a subscription to the print version also gives you a discount on a subscription to the customizable online version of the *Journal*, as well.

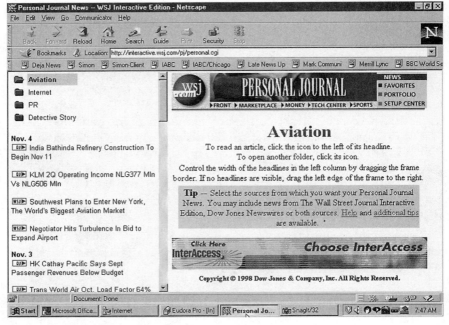

Figure 3-10. *Wall Street Journal* online.

Airline Business

Quite honestly, *Airline Business* was a new publication to me that first appeared in my mailbox as the text for this book was being written. At a glance, it appears to be a competitive magazine to *Air Transport World*, except that *Airline Business* is produced in the United Kingdom.

Recent issues were a look at Polar Air Cargo, the recent alliance between American Airlines, Canadian, Quantas, Cathay Pacific, and British Airways, as well as an excellent commentary piece on the outlook for the world economy after a recent meeting of the International Monetary Fund (IMF).

Visit their Web site at: www.airlinebusiness.com.

FAA Public Affairs List

One of the best—and least expensive methods—of staying in touch with what is happening within the aviation industry is by subscribing to the FAA's list for news releases.

Send an e-mail message to this address: listserv@listserv.faa.gov

In the body of the message, type: subscribe-faa-newsrelease (your name)

A copy of every press release the FAA sends out will be delivered automatically to your e-mail account. You'll receive an e-mail back from the FAA confirming you've been added to the list.

Some other publications to check out are *Air Jobs Digest, Air Transport World, Commuter Air, Air Progress,* and *Flight International.* Again, I'd never pass by any aviation publication without investigating the classified ads. Keep in mind that names and addresses for most publications can also be found in the *World Aviation Directory,* which you might need because so many of these publications aren't available on the street. But don't think aviation publications are the only places jobs are advertised. I've certainly found them in the "Help Wanted" section of the *Chicago Tribune,* the *Denver Post,* and even the *Wall Street Journal.*

Work Rules: How They Really Work

If you mention work rules in the airline industry, most people envision a labor contract overflowing with benefits. While a contract is one form of working rules, all the rules a pilot must live with don't evolve from standard collective bargaining relationships. When no union is present, a set of work rules is often produced by the airline, sometimes with little or no input from the workforce. These sometimes use a shoot-from-the-hip type of approach that often leaves the employee confused and floundering in the mire. Sometimes there are no work rules in the flight department other than the FARs.

According to Robert Hammarley, Executive Administrator of the Air Line Pilots Association (ALPA), "The airlines are one of the most unionized of all industries. The work rules are an important part of that organization. I think the pilots who are involved in the union see that it works; the wages and the working conditions are spelled out in the contract. And the pilots know these things are not received out of the goodness of the company's heart. I think we will see more union involvement in the future with the consolidation of companies, though. The carriers that will survive realize the last thing they need is labor fighting over how things are run."

Former Midway Airlines pilot Dave Bear said that even though many of that company's pilots were unsure about the union at first, "most felt ALPA would be there as an insurance policy to protect them against the company unilaterally making decisions about working conditions and rules. The pilots believed the rules contained in their contract made management take them more seriously" than when Midway was still nonunion.

At Southwest Airlines, Vice-President of Flight Operations, Paul Sterbenz, said the company's employees try to do the best job they can, as pilots...to support the company directives. They know that working together is what makes the entire company work. A lot of our pilots have been at less-fortunate companies that had quite different labor management relationships that went bad. Sometimes certainly, we do get into arguments over work rules, but we have methods of arbitration and board appeal here at Southwest to settle those matters too."

One nonunion pilot said that at his carrier, "the work rules were very confusing. Even after we sat down with flight management about a problem, there was nothing in place to compel them to comply with what had been agreed to. Our work rules just had no teeth." When asked about work rules, yet another pilot said the rule book "was a tough thing to define. Some people only see the rules they find important at the moment. I saw our new contract presented recently, and the first thing the pilots turned to were the sections about wages and schedules."

How much protection some rules provide employees was best described by some nonunion pilots who would not allow their names to be used in this story for fear of retaliation by their superiors. A Continental Airlines pilot, too, felt that even though there were no specific rules outlining his contact with the press, he would only talk if we did not use his name. At one union carrier, however, their employee handbook dealt much more matter of factly with their employees' expressions of their opinions; "Before speaking with the press, employees should contact their supervisors."

A good definition of work rules can be as difficult to pin down though as the rules themselves. Some pilots and managers said work rules encompassed anything that affected your daily work life at the company. John Schleder, an attorney for the Air Line Pilots Association (ALPA) though, said, "Anything that affects a pilot's wages, terms, or conditions of employment would be considered the work rules, but only when they had been established through collective bargaining."

Do pilots consider the work rules of a company before they walk through the door for that interview? John Schrage, a pilot from unionized Metro Airlines thinks they do. He thinks some of the most important rules "are the probationary pay scale. When will a pilot earn more? How long will it take to upgrade to the left seat and how much will a pilot's pay increase after they do?" Another area Schrage thought important for the new hires was the scope language of the contract that really can affect the financial future of the airline. "Can the company farm out work to cheaper pilots or start a new airline to cover routes you currently fly with cheaper labor," he said.

Don Campbell, Past President and current Secretary/Treasurer for Horizon Air Pilots, the organization that represents all of Horizon Air's nonunion pilot workforce, said," Many entry-level pilots do not consider the work rules of an airline before they hire on. All they want is a job." And what of professional pilots who have recently returned to the job market? Former Midway Airlines first officer Steve Marcum said, "I don't think I am really considering the work rules at all since I'm unemployed. I'll just get on with whom I can and take my chances on what comes later."

Mike Ballenger, a part-time Citation co-pilot, is looking for full time work, but still believed "the work rules are very important. I really try and find out what they are before I look seriously at a job. Some of the corporations expect you to work every weekend too, whether you're scheduled or not. I don't think you should have to wear a pager if you were scheduled to be off, unless something unusual came up." Former regional airline pilot Scott Janicki said, "I don't think first-time pilots think about work rules. When I first got hired at Britt (nonunion), I know I didn't care. The most important thing was getting a paycheck for what I liked doing—flying airplanes."

A contract is the major benefit to working at a union carrier most of the time, since it clearly sets down the rules both the pilots and management must work with and what is expected of each group in day-to-day operations. But one AMR Eagle pilot warned that "even with a contract these days, some companies think, the heck with the contract. We'll just do what we want and take our chances with the grievance process." Except when a pilot might be pressured to try and violate an FAR to make a flight though, it is not usually wise for a pilot to refuse a direct order from the boss. Accepting the trip, as long as it's legal and later filing a complaint is better than a direct confrontation any day. Another problem with an argument before a flight is you're liable to take the flight anyway but with your mind on the argument and not on flying the aircraft.

Surprisingly, wage scales were not the major issue to most pilots we spoke to. Scheduling won out overall as the item that more pilots wanted better control over, whether they were union or nonunion pilots. One regional airline pilot said the reason people were often willing to accept the somewhat lower wages of that type flying was to gain a schedule for their lives. Mike Ballenger said, "With a family, scheduling is very important to me." Scott Janicki felt "If the scheduling could be worked out, I could be flexible on just about everything else." Janicki went on to say though that "there is a fine line between what amount of flying hours that are legal to fly and what are safe. Since there are so many gray areas in Part 135 scheduling, you can toss those rest regulations right out the window. The FAA sets up these rules and as long as companies can work pilots the way they currently do, legally, they will. At one airline I flew for, all the shifts were legal but I was constantly tired."

At a union carrier, each contract has a specific grievance process to handle complaints. At a nonunion airline the process for handling complaints may be somewhat different or even nonexistent since technically, as long as a company has not violated any of the current civil rights or Equal Employment Opportunity regs Federal law sets down, a nonunion employee often has very little recourse in terms of work rule violations. At one on demand charter operation for instance, a pilot said he was flying King Airs with "the only restrictions being the FARs which normally only address flight time." Short flight segments were usually accented by long 14 to 15 hour days of sitting around waiting for passengers, a common day for an on-demand pilot. "I was told if I didn't like the work day, or the fact that my pager was to be on, ALWAYS, they could easily find someone else for my job." Another pilot said at his company the attitude toward work rules was "their way or the highway!"

If a company does violate the trust of its employees and changes the rules in midstream, whether they have a contract or not, the results can be devastating. Consider the predicament at Eastern Airlines. Eastern management, headed by Frank Lorenzo almost to the end, operated in a daily state of siege with labor, each side blaming the other, until the airline disintegrated in early 1991. Even though labor and management don't always agree, Don Campbell and his group realize the fine line and the importance of "not wanting to work rule a company into the ground."

How much input do pilots normally have to their work rules? In a union airline, pilots make their thoughts known to management through their negotiating team as the contract is written. Even after the contract is signed, changes are possible through a side letter of agreement. In day-to-day operations, the pilots make their voices heard on the rules they don't like by talking to the local union representative, who brings the concerns to management. At one Midwestern nonunion carrier however, one pilot said "Our Director of Operations sets all the rules. We don't have any input to those rules at all. If we happen to put together a petition, which is much like a union effort, the company seems to listen and sometimes will modify what they are doing." This same pilot said many of the pilots at his company, who are flying DC-9-type equipment, are frustrated because they ask "why should we keep putting these petitions together every few months to get the job done? Our only work rules are contained in our parent corporation's employee handbook. That's all we have for our flight department other than our ops manual."

Jeff Clark, a former Chicago Air (nonunion) pilot though took a different approach to pilot input into the work rule process. "If a guy is putting up $12M to get an airline running, I don't think it's reasonable to give the pilots a say in how he runs his airline." Clark did counter that comment though with "The work rules are important, but I think there are other issues too, like health benefits. I think there are times, that when a company is making a good profit and that company is not rewarding its employees in some way, those employees will definitely have some issues management needs to address."

You'd think at first then that union pilots would be a happy bunch and nonunion pilots an unhappy lot. Nonunion Horizon Air though certainly does dispel some of those rumors since Horizon has some of the characteristics of both union and nonunion operations. Dan Scott, Assistant to the Director, Flight Operations at Horizon Air said," We have signed a written agreement, between the company and the pilots. Additionally, we have a Pilot Representative program. This program was set up by the company 7 years ago to further enhance communication between pilots and management about work rules, pay, and other areas of concern to the pilots. Together we concentrate on the areas that are important to Horizon Air pilots as viewed by the pilots themselves." Even though the Horizon Air Pilots are not a formal union, both management and pilot representatives meet twice a month to defuse issues while they are still small.

Union or nonunion, pilots need the opportunity to discuss the work rules they must live with. Jeff Clark said "to accomplish the job of running the airline, management needs to enlist the help of the pilot group to let them understand the day-to-day problems that may cause a change in policy. Too often, when management has a problem, they try to hide things from the employee instead of talking to them about it. It makes it look like management is hiding something." While any nonunion company is free to impose a rule on its pilot workforce at any time, the people at Horizon have found that telling the pilots what is going on works much better. Dan Scott believes "historically, working as a team on establishing work rules enables us to have a better understanding of the needs of the pilots while they gain a better understanding of real-world airline operations. We speak very frankly to our flight crews about operating margins, profitability, our competitors, and what portion of our operating revenue goes to salaries. The pilots are integral to operating this airline. They really know what is going on inside the company."

At Horizon the relationship between the pilots and management is definitely healthy but not without its mountains yet to climb. One rule that causes some difficulty for both sides according to Horizon Air Pilot's Don Campbell "stems from the fact that we are salaried while the whole rest of the industry is paid on an hourly basis. What typically upsets our pilots is when the weather is bad and airplanes are all stuck in the wrong places. Dispatch calls up at some point in the day and assigns the crew to another round trip to cover for a shortage somewhere. You thought you were going to be off work at 4 P.M. and now you work until 7:30. Elsewhere in the industry, the pilots might say, `well, I'm going to be late but at least there's going to be a little something more in my paycheck to make up for the extra time.' Not here though since you're salaried."

When promanagement types read stories like this, they believe anyone hiring on at an airline has ample opportunity to check out the work rules prior to the first day and has no business taking the job if they don't like what they see. Unfortunately, human nature doesn't always work that logically. As pilot John Schrage said, "Some of the work rules will always need changing and some of the pilots will always complain about some of the rules."

4

Where Are the Jobs?

I've talked about places to train and some sources to help you find a job, but I believe a dash of realism is necessary. The number of jobs open to low-time pilots is somewhat restricted, but they do exist. However, I want you to be certain that you understand this logged-hour question because low-time is really a relative term. For this chapter, my definition of low-time is a pilot with less than 500 hours.

A factor in looking for a flying job, any kind of flying job, is letting as many people as possible know that you're looking for work. I'd call classmates, old employers, even people you've interviewed with in addition to the other methods I speak of in this book. This can effectively multiply your own job search efforts many times. You just never know when a friend might run into someone else who has an opening.

The Resumé

But before you begin looking for any flying jobs, you'll need something to announce yourself as a professional pilot in search of work. You'll need a resumé, in addition to a snappy cover letter that's specifically tailored to the company you're applying to. The thought of putting together a resumé makes some pilots cringe, but you might just as well become used to it if you intend to remain in the aviation profession, or just about any other.

There are a number of excellent books on resumés currently available in your local bookstore, but I would be careful of them. They aren't normally designed for pilots, who often have some unique talents that don't always fit into a standard resumé format. Air, Inc. and the Berliner/Schafer Group run pretty nice resumé services.

In case you've decided to go it alone and write your own resumé, here are a few tips you'll need to consider. First, neatness does count. Not simply because Human Resources people like to grade resumés, but because you're trying to sell yourself through your resumé, a document that is usually your initial contact with a company. Make it easy on the receiver to locate the information they care about. That means keep it short—one page if possible—and don't make the type size too small (see Fig. 4-1a).

Every resumé needs a stated purpose or goal. This not only offers you an opportunity to state your goal clearly, "A career pilot position," but also gives you the chance to tailor your resumé specifically to the company you are applying to, "A Career Pilot Position with Continental Airlines." Word processors make this kind of targeted marketing of your qualifications easy, but with one caveat. Be sure that you have looked at all the qualifications on the resumé before you send it out. If you send a resumé off for a corporate job, I'd remove the reference to having passed the Flight Engineer Written with a score of 98 percent. And don't forget to put it back in when you send it off to an airline that requires the FE written to apply.

The exact layout is up to you, but it must contain at least all of your personal contact information, the certificates you hold, your flight times, education, and other activities that might have an influence on the hiring decision, such as awards or membership in outside organizations, especially those that you held leadership positions in. The truly personal information—age, height, weight, marital status, etc.—is extremely controversial. By law, an interviewer cannot even ask you your age or marital status anyway. My view is to leave the rest off as well. As a final check on the finished document, why not offer it to another pilot for their comments.

If the writing is through—make certain you've verified that all the dates make sense too—plan to print your resumé on a good quality laser printer, although many of the new inkjets do a great job as well. Use only a bright, white paper as well to make the type stand out easily. No matter what, make a backup copy of your resumé and save it on a floppy disk. Try producing an electronic version as well in case a company requests you send them via e-mail. But before you send it off, try sending it to another e-mail account as a test to see what it looks like when it's downloaded. The formatting can sometimes change during transmission.

Certified Flight Instructor (CFI)

If you considered the plan I offered earlier, you might already have your CFI rating. If you've not yet earned yours, you might want to give the idea some thought. With a CFI certificate, you'll be eligible to approach a flight school about a position. I would begin by calling the local flight

Paul Allan McCartney - 291-42- 2159 - 1201 Abbey Road
Herndon, VA 20169 • **Phone:** 703-685-2901 • **Fax:** 703-685-8945

Objective: Career Pilot Position with Continental Airlines

Certificates and Ratings:

Airline Transport Pilot: Airplane, Multiengine Land
Type Rated: C-550, EMB-120
Certified Flight Instructor: Single and Multiengine Land, Instrument
FAA Class 1 Medical
Restricted Radio Telephone Permit
Flight Engineer Written - FEX, 11-98

Flight Experience: Total - 3860

PIC	2574	Turbine	1362
Multiengine	2591	Jet	541
Cross Country	3421	Night	651
Instrument-Actual	534	Instrument-Simulated	130

Experience:

Atlantic Coast Airlines, Dulles, VA   1998/present
    Captain, EMB-120 Brasilia - Part 121, Scheduled Airline

Atlantic Coast Airlines, Dulles VA   1997 to 1998
    First Officer, EMB-120 Brasilia - Part 121, Scheduled Airline

Quick Air Charter, Charleston, SC   1996 to 1997
    Captain, Cessna Citation II - Part 135 charter

Quick Air Charter, Charleston, SC   1995 to 1996
    First Officer, Cessna Citation II - Part 135 charter

Sam's Flying Service, Charlotte, NC   1992 to 1995
    Flight Instructor, Part 141 school

Education:

University of North Dakota   1988 to 1992
Bachelor of Science (BS) - Aviation Management

Extracurricular Activities:

President University Debating Team   1992
Vice President, University Flying Team   1991 to 1992

References Available Upon Request

Figure 4-1a. Sample resume.

schools in your area and simply asking for the chief flight instructor. Ask for a few minutes of their time or possibly set up an appointment to stop in and visit. It won't be a secret that you're a low-time pilot because the vast majority of new flight instructors are. Actually, many flight schools see this as a benefit. New flight instructors will not only have a great deal more enthusiasm for the job, but will also be much more familiar with the subject matter because they just completed the courses. Again, I would always try some local schools first. If the school is not interested in a new full-time instructor, ask if they'll consider you for a part-time teaching job. If you really want the job, I would try just about anything to reiterate your desire to work to the people who own the FBO. The worst they can say is no. And at that, the no means no just for right now. Try them again in a month.

If the local flight schools don't prove fruitful, you need to decide whether an out-of-town position is a consideration. In aviation, I've found that the best jobs always seem to be somewhere other than where I live. For a while, I was lucky enough to work for an airline based where I lived. Unfortunately, the airline eventually went bankrupt and I was faced with a major decision . . . to move or not to move. I decided not to move and, a year later, I was forced to change my mind. Certainly you could sit around and wait for a position to open up at your local flight school, or you could just bite the bullet and start looking elsewhere (Fig. 4-1b).

One method you could use would be to pick up a copy of *Flight Training* magazine's (see chapter 3) annual directory of flight schools. While the list is primarily designed as a guide for students seeking flight schools for their own training, it only makes sense that these schools need instructors.

Figure 4-1b. Sources of aviation work can be found in many locations. *Robert Mark*

There always seems to be a turnover in flight schools as pilots move on to bigger and better things, so the job is to locate the school that needs another CFI. The school directory is broken down by states, so if your sights are set on Florida, there are 80 schools to try. If Colorado is more to your liking, the number is less, but still a hefty 25. The directory also includes full addresses and phone numbers to ease your job search, so tell the printer to run off a few dozen more resumés, put some money in the kitty to pay for the higher telephone bill this month, and get started.

Critical Shortage

by Scott M. Spangler, Editor,
Flight Training **Magazine**

June 1998—Aviation is facing a shortage of specially trained pilots who are critical to its survival, and few people realize this shortage looms.

Aviation is facing a shortage of specially trained and certificated pilots who are critical to its survival. This frightens me, but what's even scarier is that few people realize this shortage looms. Who are these pilots that aviation cannot live without? Quite simply, they are the poorest paid and the most maligned of all professional pilots. They are certificated flight instructors.

It shouldn't have to be said that if there were no flight instructors, there would be no pilots. Yet, the aviation industry and pilot community take CFIs for granted. Young instructors are branded as "time builders" who teach only to acquire the flight time needed to get a "real" job as an airline or corporate pilot. And with more than 67,000 pilots holding current flight instructor certificates, we always have plenty waiting in the wings to replace those who do manage to hitch a ride on the airline gravy train, right?

But how many of those 67,000-plus CFIs actively teach? Reliable estimates put the number at less than 15,000, or fewer than 25 percent of all CFIs. Why the disparity between current and active? No one can say for sure, but here's a safe bet: Most of the pilots who earn a CFI see it as a waypoint—not a destination. Some CFIs have full-time jobs unrelated to aviation and teach a couple of new students a year. Far more formerly active instructors have moved onto "real" flying jobs, but they keep their CFI current as a matter of pride.

Of all the certificates and ratings a pilot earns, the flight instructor certificate is the most difficult. The fact that the FAA recognizes pilots who possess the knowledge and ability to teach the art and science of flying says a lot about their commitment to their aviation education, and letting it lapse would mean throwing away the time and effort it took to earn it.

The problem the aviation community faces is a declining number of active teaching professionals. Here are some additional indications of the growing CFI shortage. In 1997 the airlines alone hired 11,396 pilots. During the same period, the FAA issued 3,958 new flight instructor certificates. That disparity can't continue for long before we feel its effects.

At this year's Women in Aviation International convention in Denver, not one of the 150 participants at a career planning seminar raised a hand when the speaker asked how many wanted to be a flight instructor. Not one.

Can you blame anyone for not wanting to be a flight instructor? My son's piano teacher gets paid more ($20 for a half-hour lesson) than most CFIs, and he doesn't have to earn—or renew—any type of certification to teach.

Whether people in aviation admit it or not, "the industry" caused the shortage by its treatment of CFIs—low pay, and low status—and only the industry can solve the problem. How? Here are some ideas the participants (all 25 of them) touched on during a session at the recent National Air Transportation Association/Professional Aviation Maintenance Association "Super Show" held in Kansas City.

Pay flight instructors more money for the essential service they provide. Too often, flight instructors bear the brunt of general aviation's price competition. Ironically, this works against getting more people to realize their flying dreams. People aren't stupid. They pay $40 an hour for their kids' piano lessons, so they naturally assume they'll pay more for a teacher of flying. When a flight school or instructor apologetically says they must charge $25 an hour for flight instruction, these smart newcomer-students are going to wonder why it's so cheap—and whether they're getting what they're paying for.

Naturally, instructors must give good value for the money they're paid. Flight schools, the employers of most flight instructors, should encourage giving good value and support it by paying for or offsetting the cost of additional training, classes, seminars, and participation in programs such as the National Association of Flight Instructors' Master CFI program.

Paying CFIs more will encourage instructors to make a career of flight instructing, which should benefit students and the flight training industry. Let's face it, a high rate of CFI turnover costs everyone money to fund such things as training new instructors, insurance, and maintaining and building a strong customer base.

In addition to paying flight instructors a living wage, we—all of us who fly—need to treat CFIs as what they are—professional educators. Give them the respect they deserve. And we'd better start now because the shortage is real, and it's getting worse every day. Think about this. When we have no more flight instructors, our flying careers will last only until our next flight review.

POSITIONS AVAILABLE
Western Michigan University—Career Flight Instructors

WMU's School of Aviation Sciences currently provides 4-year degree courses for 500 under-graduates and conducts 12-month airline pilot training courses through its International Pilot Training Center (IPTC), for 96 cadets from Aer Lingus, British Airways, and Emirates. State-of-the-art teaching facilities at Battle Creek's fully equipped W.K. Kellogg Airport and 60 top-of-the-range aircraft contribute to their training environment.

The school also invites applications from Career Flight Instructors to instruct in its IPTC. Applicants should possess an FAA CFII and ME Ratings with 800 hours of previous instructing experience. Candidates who successfully complete the WMU instructor standardization course will qualify for the salary of $45,000.

Information on WMU may be found on our Web site at *http://avs.wmich.edu*. To apply for these positions, submit a letter of recommendation including resumé with a breakdown of your previous flying experience to: Bonnie Sleeman, WMU School of Aviation Sciences, 237 Helmer Road, Battle Creek, MI 49015, USA. FAX 616/964-4676, e-mail: bonnie.sleeman@wmich.edu

Figure 4-1c. CFI pay is changing.

How much is flight instructing changing? A recent ad that Western Michigan University posted tells exactly how the school views its teachers. (Fig. 4-1c.)

Profile: Neal Schwartz, CFI, 340 Hours Total Time

Senior at Duke University, Majoring in Economics

When *Flying* magazine editor Richard Collins called Neal Schwartz on the phone, Schwartz was shocked. But shock was quickly transformed into excitement as Collins told the college student and licensed private pilot that he'd won the Sporty's Catalog Scholarship that would bring him $15,000 to help pay for his additional ratings.

Right from the start, Schwartz attacked the flight school search problem like any good student of economics would. "I didn't want to spend the money inefficiently, so I did tons of research on flight schools," Schwartz recalled.

"I found that Part 141 schools were not necessarily better than Part 61, just more expensive," Schwartz said. "You're really only as good as your own personal study habits. I happen to think that pilot education is more a function of the pilot themselves than the school. The airlines really don't care where you got your ratings from. I went to a Part 61 school and felt it was a better education because I was much more involved in the decision process from the beginning."

Schwartz knew he wanted to fly almost from the beginning. "I knew how fast an F-14 flew when I was 10. I wrote my college acceptance essay on learning to fly since I picked up my private when I was 17. I paid for that by being a lifeguard during the summers." Schwartz now dreams of flying an F-16 for the U.S. Air Force.

For a man who is only 21, Schwartz has many valuable insights to share with other pilots. "Most people completely underestimate the value of networking. Pilots often don't want to be involved in the business end of looking for a job. They feel they've spent their money and expect a job to just come to them. I've met pilots by just walking up and talking to them. Sure it's a little intimidating, but pilots love to talk about themselves."

Schwartz also felt that the Internet has played a key role for him in staying in touch with other pilots. "AOL and its message boards have been incredibly helpful," he said. "I got a job flying an airplane across the country from a message I found online. I'm very comfortable and believe you can weed out the goofs online pretty easily by the questions you ask and their responses. I've found the Student Pilot Network, AVWeb, and the Landings sites (all mentioned in the chapter on useful Web sites) to be very helpful as well."

But Schwartz differs from some pilots when it comes to the issue of a formal education. "I think an aeronautical science degree is a copout," he said. "It's important to have something to fall back on. What can you do with that degree if you need to? You need to have the whole college experience. I think you can fly on the side and read the books and be just as good as someone from Embry-Riddle."

Finally, Schwartz tells aspiring pilots to "get a job at the local airport. The money is terrible, but you're always surrounded by pilots, people who are already doing what you want to do."

A CFI'S Job Is An Important One

On the first day of work as a CFI, you could easily be saddled with three new primary students and a commercial one as well. Obviously, the job is to prepare these students for their ratings, but there's certainly more to your job than that. No one should begin instructing without being aware of the awesome responsibility they hold. Because of you, or in spite of you, a student will eventually take to the skies alone, based on what you've taught them. Teach them well. Make certain your student's brain contains the knowledge they need to keep them out of trouble. Think about them before you send them up solo. Can this student handle an engine failure? How about 360s on downwind from the tower? A good crosswind that suddenly appeared from nowhere? You are your students' role model when it comes to airplanes. You can make or break a career with your attitude and your style.

I've never told this story to a soul because I was too embarrassed, but many years ago I wanted to be an airline pilot and attended a large state university with a flight school program to begin the work to make my dreams come true. I was 17 years old. Perhaps you'll keep this story in mind as you start off on your adventures as a flight instructor.

My flight instructor was a young man of only about 22, but he obviously knew a great deal more about flying than I did, so I settled myself down to learn. Dick was what they called a screamer in the military. (No, I didn't change his name. Perhaps he'll read this someday and realize what an idiot he was.) He didn't teach by presenting and reinforcing; he taught by yelling until students either really understood or were intimidated enough to say they understood. In the airplane, a 7FC Tri-Champ with one seat in front and one in back, Dick taught the practical portion of flying the same way. He yelled! Even worse, however, in the cockpit, Dick would hit me from behind if I didn't perform correctly. I still remember my first landing with him; it's like it was yesterday instead of 25 years ago. On final, I guided the aircraft fairly well, but he screamed at me the last few feet, "Flare . . . Flare . . . Flare!" Wham! We hit the runway. He just kept yelling flare on final, but he'd never bothered to tell me what a flare was. The man was a class A jerk. Unfortunately, at 17 years old, I could only cope with this for a short period of time. I finally left school, convinced that I was an idiot. I didn't fly again for five years.

You heard about the bad instructor. Here's how another instructor saved a career that had not yet begun. As a sergeant in the U.S. Air Force, I was assigned to a Texas air force base that just happened to have an aero club. At this point in time, I had not been inside a small aircraft for nearly five years. Looking back on this, it was pretty obvious to most everyone, except me, that I really did want to learn to fly . . . badly. I began hanging around the aero club but not going in. I just stood around outside, like some lonesome pup, looking at the three Piper Cherokees the club owned.

One day, an instructor happened to come out as I was walking around one of the planes. "You a pilot?" he said. "No," I said. "No, I'm not." "Sorry, you just sort of looked like a pilot." He smiled and walked away. About a week later I was back. Just snooping. The same instructor walked out of the hanger. "Boy. For a guy who isn't a pilot you sure hang around here a lot. Why don't you just join the club?" "No. I don't think I'm smart enough to do this," I said. He just stared at me for a minute. "Who told you that? Some really dumb instructor?" "Well, not exactly," I replied. "Hi," he said. "My name is Ray." I introduced myself and he told me he was about to fly the Cherokee out to check the VORs. "Want to go along?" I only waited half a second before I said yes. Ten minutes later we were climbing westward out of Austin, Texas. "Want to fly it a while?" I took the controls and felt the airplane move to my inputs at the control wheel. I turned, I dove, I climbed and turned, and climbed again. When we landed, Ray asked me if I wanted to join the club now. I did, and I never stopped flying again. Thanks, Ray, wherever you are. You saved a flying soul that was almost lost.

Good teachers not only teach; they influence lives. Use your power wisely.

Time Builders

Once you have your certificates, don't be terribly surprised if some potential employer checks out your logbook and says, "Thanks. But call us when you have a little more time." For the most part, I can guarantee you that this is going to happen. Plan to grin and bear it, but take a look around for some of the flying jobs that perhaps aren't as glamorous as you might think at first. Let me share a bit of personal experience with you on a couple of the ways that I built quite a bit of my time over the years.

Ferrying Aircraft

The first method is by ferrying aircraft, and the second is by towing banners. I still ferry airplanes around the United States. These very words are being written from a splendid Florida hotel where I've arrived after bringing one twin-engine aircraft down to trade it for another to take me back to Chicago.

Ferrying airplanes began for me about 15 years ago, just after I'd received my flight instructor rating. I didn't need the flight instructor certificate for this kind of flying, but the opportunity just happened around that time. I was instructing at an airport near Chicago, and after many months of hanging around everywhere on the airport, looking for a break, I got one. At the airport restaurant one afternoon, another pilot pal of mine introduced me to a young lady who just happened to run an aircraft rental firm (that, unfortunately, has long since gone out of business). During lunch, I mentioned that I had recently picked up my CFI and hoped for a professional pilot's job someday if I ever got lucky and could make the flight time requirements. She looked over at me very casually and said, "I need someone to ferry some airplanes around for our company. Would you be interested?" I almost choked on my lunch!

The first airplane I flew was a Piper Arrow. I think I must have had all of about 10 or 15 hours of retractable gear time, but they didn't seem to mind. I flew to St. Petersburg, Florida, on the airlines, got a checkout and made my plans for the trip back to Chicago. The experience was valuable. Not being too weather smart about Florida, I didn't know that leaving Florida in the middle of a hot summer afternoon was not the greatest of decisions, but I checked weather and left anyway . . . VFR, because the airplane had only one VHF radio and an ADF.

I spent the next three or four hours dodging showers, thunderstorms, and ever poorer visibility. By the time I reached the Atlanta area, the weather was terrible and I finally managed to land at Charlie Brown airport, thanks to a radar steer from a kind Atlanta Approach controller. I landed just in time to learn that the field had gone IFR in rain and approaching thunderstorms. The next morning the field was clear, with two miles visibility and fog, so I asked for and received a special VFR clearance from the tower and departed northwest bound. Once clear of the Atlanta TCA, I called Atlanta Center for VFR advisories. The rest of the trip offered me the chance to talk to more towers and centers on the rest of the trip back. All totaled, I put in about 8.5 flying hours. I gained experience as well as the time, which helped to make up for the fact that the pay was pretty poor. At the time, though, I felt that the low pay was more than worth what I received.

Before that month was out, a conversation with another man gained me another ferry trip. This time I was off to Miami to bring a Cessna 150 back to Chicago. From that trip, I gained 11.5 more hours in my logbook.

What made this whole venture really great was the way it sparked my enthusiasm to find more ferry work. Now, with two long trips out of the way, I was starting to feel more confident about my abilities. I began circulating to other operators at other airports. I had some business cards printed up with my phone number on them. I added an answering machine to my phone just to make sure I didn't miss any possible trips. Another month later, I made a trip to Colorado Springs in a single-engine airplane and came back in another. I ended up flying this trip twice. One trip, I logged nearly 20 hours round trip, the next about 16. It all counted toward increasing those total hours (Figs. 4-2 and 4-3).

Just in case you think this is all glory and fun, though, let me give you a real-life example of just how a typical ferry trip ran just recently. It began with the phone call. "Hi, Rob. It's Jan. Are you free for a two-day trip tomorrow?" There's seldom a great deal of notice in this game. "Sure," I said. "What's the trip?" "First of all," Jan said, "You're going to catch a ride with Tim in the Mooney over to DuPage to pick up a Cessna 421. You have 421 time don't you?" She was relieved when I said yes, because I possibly could have been one of a dozen pilots she called who were either busy or not interested in the trip. "We'll want you to leave as early in the morning as possible. Actually, you'll be taking the 421 to Naples, Florida to drop off Chris, who's going to look at a Mooney. Then you'll be flying the 421 back up to St. Pete." "Sure, no problem," I said. We hung up and I called my wife to tell her what was happening.

Before the phone call was complete, my call waiting beep told me something had changed. "Hi Rob, it's Jan. Change of plans. Why don't you get here around 10 A.M.?" "Fine," I said. The next morning, all packed and ready to go for a two-day trip, I checked my bag. All my IFR charts were current. I don't carry Jepps anymore because the NOS are easier. When

Figure 4-2. A ferry pilot could fly a single engine one day and . . . a turbine aircraft the next. *Hiser Helicopters*

Figure 4-3. A ferry pilot could fly a single engine one day and . . . a turbine aircraft the next. *Beech Aircraft*

they expire I just buy new ones like the VFR charts. If I don't fly IFR for a month, I don't end up spending money on charts I don't use. Next item is a book. Never take ferry flights without a book. You never know when you'll find time to read. Finally, clothes. It's a two-day trip, but I pack for three . . . just in case.

Before I get ready to leave the house for the airport, this morning I call in. "How's the trip coming? Everything okay?" "Hang on just a minute, Rob," Jan says. "Be here around 11 A.M."

"Elevvvven . . . Ah. We're going to have a tough time getting all the way to Naples and back up to St. Pete if we don't get going 'til almost noon. I have to be back here Friday morning." "Okay," she said. I show up and we head out for the ride to DuPage to pick up the 421. After my ferry pilot pays the bill, I do a thorough preflight; this is important! Never fly a ferry flight without a thorough preflight. This is where ferrying airplanes becomes serious work. Since most ferrying work is to transfer aircraft from one place to another so they can be sold. Sometimes the aircraft will be sold before you take off; sometimes you'll be the first contact a potential buyer has with the sales company and the aircraft. Always remember that it's very much like buying a used car. Some people take meticulous care of their machinery, while others are lucky they change the oil . . . ever.

With aircraft, you take a look at the logbooks to trace the history of the machine as well as whether or not the aircraft is legally capable of IFR flight. Your duties as a pilot involve your being sure of the aircraft you fly. Don't take someone else's word for it. Check the aircraft for oil or hydraulic leaks. What shape are the tires in? How close is the aircraft to its next annual? Many sales occur as the airplane approaches the annual. The current owner might just not want the expense, but there could be more problems.

When I preflighted the 421, it looked fine, as did the logbooks. I noticed, though, that the right engine was fairly high time. It had just about reached its TBO. That was definitely something I wanted to keep an eye on, but the aircraft had passed the last annual, so it was legal to fly. After I sat down in the cockpit I took a few minutes to refamiliarize myself with the 421 cockpit before I started running the before-start checklist. When I reached engine starts, I hit the button for primer and starter, and the left prop growled for about a half turn and stopped. The battery was dead. I called for a GPU. The first GPU didn't work, so a second was brought as a replacement. Once the aircraft was started, I taxied slowly, checking ground steering and making sure that all the electrical systems were up. Remember, I still didn't know what had flattened the battery.

After a VFR takeoff (I wouldn't have left IFR with an almost-flat battery), I watched the engine gauges closely. Things looked normal during the short 15-minute flight to the next airport at DeKalb, where I arrived at about 1:30 P.M. So much for a crack-of-dawn departure! At DeKalb I waited for the other pilot, who was supposed to be there when I arrived. He wasn't. I checked weather towards Florida while I waited, and I learned that thunderstorms were building near Chattanooga and Atlanta. There had already been a few funnel clouds at St. Petersburg, thanks to a stationary front about 40 miles from there. I just shook my head and sat down to wait. If my passenger didn't show up soon, the chances of making that kind of distance would be slim, not even counting the state of the weather, which was becoming more exciting by the hour.

The problem with just waiting while the destination weather gets worse is that your anxiety level tends to rise. My passenger didn't arrive until 4 P.M., 5 P.M. Florida time. We launched by 4:20 and made it as far as Birmingham, Alabama, where we stopped for fuel and a potty break. We would have been a bit closer to south Florida if the thunderstorms had not made us deviate west. Along about Chattanooga, where the cumulo bumpus really began, I realized that the aircraft's radar was out to lunch. As I did the walk around at BHM while the fuel was being added, I found an oil leak in the right engine. Not just streaks, but a fair amount dripping off the inboard side of the cowling. It was now 8:30 P.M. Florida time, and the flight was definitely ending right here, right now, until a mechanic could check out the leak, which would not be until the next morning. Don't fool around with this kind of thing. If you're going to ferry airplanes, you must know when to cry uncle. To tell you the truth, the fact that the radar had already rolled over and died had already made the decision for me that I definitely wasn't flying after dark. There would be no way to see what was ahead unless I only tried to avoid the areas of lightning, and I'm not that brave!

The next morning, the mechanics found a loose fitting; they tightened it up and replaced some of the oil we'd lost. By 10:30 A.M., I was on my way. Notice I said I; my passenger had left. He needed to be in south Florida earlier, and he caught an airliner out, so all I needed to do was fly to St. Petersburg. That took about 2 plus 20 right up to the hangar of the waiting buyer. Of course, by this time, I was almost a day later than I had planned. Good thing I had two day's worth of clothes. The mechanics would be inspecting the aircraft before the buyer decided to buy or not and said they might be finished by noon the next day. Other commitments were now conflicting with this late time frame, and I called the sales office to let them know it could be a problem. They weren't happy about it, but it happens. These days, there are a lot of freelance ferry pilots, with very few on salary to anyone. This is a benefit to the sales company because they only pay for you when they need you, but it does sometimes cause headaches.

By the next morning, I was starting to get concerned because of my commitments. If I brought the buyer's aircraft back (a Piper Seneca) it would take seven hours from St. Pete back to Chicago, but there were a ton of thunderstorms between Florida and Illinois. The two-day trip was now in its third day, looking at a possible fourth. Luckily, the salespeople realized the dilemma before I even mentioned it again. The 421 stayed in St. Petersburg, as did the Seneca for the next pilot. I took the airlines back to Chicago and made my meeting with an hour to spare. Of course, I did manage to add almost six hours of C-421 time to my logbook.

Is ferrying airplanes for you? Well, in this little scenario I've tried to give you a look at some of the bad along with the good. Ferrying aircraft can be a great adventure and a heck of a lot of fun. I've gotten to fly many different kinds of aircraft into places I'd often only heard of. Becoming involved in this kind of work tends to have a cumulative effect, too. Once people know you're around and available for a trip, they tend to use you. More than once, I've been called by more than one firm to fly trips on the same day. Then there might be times you might not fly any for a month, so while the work can certainly be interesting at times, it definitely is not steady. A friend of mine and I both ferry for the same company, but I seem to be called more often than him. This is where being assertive helps. I call the scheduler about once a week, just to say hi. I'm convinced that's why I'm called out more often.

The pay for ferrying aircraft is hardly union scale. The pay is based pretty much on a couple of things: how badly the company needs you or an airplane somewhere else, and how good a negotiator you really are. I remember ferrying airplanes years ago for $50 a day plus expenses. Today the rates are higher, but you'll have to stand up for what you think you're worth. Just be aware that some other people might not value your services as highly as you do.

The only regrets I have are having missed some of the really great trips. Last winter, I was set for a trip from Chicago to Tacoma, Washington, in a C-172. It probably would have been about 12 or 15 flying hours, but the scenery would have been great. I was weathered out. The other trip I really wanted was ferrying a couple of C-421s from London, England across the Atlantic back to the Midwest. Stops in Iceland and Greenland and Labrador would have made for quite an adventure.

I don't think I would ever call ferrying airplanes boring, as long as you realize your limits—just how far you're willing to go in what kind of an airplane and into what kind of weather. If you ever run into a ferry job where the contractor seems to care more about his machines than about your life, there's only one solution. Run—fast!

Banner Towing

Banner towing actually turned out to be more interesting than I first expected. Let me tell you how I found the job. This might have been just luck; I don't know, but as I was wandering around a local airport, I happened along on an airplane of a slightly different model than one I had once owned, a 1968 7ECA Citabria that I logged some 600 or so hours in during the few years I owned it. As I was wandering around this airport, just being generally nosy, which means looking and talking to people, I happened to see another Citabria painted in the same scheme as my old one. Because taildragger aircraft are such a rarity these days, I walked over to look at it and found the aircraft to have an unusual array of what looked like chicken wire strung beneath the aircraft from wing to wing. The tail also had some sort of unusual hardware attached near the tailwheel. As I found myself wondering what it was all for, the owner walked up and we started talking. I mentioned I had owned a similar aircraft.

He told me the hardware on the back was for the tow hooks to grab aerial banners, while the chicken wire arrangement was a night sign that from the air would look much like the moving marquee at a bank that tells you the time and temperature. The contraption that ran the night sign was a big steel box that sat between the pilot's feet and in front of the control stick. I guess the owner figured out I was okay when he heard how much taildragger time I had logged. All totaled, I probably had about 1,000 hours at that point in time. He told me he'd be looking for a new tow pilot in a month or so and to give him a call if I was interested. I must say I made a regular weekly pest of myself. I called just often enough to let him know I was interested. He finally called back to say the training would be in a week, and he asked if I could make it. I was there in a flash.

Because I already had time in the airplane, the training process was greatly reduced. The ground school lasted an hour, I think, while Barry explained the main parts of both the underwing night sign and the tow system. Considering what the job was, aerial advertising, the equipment was really quite simple. For a banner tow, the pilot connected three long lines, each with a hook on one end, to the hardware at the back of the tail-wheel. The hooks as well as the rest of the rope were pulled back inside the cockpit. As the pilot needed another hook and line for a new banner, he or she simply opened the door in flight (better be buckled in), and tossed the rope and hook clear of the aircraft. It then swung free about 15 feet below and behind the plane. Each time you were through with the banner, you'd pull a lever in the cockpit, and both the banner and the rope would drop free back to the ground. Sounds easy. (Figs. 4-4, 4-5, 4-6.)

Figure 4-4. A good banner tow is a three-step process. *Ad Air Lines Inc.*

Figure 4-5. Message viewed from the ground. *Ad Air Lines Inc.*

Figure 4-6. Banner tow on its way to work. *Ad Air Lines Inc.*

"If there was a tough part, it was picking up the banner. But I'm going to let my friend Sark Boyajian—another banner tow pilot—tell you how that all works."

Profile: Sark Boyajian, Banner-Tow Pilot

Banner towing is basically outdoor aerial advertising that sends personal or business messages. I am authorized to fly anywhere in Chicagoland and surrounding states.

The banner tow behind the aircraft can be as long as 125 feet. Each letter attached to that sign is 5 feet high and 3 feet wide. The maximum sign I will tow, then, is about 40 characters, although some banner-tow companies, like those in Florida, might tow a sign with as many as 60 characters. I've never had to refuse a message, because there always seems to be some way of abbreviating the words.

The aircraft I fly, the Cessna 150, is the most efficient and least expensive one to fly in a banner-tow operation. To pull signs longer than 40 characters, I'd need an aircraft with a more powerful engine because of the drag the big banner produces on the aircraft, and such an aircraft could be much more expensive to maintain.

The aircraft towing the banner is normally flying about 50 to 60 m.p.h. If I fly much faster than that, the nylon banners behind the airplane can rip. The banner itself is weighted on the bottom to make certain it flies correctly behind the airplane. There's also a small tail chute on the end of the banner. Without that, the banner would continually twist and spiral, making it impossible to read from the ground.

The method that I use to pick up the banner is rather unique since we never take off with the banner already attached to the airplane. It would drag on the ground. Imagine the sign lying on the ground flat. A short pole at the beginning of the banner, where the message starts, has a 250-foot rope attached to it. That rope is then stretched across the ground and eventually strung between two 10-foot-high poles, much like a line strung between two small goalposts. Attached to the tail of the airplane is a 30-foot steel cable with a grappling hook on it.

Once the aircraft has taken off, I release the grappling hook, and it trails along behind the airplane. I try to fly the airplane close to the ground, normally about 20 feet high, with the hook trailing in back, and aim between the goalposts. Just before I reach the posts, I pull the nose of the aircraft up very quickly. The hook in back swings much like a pendulum on a clock and, hopefully, grabs the rope from between the posts; if so, I fly away with the banner attached.

I know when the banner is attached because I'll feel a bump and actually feel the airplane slow down as I'm pushed forward from the drag that's been added by the banner. I climb the aircraft out steeply to prevent the banner from dragging along on the ground. Eighty percent of the time I pick up the banner the first time, but on really windy days it may take three, four, or even five times to pick the banner up because of the way the hook swings around.

I also never land the airplane with the banner still attached. Again, it could get torn up on the ground. I tell the tower I'm coming in to drop the banner, and I begin a gradual descent over the airport to around 100 feet. When I am near the ground crew, I pull a lever in the cockpit that releases the tow cable, and the banner floats down to the earth.

Weather is also very important when I tow a banner because sometimes the forecasts are not correct. The federal regulations say we can't be closer than 500 feet to any clouds, nor any closer to the ground than 1,000 feet in a congested area. When I look out the cockpit window, the visibility must also be a minimum of five miles.

We will obviously not fly in thunderstorms or freezing rain, but we will fly when it's raining or in light snow or even when the winds are strong. The wind doesn't affect you as far as getting up into the air, but when you are up there on a windy or gusty day, I am going to be tossed around because the airplane has this long rudder and the wind just pushes me up and down and sideways, particularly in the Loop area. There have been a couple of days when I was flying that I had to go back to the airport because the airplane was almost uncontrollable because of the way the winds blow around the downtown skyscrapers. As far as the readability of the banner to the customer, it might be twirling a little more in the wind, but it's still quite stable.

During the evening hours, when you would not be able to see a banner, I can advertise with my electric aerial billboard from about 1,000 feet above the ground. It's a wire grid strung beneath the aircraft from wingtip to wing tip. When it's extended, it measures about 40 feet long and about 8 feet wide. The grid is wired with between 200 and 300 light bulbs, about a foot apart. The machine with the message is in the cockpit, and when I run that, the message appears on the grid, beneath the aircraft, like the moving marquee sign at a bank. I can also vary the speed that the message moves at. The advantage of the billboard over a banner is that I can change the message while I'm in the air and stay over, say, Soldier Field, for three hours at a time with different messages.

The maximum I will fly is eight hours a day, but that long a day is rare. For instance, when the Bears are playing, the flying time is about $2\frac{1}{2}$ hours or possibly $3\frac{1}{2}$. This gives me time to pull about two banners in an afternoon.

There is a lot of pressure in this job. From the first moment I speak to a customer until the job is completed, I have to be aware of the weather. I am always looking at my watch wondering if I will take off on time. Will there be a problem? Will I be able to get back home safely? But it's still a great job.

I fly almost everywhere. I'll fly around a house, a boat in the lake, a ballpark, or even the Hancock building. The banner plane circles the target—the Hancock building, for example—in a counterclockwise direction, since that's the way the banner is written—left to right. The normal distance I fly is about a half a block away from the building. If I fly farther away, the letters are too small to read.

I flew past the 95th restaurant there once with a banner that said, "Jane, will you marry me? I love you. John." My policy is if the woman says no, the guy doesn't have to pay.

Pipeline Patrol

Pipeline patrol jobs are tough to locate, but they certainly can help you build time—quickly! The days are long, and the airplanes or helicopters are sometimes small (Fig. 4-7). One company I know of employed a pilot in a Mooney to begin in Minnesota on Monday and follow a natural gas

Figure 4-7. Helicopter on pipeline patrol. *Helicopter Association International*

pipeline (visible from above by certain markings unique to pipelines) and follow it for almost 800 miles down through Louisiana. To avoid the risk of missing something on the ground, the flying couldn't be performed at high speed, nor at high altitude. I've seen pipeline patrols that finished five states away from where they began, only to take the pilot off on some tangential course in some other direction three more states eastward.

The flying speed in the Mooney for this work was around 80 knots, so the flight took a very long time. Altitudes were often at 500 feet or even less in uncongested areas. Realize too, that pipeline or powerline patrol is not always flown in good VFR weather. One powerline pilot told me that special VFR can be a way of life for this kind of flying. It can be dangerous, but you certainly will see a great deal of the country. When the pilot in this flight finished his patrol, he'd turn around and fly the reverse track along the route to his starting point again. One of my friends found his pipeline job by word of mouth, just hanging around the airport and telling everyone he met that he was looking. You might also try calling local utilities. Their public relations departments should be able to tell you if that company uses aircraft to patrol their lines. If not, they might be able to steer you in the right direction.

Freight Flying

In case you're wondering why I'm bringing up freight flying as a somewhat alternative job when everyone knows the pilots at Federal Express or UPS are flying big aircraft, DC-10s, and Boeing 747s, let me say that those jet carriers represent only a small portion of the freight carried in this country. There are literally hundreds of small charter companies flying mail and small packages or even boxes of nuts and bolts to where they need to be—usually to small towns or medium-sized cities. Often these freight routes are flown in older aircraft specifically selected for their ability to haul large amounts of cargo, but not necessarily in conditions some pilots might find ideal.

A company here in Chicago flies night freight in Beech 18s and DC-3s, both old radial engine aircraft from the era of World War II. The reason they use these aircraft is they're like flying Mack trucks. They'll fly with just about anything that can fit into the fuselage cargo doors. I recently had an old freight pilot, Mark Goldfischer (now a DC-9 first officer for a national airline), tell me what he recalled about his time flying cargo.

> I was employed at Zantop International for almost four years (May '85–March '89). Believe it or not, I first learned of Zantop from an old magazine story, while flying canceled checks in Florida and California. Through the article, I realized that they would consider me with my 2,000 hours total at that time; I was 25. I was also parking next to

Zantop at the Atlanta freight ramp most nights. Using this opportunity, I spoke to a couple of pilots, and they said to get the Flight Engineer Turboprop written out of the way, then call Zantop. I did, and I landed an interview and the Flight Engineer job on the Electra (L-188).

I spent my first year and a half flying out of Willow Run Airport in Ypsilanti, Michigan. The operation involved transporting civilian freight to most of the major cities in the United States.

Zantop had a fleet of 23 Electras at the time, the largest Electra operator in the world. The Electra is a wonderful airplane from a pilot's perspective. To this day she's my favorite. Lockheed made her as strong as a workhorse, with plenty of power. I've talked to many pilots throughout this industry who have flown the L-188, and I always see a twinkle in their eye and a warm smile on their face when they think of the days when they used to pilot her around the skies. Whether flying people or boxes, she's a real treat. Zantop also operated DC8s and Convair 640s.

Flying freight takes a toll on your body and health, though. Whether you make $25,000 or $125,000, it doesn't matter; your body doesn't know the difference. I can honestly say that 50 percent of the time in the saddle, I was fatigued, and 25 percent of the time I was too tired to be flying. That was the worst part of the job. Inevitably, when your body's clock struck bedtime, it was time to go to work. Of course, I've said nothing of all the time spent eating out of machines, trying to stay awake by gulping coffee in absurd quantities, and all the other wonderful luxuries that come with the freight industry. (See Fig. 4-8, "The Sign of the Freight Dog.")

On the other hand, direct clearance to your destination, less weather to worry about, and uncluttered radio frequencies and airways are part of the good side to hauling boxes when the rest of the world is sleeping. Hotel time? Yes, plenty of it. For the most part, package pilots spend all too much time in the "pilot prisons." One year, I spent 250 out of 365 days in hotels! Ouch! Lucky for me, I didn't have a family back home.

Yet, through all the sham and drudgery, somehow I still managed to have a lot of fun. What do I miss the most? The guys. My fellow freight dogs.

For some reason (and I'm sure I'm not being totally impartial) freight pilots seem to have more varied and vivid personalities than their people-pushing cronies do. Freight pilots are more humble and likable as people. For some reason, their egos don't get overinflated with huge salaries and flirting flight attendants. They do their job silently and safely and go home without much ado about anything. They don't get enough credit for a job that's done as well as any passenger pilot who works for some major airline. I miss their kind.

The schedules vary, depending on your company, base, and equipment. At Zantop in Ypsilanti, I worked three to four days a week. Weekends were mostly free because most freight companies didn't move freight the whole weekend. Zantop played the game a little differently, though, because, on Saturday, they'd spend the money to

Figure 4-8. Freight Dogs Anonymous. Order of the Sleepless Knights. *Mark Goldfischer*

commercial flight you back and out to whatever city on Monday in time for the Monday night hub. Most companies don't foot this bill. A typical showtime would be 4:00 P.M. If you had to come in and go out on the same hub, that would really take the wind out of your sails. The wait time was about three hours. You could either go to the Lazy-boy lounge for sleep or to Denny's for eats. Needless to say, you get to know Denny's menu real well. And sleep was always that kind of unsatisfying, dirty sleep.

Military Contract Flying

Logair was different. I spent my last year as FE in Warner Robbins, GA, flying freight for the Air Force. Schedules were a week on, and a week off. Actually, Logair was where you wanted to be. You were home almost every night.

I left for two reasons. One was, of course, to further my career at Pan Am, and two was because I had simply had enough of freight flying. It was starting to affect my health, which I couldn't accept. I would definitely recommend a freight job to build time; it's a neat side of the industry to experience and learn about. As a career? Well, for some, maybe. It's up to the person. I can only speak for myself in that it wasn't meant to be. For some, it could be very rewarding. Incidentally, my qualifications at time of hire were CFII, MEL, A&P, turboprop FE written, and a four-year B.S., Aviation Technology, from Embry Riddle.

Flying for AirNet: A Time Builder or a Career?

An Interview with Craig Washka, Director of Training

In April, 1974, Jerry Mercer became AirNet's first pilot, in addition to being the sole team member. Little did he know then that AirNet—originally called PDQ Air Service—would grow from a single pilot, single airplane company, into one 25 years later with 118 aircraft and 150 pilots and 1,200 employees nationwide. That original aircraft—N1814W, a BE-58 Baron—still flies for AirNet.

The concept was simple. Banks needed canceled checks moved between cities as quickly as possible to reduce the interest they'd pay on those checks—the float, in banker lingo. Mercer provided that opportunity with an ever-increasing fleet of piston aircraft—Navajos, C-310s, Aerostars, and Barons—and jets—Lear 35s. AirNet is now the largest transporter of canceled checks in the world. "Our niche market is delivering time," said Craig Washka, AirNet's director of training.

One of the best parts about flying for a company like AirNet—Starcheck on the radio—is that pilots will be out building PIC time almost from the first. They can move up rather quickly too. "In a year and a half, a pilot can move from piston captain to first officer on the Lear 35," said Washka. "In fact, we have the largest civilian fleet of Learjets in the world. Upgrades are based on a seniority bidding system. Because of attrition, (about 6-8 pilots per month) AirNet is always looking for new pilots. "We have never had a furlough either," Washka added.

Pilots will log 1000 hours or more a year at AirNet, flying about six hours per night, four nights per week. They could be based at any of AirNet's 70 bases scattered around the continental United States from which they fly some 585 flights each night during the week and 150 on weekends. AirNet pilots are all salaried based upon their position—piston captain, Lear first officer or Lear captain (see insert). All salaries are based upon the number of years a pilot has worked for the company. A piston captain starts at $18,000 per year, with a $3,000 raise after one year. Washka adds that "there is a $1,500 sign on bonus as well." All training is paid for by the company, but pilots do sign a training contract—a promissory note actually—to help AirNet recoup some of their training costs if the pilot leaves the company early. For a closer look at AirNet's pay scale, see Appendix D.

"We know that many of our pilots will use us as a stepping stone to the majors, although many of our pilots are career-changers." Washka said. "We just expect them to be very involved in a challenging career on the back side of the clock. Short of going to war, this is the most challenging flying around."

A typical day—or night actually—for an AirNet pilot begins around 9 pm, like run number 501 for example. The pilot leaves AGC (Pittsburgh, Allegheny County) at 9:10 and flies one leg to CMH (Columbus, Ohio), where they'll help unload and reload the aircraft and be off again by about 1 am for CLE (Cleveland, Hopkins). There they layover until about 4:15 am when they depart again for CMH and turn the aircraft around to depart by 5:30 am for AGC. Upon landing at AGC, the pilot signs out for the night about 7 am—10 hours of duty and about 3.6 hours of flying.

Interested? Here's what you'll need to be considered—A minimum of 500 hours total time, with a multi-engine and instrument rating. An ATP is not necessary. "Initially, our interns screen the resumes," says Washka. "That resume must be put together well. Those interns are like hounds and look for spelling errors. More than two pages and you can forget it too. We like to learn about a pilot on a single piece of paper and respond to every resume we receive, currently about 250 per month."

"If the pilot meets our minimums and the resume is well laid out, they receive an application. If an applicant lives close to an American Flyers location, we send them there for a preliminary assessment that we call PASS (Preliminary Assessment Selection Service) Phase 1. There, they take a written test to learn their basic knowledge level. If they score 70 percent or better, they go on to a simulator check in a Frasca 142 to see if they can fly to commercial pilot standards. 80 percent of the hiring decision is weighted toward the simulator ride. They also take a PPS Personality test. It is a tool to help us make a decision, but it has been deadly accurate. If they pass all of this, we invite them to Columbus for an interview."

The interview lasts about 20 to 30 minutes and we do it in the middle of the night. We are trying to learn who they are. Do they have knowledge of the aircraft they're currently flying. We also ask them some situational questions and whether they'd like flying at night. We'll tell them on the spot, or certainly within a few days, if they're hired. When we give them a class date, they give us a $300 deposit to hold their spot. We give them the money back after training is completed through bonuses, moving expenses, per diem, etc."

Training at AirNet—like flying there—takes place in the middle of the night. Ground school runs from 8 P.M. to 6 A.M. for a month. "This is where some pilots might decide that flying all night is not for them," Washka said. "Pilots are not actually hired until they pass their Part 135 checkride at the end of training." All initial AirNet training is accomplished in a Baron and the Frasca 142 and AST 300 simulators. AirNet pays for a pilot's lodging in a large apartment complex near the airport where they learn to work together in training teams.

"While they're in training, we observe how well they work together," Washka adds. "We expect the strong ones to pull the weaker ones along. The pilots who go home every weekend usually don't make it." After training, pilots will fly together with a line captain before being set free to fly alone. That could be two days later or a week. Most PICs are on their own within a week. "It's like a whirlwind, night PIC and IFR right off the bat," said Washka.

Want a few tips for getting hired at AirNet? "Be prepared and research the company," Washka says. "Make sure you review the regs, weather, multiengine procedures and the AIM. You'll need to fly a simulator before the interview too. If a pilot is not current on instruments, they're kidding themselves to think they'll get through. "Here, attitude is everything," he added. "Some applicants we've met are great pilots but don't have the right attitude. Remember too, that part of the interview is me asking whether you have any good questions about the company or the job. A pilot with the right attitude and good stick skills will succeed here."

"The major airlines realize our pilots are very talented," Washka concluded. "We hire so many."

5

The Regional Airlines

Now is the time to set the record straight. "Commuter" pilot is out, and "regional airline" pilot or simply "airline" pilot is definitely in. If you're flying a B-737 or an Airbus, you might wonder why anyone cares, but to a regional pilot, "commuter" pilot just doesn't tell the right story anymore (Fig. 5-1).

In years past, a commuter pilot was someone who drove little airplanes—Aztecs, Navajos, King Airs, and the like. Those eventually evolved into Metros, Dorniers, Shorts, and Beech 99s, hardly little

Figure 5-1. New-generation regional airliner. *Courtesy Canadair*

airplanes, but still the name "commuter" pilot stuck. With that name came an impression of the pilot—someone young and inexperienced and flying little scooter airplanes. Certainly not something to encourage much respect in the major airline circles. But, as Bob Dylan once said, "The times they are a'changin'."

At the time, the norm for regionals was a loud, cramped, 19-seat turboprop. Today, only a few companies produce 19-seat airplanes. New regional aircraft appear today with glass cockpits and sophisticated flight management systems to squeeze out every ounce of bang for the buck. Even the level of regional airline cabin service is improving: On some flights, instead of packs of peanuts, meals are served. Because the vast majority of regional airlines now code-share with a major airline, the overall trend of the '90s is to provide a class of service that closely resembles that of the regional airline's major partner.

A current market forecast says, "Regional airlines are already a key element in the strategy of majors and flag carriers, and will continue to demonstrate strong growth in the long term." The forecast, recently released by Bombardier/de Havilland, also states that, "The number of seats offered by regional carriers is expected to grow at an annual average rate of 3.7 percent . . . with delivery of 7,420 aircraft in the 15- to 90-seat range . . . over the next 20 years."

Along with increased seating capacity comes increased range as regionals opt for small, pure jet aircraft. Certainly a less well-publicized aspect of pure jet aircraft is curb appeal. Some passengers think a jet is safer, more comfortable, and overall, a better machine to fly on. A recent Bombardier study says that as many as 100 new city pairs might materialize from agreements between the United States and Canada, routes that would be perfectly suited for the regional jets like the Canadian RJ, EMB-145, and the Dornier 328 jet.

Business Express, recently purchased by American Eagle, now flies jets, while Comair, a Delta Connection company, expects to be an all-jet regional soon. But the new turboprop regionals soon to be showing at major airports everywhere will give the jets, and even the majors, a run for their money when it comes to treating passengers in style and safety.

Some of the new regional aircraft are brand new, while some evolved from earlier models. And some fall somewhere in between. The Saab 2000 looks like a superstretched version of the Saab 340. The Fokker 50 looks like an F-27. The Canadair RJ looks just like a stretched Challenger business jet. Yet, while these aircraft resemble earlier models, their performance places them in quite another category. The RJ seats 56. The Saab 2000's 360-knot cruise is almost 80 knots faster than the smaller Saab 340 and the fastest of the regional turboprops. The Fokker 50 slices through the sky nearly 40 knots faster than the F-27.

The Dornier 328, the ATR, and the Embraer Brasilia are all new designs, while the Beech 1900D is a new version of the 1900 first introduced in 1984.

If you've flown a Boeing 737-300, a B-757, or even an Airbus, you'd feel right at home in a new-generation regional airplane. Almost every regional airliner leaving the factory these days incorporates a glass-cockpit design that brings to these aircraft the efficiencies of advanced flight management, engine and avionics control systems, which until recently were the domain of only large air carrier aircraft.

Regional carriers operating the de Havilland of Canada Dash 8, for instance, expect to begin certification of crews to Category II landing standards in the near future, too. The new 35-seat Dornier 328 will use a sophisticated Honeywell SPZ-8800 integrated avionics system to give the crew a simple, yet precise, answer to whether climbing to FL 250 even with a headwind or staying low is more efficient for the best fuel burn.

This all means that much of the back-breaking work involved in regional operations, like eight or more instrument approaches a day, will be reduced. At some regional carriers today, large aircraft—like the Shorts 360, which weighs in at 26,000 pounds, or like the 46,000-pound F-27—are being hand-flown on 12- to 14-hour duty days because the aircraft have no autopilot. Many of the 19-seat aircraft, such as the BAe Jetstream 31 and the Dornier 228, have the more common two VORs, an ADF, and a transponder for continuous operation. When was the last time you flew a 12-hour day by hand in a B-737?

As cruise speeds for turboprops continue to rise and as more and more jet aircraft join regional airline fleets, the typical trips that regionals fly are changing dramatically. No longer will all regional airline pilots be found banging around in the bumps and weather at 6,000 feet and 200 knots. Many regional pilots are moving into the big leagues with the aircraft they fly. But these new airplanes are going to force changes in the way people—such as airline managers, FAA officials, and fellow pilots—view regional operations.

BAe Jetstream 41

(See Fig. 5-2.) The British Aerospace Jetstream 41 that rolled out of the Prestwick, Scotland, facility in March 1991 might look very much like just a stretched version of the Jetstream 31, but it's not. One of the most obvious differences between the 31 and 41 is the two-piece windshield of the 41 and the wide scan visibility it now provides. A four- or five-tube Honeywell electronic flight instrument system (EFIS) dominates the instrument panel of the 41, while a Honeywell Primus II system controls the radios. Although the aircraft will also carry a flight data and

Figure 5-2. Two Jetstream 41s. *Courtesy Jetstream Aircraft, Inc.*

cockpit voice recorder, the autopilot is again an option. Most North American operators are expected to order the aircraft with this option, however.

Although the cabin of the Jetstream 41 will seat as many as 30, most airlines are expected to fly the aircraft with just 29 passengers. British Aerospace will continue producing the 19-seat Jetstream 31 even after the company begins deliveries of the first of the 114 Jetstream 41s already ordered. The aircraft should pass its FAA certification tests by fall, 1992, with deliveries to begin in mid-1993.

British Aerospace RJ70

(See Fig. 5-3.) A derivative of the successful BAe 146, the RJ70 is poised for a head-to-head fight with the smaller Canadair RJ. To expand service as well as gain operating experience with a jet in the regional airline marketplace, Business Express began operating five leased BAe 146-200s while the carrier waits for its RJ70s to arrive. Air Wisconsin, and other ALPA regional airlines have been operating BAe 146s for many years.

Figure 5-3. British Aerospace RJ70. *Courtesy British Aerospace*

The RJ70 is expected to operate with essentially the same reliability as the proven BAe 146, which currently shows approximately 99 percent dispatch reliability. With five-abreast seating, the RJ70's cabin, which will seat 70 passengers, is only four inches narrower than a B-737. The RJ70 is also compatible with most airport jetways. While the early BAe 146s were equipped with standard analog flight instruments, the RJ70 will incorporate a full four- or five-tube Honeywell EFIS system designed especially for this aircraft. Full major airline cabin service will also be available.

A new two-engine version of the RJ70 is on the drawing board, too, featuring seats for 136 passengers and a fly-by-wire control system. The RJ70 is one of the largest of the regional aircraft available. (See Table 5-1.)

BAe ATP

(See Fig. 5-4.) While airlines operate fewer British Aerospace ATPs than either of BAe's two other regional aircraft, the BAe ATP (for advanced turboprop) is a massive aircraft when viewed from the traditional scope of 19- to 29-seat airplanes; the ATP seats as many as 72 passengers.

Table 5-1. *BAe RJ70 specs*

Max. takeoff weight	84,000 lbs
Payload	19,000 lbs
Range	1,135 nm
Engines	Lycoming LF 507s @ 7,000 lb thrust
Max. passenger load	70
Mmo	0.73 Mach

Figure 5-4. BAe-ATP. *Courtesy British Aerospace*

In the United States, United Feeder Service in South Bend, Indiana, is the only airline operating the ATP, with 10 currently in service. They're configured for 64 passengers.

The ATP also features a Smiths' glass cockpit for the primary flight instruments system and the navigation display. The ATP operates with two Pratt & Whitney PW 126As, which produce 2,653 shaft horsepower each. Hamilton Standard division of United Technologies has designed and is manufacturing a new-generation six-blade propeller for the ATP,

Table 5-2. *BAe ATP specs*

Max. takeoff weight	55,550 lbs
Payload	15,200 lbs
Range	600 nm
Engines	P&W 126As @ 2,653 shp
Max. passenger load	64
Vmo	227 kts

which gives the aircraft an extremely low operational noise level. (See Table 5-2.)

De Havilland Dash 8

(See Fig. 5-5.) The original de Havilland of Canada Dash 8-100 first flew in 1983. It's configured for 37 to 40 passengers, but a larger model 300, arranged for 50 to 56 passengers, will soon be available. In March, 1992, de Havilland announced the launch of the Dash 8-200 model, which essentially is the airframe of the —100 model with the engines and propellers of the —300 model, giving the aircraft a 30-knot increase in speed

Figure 5-5. Dash 8-400. *Courtesy Bomardier.*

Table 5-3. *Dash 8-100, —300 specs*

Max. takeoff weight	41,000 lbs
Payload	11,500 lbs
Range	840 nm
Engines	P&W 120As 2,150 shp
Max. passenger load	(-100)40
	(-300)56
Vmo	271 kts

as well as improved single-engine capabilities. A glass cockpit is the standard in new models of the series.

U.S. Air Express is the largest operator of Dash 8 aircraft in the world, currently flying 43. In all, de Havilland has delivered 304 Dash 8 aircraft and has an order backlog of 80. (See Table 5-3.)

Saab 2000

(See Fig. 5-6.) People want to fly to their destination quickly. With a 360-knot top cruise speed and a service ceiling of 31,000 feet, the Saab 2000

Figure 5-6. Saab 2000. *Courtesy J. Lindahl*

Table 5-4. *Saab 2000 specs*

Max. takeoff weight	48,500 lbs
Payload	13,000 lbs
Range	1,200 nm
Engines	Allison GMA 2100s @ 4,152 shp
Max. passenger load	58
Vmo	360 kts

will accomplish that goal easily. Designed to carry 50 to 58 passengers, the 2000 will complement the 34-seat Saab 340; the company also markets to the regionals. The 2000 is outfitted with a Collins Pro Line 4 avionics system, which includes a six-tube CRT display system as well as electronic engine indication and crew alerting system (EICAS). The Saab 2000 is expected to be one of the first turboprop regional airliners certified for landings down to Cat IIIa minimums.

Saab has delivered 266 of the smaller Saab 340s and has a backlog of more than 100 aircraft. At present, the manufacturer has nearly 200 firm orders and options for the Saab 2000. (See Table 5-4.)

Beech 1900D

(See Fig. 5-7.) The Beech 1900D is one of the last remaining 19-seat aircraft in production. The D model features a stand-up cabin 5 feet, 11 inches tall, an unheard of height in a 19-seat aircraft. The cockpit contains a four-tube EFIS with full flight director and autopilot systems. Three different airlines have 89 of the 1900D models. The original Beech 1900 was first introduced in 1984. (See Table 5-5.) (See Figs. 5-8 and 5-9.)

Dornier 328

(See Figs. 5-10a, 5-10b, and 5-11) Until the Saab 2000 was announced, the Dornier 328 hoped to be the fastest turboprop regional airliner around. Now it's the second fastest. But at 345 knots, it's still a quick way to shuttle 30 people around by most anyone's standards. The near supercritical wing is similar to that used on many large jet aircraft. The 328 is also made from a number of composite materials to save weight and speed the construction process.

The advanced avionics of the Dornier 328 cockpit equipped with a Honeywell SPZ-8800 are what will attract pilots, though. The initial aircraft, destined for Horizon Air, was certified with a head-up display (HUD), like

Figure 5-7. Beech 1900 D. *Courtesy Beech Aircraft*

Table 5-5. *Beech 1900D specs*

Max. takeoff weight	16,950 lbs
Payload	6,510 lbs
Range	700 nm
Engines	PT6A-67Ds @ 1,279 shp
Max. passenger load	19
Vmo	289 kts

that on Alaska's B-727s. With the instrument panel in the Dornier dominated by as many as five 8-by-7-inch CRTs, the cockpit of the 328 looks much like a foreign car with a clean, smooth appearance. Operators have a full range of options for the 328, including full flight management systems (FMS) and a laser inertial reference system (LIRS). Dornier still produces the unpressurized 19-seat model 228. (See Table 5-6.) Dornier has also begun certification work on an entire new family of jet aircraft as well. The smallest, the 328 Jet, is based upon the turboprop 328 fuselage. Other versions include a 70-seat, Dornier 728 jet.

Figure 5-8. Embraer ERJ-145 regional jet.

Figure 5-9. EMB 145 cockpit.

Table 5-6. *Dornier 328 specs*

Max. takeoff weight	27,558 lbs
Payload	7,606 lbs
Range	700 nm
Engines	P&W PW119s @ 1,815 shp
Max. passenger load	30
Vmo	345 kts

Figure 5-10a. Dornier 328 jet.

Figure 5-10b. US Air Express Dornier 328. *Courtesy www.aeroimages.com.*

Figure 5-11. Dornier 328 cockpit.

Canadair RJ

Anyone who has ever flown the corporate Canadair Challenger jet will like the 50-passenger Canadair RJ. Those who have never flown the Challenger will certainly like the RJ anyway. Basically a stretched version of the Challenger, the RJ competes with the EMB-145. (See Table 5-7.)

Table 5-7. *Canadair RJ specs*

Max. takeoff weight	47,450 lbs
Payload	13,878 lbs
Range	1,435 nm
Engines	GE 34 3A-1s
Max. passenger load	50
Mmo	0.80 Mach

ATR 42 & ATR 72

Produced in Europe by the partnership of French Aerospatiale and Italian Alenia, ATR's two models—the 56-passenger ATR 42 and the 74-passenger ATR 72—offer something for each end of the regional spectrum. Currently in operation with American Eagle and Continental Express Airlines, the ATR has a cockpit that was designed with the technology that Aerospatiale used in its partnership designs on the Airbus A310.

The ATR uses composite materials extensively in the wing and propeller structures. Up front, the ATR uses a four-tube EFIS connected with a King Gold Crown III avionics package. The ATRs can be certified to Cat II standards. (See Table 5-8.)

Embraer Brasilia (EMB-120)

(See Fig. 5-12.) The Embraer Brasilia (EMB-120) still reigns as one of the quickest turboprop regional airlines. Capable of 300-knot airspeeds, the EMB-120 carries 30 passengers at altitudes up to 32,000 feet. The large, roomy cockpit is dominated by a standard five-tube EFIS panel and a system layout that most Brasilia pilots agree is the best training around for anyone who is planning to move up to jet aircraft. But most of all, the EMB-120 is fast. Even down low, Brasilia pilots are used to being asked to slow down for the B-737 ahead. (See Table 5-9.)

Code Sharing

So now you've had the opportunity to learn a bit more about what the regionals are all about and to take a look at some of the aircraft they fly. In 1992, the major airline industry was still attempting to pull itself out of one of the worst economic times in its history. They'd lost about $10 billion, more than all the combined profits of all the airlines, since commer-

Table 5-8. *ATR 72 specs*

Max. takeoff weight	47,400 lbs
Payload	15,870 lbs
Range	1,220 nm
Engines	P&W PW 124s @ 2,160 shp
Max. passenger load	74
Vmo	284 kts

Figure 5-12. EMB 120 Brasilia. *Courtesy www.aeroimages.com.*

Table 5-9. *Embraer Brasilia (EMB-120) specs*

Max. takeoff weight	25,353 lbs
Payload	6,700 lbs
Range	550 nm
Engines	P&W PW 118s @ 1,800 shp
Max. passenger load	30
Vmo	300 kts

cial flying first evolved from the old biplane mail-carrying days. There has been one bright spot in the airline industry, however, and that's at the regional level. If you were to take a look at the financial profitability of three of the most successful airlines, you'd find one of them to be Southwest Airlines, which flies Boeing 737s. The other two, Atlanta-based Atlantic Southeast Airlines (ASA) and Cincinnati-based Comair are regional carriers. And these two regional carriers were not simply meeting their payroll; they were making money hand over fist. Another carrier on the East Coast, Atlantic Coast Airlines, is also making a tidy little profit as a United Express Carrier. Predictions are that more and more jobs for pilots will be appearing at the regional level.

One major change to the airline industry has been code-sharing. In the early days of regional airlines, back when these airlines really were flying fairly small, poorly equipped aircraft, the commuters, as they were called then, became involved in agreements with major airlines in a kind of "I'll scratch your back and you scratch mine," kind of deal. The major airlines, like American, for instance, contracted with a small carrier to provide a feed from the smaller cities to American's jets at a hub location like Chicago O'Hare. It was much cheaper to run a 19-seat turboprop from Peoria to Chicago than it would have been to fly that route with one of American's DC-9s.

To make sure that the turboprop aircraft were flying as full as possible, American allowed these turboprop airlines to use American Airlines flight codes in the massive SABRE reservation system. A travel agent could book someone through from a large city to a smaller one and fly on American all the way, or at least what the passengers thought was American Airlines all the way (Fig. 5-13). Quite a few passengers were shocked to exit a large American Airlines jet to learn that the remainder of their trip would be aboard a turboprop that only looked similar to an American Airlines aircraft.

For the airlines, at least, it seemed a match made in heaven, for a while. With the profit margins in the airline industry being as tight as they are, some of the commuter carriers were not able to survive, and carriers like American would often awaken one morning to find out that one of their code-sharing companies had closed up shop the night before, leaving hundreds of American Airlines passengers stranded. The airline knew this couldn't continue, so American began buying the code-sharing regionals themselves. At least by owning these carriers, the airline could be certain of controlling its partner airlines.

This all might sound pretty enlightening and, indeed, you might wonder, then, why they're so successful when the other airlines are losing their shirts. The reason that some of the regionals are so successful is that their cost structure is considerably less than the major airlines. As you saw earlier, while the price of regional airliners is certainly in the millions, that's a drop in the bucket compared to $200 million price tag on a Boeing 747-400. The other major factor, besides the relatively low cost of purchasing these regional aircraft, is that the salaries paid to regional pilots is much lower than at a major airline.

The recent edition of Air Inc.'s *Pilot* salary survey tells the entire story, so hold on to your hat. Some representative salaries read like this. At American Eagle Airlines, a first officer will begin at $1,453 per month, based on a guarantee of 72 flight hours. A captain on the 42-passenger ATRs begins at $2,419 per month. When maximum pay for a captain is reached, when there are no longer any annual longevity raises, the annual salary in the ATR is $43,000.

Figure 5-13. The interior of a regional aircraft now looks like that of a major airline aircraft. *Courtesy Canadair*

Continental Express, owned by Continental Airlines organizes their pay somewhat differently. All first officers are paid the same amount, no matter which aircraft they fly, as are the captains. Other carriers normally pay more money for larger aircraft. At Continental Express, new first officers begin with the same $1,040 as the American Eagle pilots. By the second year, the pay has risen to just under $1,300 per month. A captain begins at $1,800 per month, rising to $2,040 the second year, and a maximum cap of $56,640 after at least ten years on the job. Continental Express, however, expects its pilots to fly a minimum of 80 hours each month. If there's a secondary benefit to flying for Continental Express, it's a seniority number at Continental Airlines as soon as you hire on with the Express. The airline believes you'll stay with them longer if they offer you the incentive of being able to move up into the jets if you just hang around long enough. When Midway Airlines and Pan American were still in business, they too had a similar arrangement with their wholly owned regional carriers.

Kit Darby, president of Air Inc., added this to the discussion of flow-through agreements. "Most of these agreements are designed to work when the flow is up. That's when they're easy to embrace. But when the economy is bad and people at the regionals are getting kicked backed down, it causes major dissension. It can get ugly. Pilots should not count on a flow-through to make their career for them. If this is your only plan, you need to get back to a regular job search agenda."

Finally, let's take a look at USAir Express, most of which, but not all, are wholly owned by USAirways. At USAirways Express Henson Airlines (recently renamed Piedmont), a first officer begins at $1,607 each month based on an even higher 85-hour guarantee. In year two, this rises to $1,831 monthly. A Dash-8 captain is being paid $3,885 per month in year ten, and the salary caps out at $50,102 after at least ten years service. While most of the regional carriers offer some type of pass benefits in addition to standard health insurance, the retirement programs at regionals leave a great deal to be desired. Most only offer pilots the chance to participate in a 401K program, and nothing more.

Regional Jobs

Earlier, I mentioned that you might have better luck hiring on with a regional than a major carrier. This is not a suggestion of the only way to win an airline job, if that's your goal, but it's certainly a method toward reaching that goal. A presidential panel that met in 1992 believed that the demand for pilots at the regionals would be up just slightly over that at the majors, about a 28 percent increase at the regionals versus a 25 percent increase at the majors by the year 2003.

Is finding a job difficult at a regional airline? As with any other job, the path to the cockpit door is similar to other flying jobs. Through some of the sources I've already mentioned, find the list of who is hiring, then begin sending in your application and updates to that application on a regular basis. I can't emphasize this enough. Merely sending in your application and waiting for the phone to ring won't get you an interview. I currently have applications on file with two regionals and two major carriers, and I make certain I update them on a regular basis, unless, of course, the airline requests something different. The folks in personnel at United seem content with an annual update. One of the regionals wants to hear from me every six months. I'm not saying you need to make a pest of yourself, but you do need to let the airline know of your continued interest. If you have a friend already working for the carrier you'd like to become employed with, ask if they'll write a letter of recommendation. I've never heard of a company yet that didn't like to see a note from one of their own employees, a known entity, talking about a potential employee. It just makes good sense. If you don't have a friend at an airline, make one. But most importantly, let the airline know you're interested.

Major and Regional Airline Relationships

Airline	Feeder Name	Affiliates *8/98*
Alaska	Alaska Airlines Commuter System	ERA Aviation, Harbor, Horizon, Peninsula Airways, Trans States, Wings West
America West	America West Express	Mesa Airlines
American	American Eagle	Aspen Mountain Air/Lone Star Business Express
	American Eagle Airlines	Executive (wholly owned) Flagship (wholly owned) Simmons (wholly owned) Wings West (wholly owned)
Continental	Continental Express	Colgan Continental Express (wholly owned) Gulfstream International
Delta	Delta Connection	Atlantic Southeast Airlines Business Express Comair Skywest Trans States
Northwest	Northwest Airlink	Business Express Express Airlines 1 (wholly owned) Horizon Mesaba Trans States
Trans World	Trans World Express	Trans States
United	United Express	Air Wisconsin Atlantic Coast Airlines Great Lakes Gulfstream International Mountain Air Express Skywest Trans States United Feeder Service
USAirways	USAirways Express	Air Midwest Allegheny Commuter (wholly owned) Chautauqua Commutair Mesa Airlines Piedmont Airlines (wholly owned) PSA (wholly owned) Trans States

Chicago Express—The ATA Connection

If you stand on any ramp at Chicago's Midway Airport on a clear day and face northeast, the view of downtown Chicago's monoliths of commerce is breathtaking, with the Sears Tower, the Amoco building, and the John Hancock building being some of the more prominent. As aircraft after aircraft taxi by, however, you vaguely begin to realize you've seen that cityscape somewhere before. But where? Then a Chicago Express Airlines' Jetstream 31 crosses a taxiway headed for the B concourse and you make the connection. There, painted in teal and blue on its tail, is that same image of downtown Chicago that has served the company as both a logo and marketing tool since Chicago Express's first flight in August 1993. The airline's radio callsign "Windy City" seems appropriate too.

Chicago Express is the brainchild of Michael J. Brady, the airline's chairman. The company is privately owned by the Brady family, as well, so no financial data is available. Brady is the former boss of Memphis-based Express One, the Northwest Airlink regional he sold in 1997. Chicago Express was a subsidiary of Phoenix Airline Services that also ceased to exist at the time of the sale. Brady has now taken a much more active role in the airline's operation since it is the only aviation company he owns.

The idea for Chicago Express evolved in 1993 from the void in service left when the original Midway Airlines went bankrupt in November 1991, transforming Chicago's second airport into a virtual ghost town. It also left many smaller cities with no air service to Midway Airport, a place that had emerged as the airport to fly from when price was a major travel consideration. While Southwest Airlines captured most of the longer haul markets after 1991, the 200- to 300-mile segments were virtually abandoned. Chicago Express was developed to fill that gap. The company is now also directly linked—through a code-sharing agreement with American Trans Air. ATA recently announced a major ramp-up of service at MDW from six gates to 12, with flights increasing to 36 daily in 1998 and to 90 by 2004. Doug Abbey, president of AvStats & Associates in Washington, D.C., believes that "American Trans Air is committed to growing at Midway, and that can only help Chicago Express."

Scott Hall, Chicago Express's vice president of operations, said, "We have a long-term contract to provide connection service with ATA in Chicago and perhaps Florida at some later date. We provide the service and get paid a flat fee per flight operation. This takes a bite out of the cyclical nature of this industry." Chicago Express flights are listed alongside ATA's own, essentially making the smaller carrier an invisible partner to the larger. Code sharing is not new to Chicago Express Airlines, however. The company had a limited code-sharing agreement with the new Midway Airlines. In fact, the plan called for moving the entire Chicago Express operation to Raleigh Durham in 1996. "We were spooling up, thinking about taking on Midway at RDU," Hall said. "We even had 10 pilots in training on the Saab 340. Then it all came to a grinding halt." That halt precipitated the first furlough in Chicago Express's short history as 10 pilots on the bottom of the seniority list lost their jobs. All were quickly offered work at Express One airlines.

Hall believes his company's bond with a major carrier was inevitable. "It is very difficult for an independent carrier to exist without a mainline partner. The public believes that our link with ATA is an endorsement [of our company]. The switch to all Part 121 carrier operations has also helped our company's image." Doug Abbey believes that, "with American Eagle going to Jets at O'Hare and United Express somewhat up for grabs there too, Chicago Express may have some unique opportunities at MDW. Midway is not slot-constrained like O'Hare."

Chicago Express operates 12 BAe-3100s. "Chicago Express has gone against the commonly held dogma that the 19-seat business is dead," Abbey added. They believe there are still good markets out there. It is incumbent upon all regionals to raise the bar on passenger service."

Hall added, "People are demanding cabin-class service and our load factors are up."

These factors have the company looking at larger aircraft that they hope to be operating by the end of 1998. That decision—possibly on SF-340s, EMB-120s or Do-328s—will be made jointly with ATA management. Chicago Express had a short stint with two SF-340s a few years ago as the plans with Midway Airlines took shape, but it returned those aircraft to the lessor in 1997 when the deal fell apart. The 340s could again be the logical choice for upgrading Chicago Express since, as Abbey said, "Brady was one of the largest Saab operators, and Saab is very much in the used aircraft market."

During early 1998, Chicago Express had 56 pilots on the payroll, but that number rose during the rest of 1998 as fleet size increased. Hall added that "we will hire 25 to 50 this year," a number he believes to be quite conservative. "That number could go higher if we add more cities and larger aircraft," Hall said. "We're after pilots who are outgoing, who will pitch in and are very professional on procedures," reflecting the very necessary one-on-one customer service aspect of 19-seat aircraft crews. Despite the upbeat nature of the Chicago Express-ATA link, Hall says, "we have not yet decided to change our paint scheme."

Opportunities for Chicago Express through their connection with ATA also spell opportunity for pilot applicants, especially since, Hall said,

"Our attrition is high. Of the 56 currently employed, those below seniority number 16 have only been with the company since April 1997. We are losing two to three pilots per month right now, mostly to United, Delta, and ATA."

The company has a gentleman's agreement with ATA to interview Chicago Express pilots who have at least three years' experience with the carrier. That agreement may be formalized at some point in the future, but has already succeeded in a move to the majors for nine pilots in the last year. The tenure of pilots at Chicago Express is about three years before most leave for a major airline position. Currently, the pay scale does not extend past the four-year point anyway. Matt Guyso, a Chicago Express captain said he particularly liked the upgrade training when he was a first officer. "As you progress in experience, the captains would definitely expect more and more from you."

Chicago Express's only hub is Midway Airport from which they fly to Des Moines, Iowa, Madison, and Milwaukee, Wis.; Grand Rapids and Lansing, Mich.; Indianapolis; and Dayton, Ohio. Every other leg includes a landing at Midway Airport. Crew domiciles are located at Chicago-Midway, Dayton, Des Moines, Grand Rapids, Indianapolis, and Milwaukee. The company performs all maintenance—except for small line items—at their base in Grand Rapids as well. There is essentially no flying on Saturday, except from the Indianapolis and Milwaukee bases. Each outstation is staffed by three crews with no reserves. "Everyone knows they need to be at work," Hall said. "If they think they are going to miss a trip, we expect them to call ahead," so the company has time to make other arrangements.

The company puts a great deal of faith in their pilots and makes attempts to keep them happy, especially, as Hall said, "since, you'll work hard here, for not a lot of money." Crews get together at the outstations each month and build their own schedules. Management's only concerns are that the lines are legal and that all trips are satisfactorily covered. A failure among outstation crews to agree on a schedule would cause the company's default schedules to take over, something Hall said, "Hasn't happened yet." Management is said to be quite flexible on trip swapping as well.

Another major benefit for Chicago Express pilots is the aviation education they receive operating at an airport as busy as Chicago Midway. Toss in the Chicago weather environment that changes a couple of times a day and it is little wonder that Chicago Express pilots are being scooped up by the majors. Tom Collins, a first officer felt his education was also enhanced by the captains he has flown with. "Circumnavigation of thunderstorms is the most interesting. The captains know what needs to be done, but want to see if I know what to do. They give me a lot of leeway in my decisions." Collins also believes that his experience dealing with passengers has been important to his career. "Smiling and helping them relax is important. I like to deal with the people."

Chicago Express is nonunion, although Hall said pilots may have been approached by the Teamsters. "We believe we are doing a good job for the pilots right now. A lot of our success comes from the fact that we listen to our employees." The company offers pilots a 401(k), although the airline does not match those contributions, and also provides passes and jumpseat privileges on other carriers. Major healthcare is a part of the working environment at Chicago Express. A new crew lounge at Midway was also opened recently.

Pilots typically fly about 70 to 75 hours per month. Reserve line holders will see about 10 days off per month, while regular line holders can expect anywhere from 12 to 16 days off. Depending upon their domicile, pilots might sit on reserve for three to four months before getting a regular line of their own. Pilots are all salaried at Chicago Express, with first officers starting out at $14,941 per year and a captain at $20,950. Any flying time over 80 hours makes crews eligible for bonus pay. Pilots may choose to pick up additional flying time for anywhere from $60 to $160 per day extra, depending upon their cockpit seat. They also receive $15 a day in per diem when on an overnight.

What does it take to work for Chicago Express? Hall said the minimums for pilots who do not come through a FlightSafety ab initio-type training program are 1,100 total time and 100 multiengine time. That is down from the 1,200 and 200 that was the norm just a year ago. A commercial pilot's certificate with multiengine and instrument ratings is also required. Hall added that "We've taken people from the Vero Beach [FlightSafety] program with as little as 300 hours total time. The average has been about 320 hours. We've taken about 40 pilots from there so far." First officer Pete Beckmeyer arrived at Chicago Express "as a former company intern with only 280 hours total time, 70 of that in multiengine aircraft." Richo Vergara, now a captain, had "1,200 hours of total time and just under 100 hours of multi," when he started with Chicago Express, while Matt Guyso, also a new captain, had "1,400 [hours] total and 250 multiengine."

While an ATP is not a requirement, having the written out of the way would be a plus, since it is necessary for the upgrade to captain. Upgrade time is currently a mere three to four months at Chicago Express. Hall said that "a recommendation from another pilot carries a lot of weight," in the hiring process. "It's always easier to hire someone whom you know something about. We've even had recommendations come in from pilots who have moved on to other airlines."

Since all of the initial screening—as well as initial new-hire training—is performed by FlightSafety, on behalf of Chicago Express, resumés should be sent directly to FlightSafety International's New-Hire Program, 600 S. Clyde Morris Blvd., Daytona Beach, FL 32120-1527(Additional information is available from FlightSafety by calling 888-789-4473). Higher-time pilots have also been applying to Chicago Express through FlightSafety, said Beth Thornton, FSI's new-hire manager. "We've had captain-qualified candidates go through." Additional programs to funnel new pilots to Chicago Express have also been initiated with American Flyers. The company is also investigating another program with the Prescott, Arizona, campus of Embry-Riddle Aeronautical University.

When FlightSafety receives a resumé, they will scan it to be certain it meets the minimum requirements. If it does, the applicant will be contacted to arrange the next step in the process, the evaluation at a nearby FlightSafety training center. Thornton said, "There are three major parts to the evaluation. The background history, the simulator evaluation, and a written job assessment." The background history collects all of the pilot's certificates, past job record, and letters of recommendation. The written job assessment measures a pilot's general abilities and personality traits and takes about an hour and one-half. It does not require a psychologist to administer or evaluate the assessment.

The simulator evaluation "looks for a proficiency level on instruments equivalent to ATP standards." Thornton said. "Pilots are briefed before the ride and on the results immediately after. They fly in both the right and left seat for the evaluation. We pair them up two pilots per session because we want to see how they interact with other crewmembers." Thornton said the entire simulator evaluation process takes about three hours. She believes one of the keys to a successful simulator ride is "being instrument proficient," before the applicant takes the exam. Pilot applicants are responsible for their expenses to and from the FlightSafety location they choose and should plan one full day for the evaluation. Arriving the night before would be a good bet since the work begins promptly at 8 A.M.

Applicants will find there is a wide range of full-motion Level C or D simulators they could be asked to fly. Some of these include the King Air 200, the Metro 2, or the Beech 1900D. Captain-qualified candidates may be asked to fly a Citation, Gulfstream 4, Boeing 727, Lockheed Jetstar, or Learjet. Thornton added that applicants are not expected to have a thorough understanding of these aircraft, but they are expected to be able to demonstrate the basics of good instrument flying. Successful completion of the evaluation does not guarantee a job; however, Vergara was made a conditional offer of employment when he satisfactorily completed his simulator ride.

When a pilot successfully completes the FlightSafety evaluation, his or her files are forwarded to Chicago Express for the final in-person interview. Hall will often conduct these interviews with someone from human resources as well as the company's chief pilot Don Terrell. "We sit down and review the pilot's resumé and their work history to begin with," Hall said. "We ask a number of questions about Jepp plates for the technical interview and discuss a number of crew scenarios to check their judgment. We hire about half of the pilots who make it to the interview." Collins recalled that the interview also included questions about FARs and the AIM, as well as human resources questions such as "Why do you want to work for this company?"For him, the process of getting hired "took two and one-half months from start to finish." The airline expects the more experienced pilots to explain the regulations and operational questions with considerably more depth than less experienced first officer candidates.

Once hired, a pilot heads for FlightSafety's St. Louis facility for groundschool and simulator training. Chicago Express is a pay-for-training airline, so pilots should expect to receive a bill for approximately $9,450 for the Jetstream course, not including living expenses. Financing can be arranged through banks that already have an agreement with FlightSafety, but there is no guaranteed financing. Matt Guyso financed his training through "Wells Fargo Bank at a rate that was much better than a credit card." While Guyso made a down payment to lower his monthly repayment costs, he added that "they would have financed the entire thing."

FlightSafety training includes about six days of indoctrination and crew resource management, two weeks of systems, and two weeks of cockpit procedures and simulator training. Training also includes eight simulator sessions of four hours each, where each student receives time in the flying and the nonflying pilot position. There is normally an hour or two of training in the aircraft before the pilot reaches the line to begin their new job. Guyso completed the flight training portion of his education by traveling to a Chicago Express outstation one night and waiting for that station's aircraft to return from the day's work. He and a Chicago Express instructor then went out and "flew one session of three hours." Initial operating experience is comprised of about 10 to 20 hours before the pilot begins their first regular Chicago Express trip. Guyso added that "FlightSafety had really excellent systems instructors. They would work with you if you were having trouble. They had a very positive attitude. We briefed before we started each day and at the end of the day as well. We also worked in the cockpit procedures trainer every day." Vergara said, "The training at FlightSafety is intensive. Everyone parallels it to sticking a fire hose in your mouth. I suggest being highly instrument competent."

A typical schedule for a pilot based at an outstation such as Milwaukee, might run like this. Showtime for the morning shift is a sleepy 4:30 A.M., with the first trip leaving Milwaukee at 5:15 A.M. for Midway. A quick turn finds the crew returning for three more round-trips to Milwaukee before they go off duty at 2 P.M. Other versions might include a two-hour layover after that first leg into Midway, followed by a round-trip to Indianapolis, and then a return to Milwaukee for the day's end. A typical afternoon shift would show at 1:40 P.M. for the first leg to Midway. That's followed by a round-trip to Dayton and then back to Milwaukee to sign out about 8 P.M. Out of Chicago, some crews fly standup overnights that show in the early evening. They then fly one leg out, as the last flight to an outstation, only to be the first trip out in the morning back to Chicago. These can make for a night that is pretty short on rest.

If you're looking for a career as an airline pilot, nothing can be quite as handy as a type rating and a thousand or so hours of turbine PIC time. The management team at Chicago Express understands that they are not yet at a point where the carrier will be a career for most pilots. Collins said, "I can't say enough good about this company." Guyso added, "I would recommend Chicago Express Airlines as a place to work. I think you're going to see good things happen here. Management understands that most of the pilots will eventually want to move on."

Current Chicago Express pilots had a few tips for others who might want to follow in their footsteps. Collins said, "There just is no way to prepare for the written exams at FlightSafety. Also, things happen in groundschool pretty fast. You're thrown into this transition from pistons to turbines that I thought was hard to get through. I recommend buying *The Turbine Pilot's Flight Manual*, by Gregory Brown and Mark J. Holt. It made it all much easier to understand." Pete Beckmeyer says for him, the best part of flying at Chicago Express is that "The captains still remember when they were first officers. The captains are always encouraging me to do better each day."

Profile: Pete Beckmeyer, In the Right Places at the Right Time

Pete Beckmeyer knew he wanted to fly when he was eight years old. That's when his dad used to take him to the Dayton Airshow. The family would pull the back seat out of their old car and sit on the ground and watch. Beckmeyer recalls "things flying fast and making lots of noise."

As a teenager, he talked to people about getting into one of the service academies, believing the military to be the only real way to become a pilot.

"I went to Ohio State and joined the Air Force Reserve Officer Training (ROTC) program. When I was a freshman, I learned that six seniors who were promised flight training were not going to get their wish. When I later transferred to the University of Cincinnati, I decided to forget ROTC."

Beckmeyer stayed involved in aviation by working a summer as a gate operations employee with Delta Airlines at Cincinnati. "That's where I heard about Embry-Riddle. Friends told me if you went there, you could pretty much write your own ticket in aviation. A lot of the gate agents at CVG jokingly told me to get out of aviation while I still could. I didn't and graduated from Embry-Riddle in 1996 with a degree in Aeronautical Science." But it was what happened to Beckmeyer in the summer before graduation, in 1995, that became significant to his aviation career.

I interviewed for an internship at Chicago Express Airlines at Midway Airport and got the job. I did a lot of word processing and worked on manuals. I worked with Scott Hall (now VP of Operations) or anyone who had anything for me to do. I got to do some jumpseating and visit the maintenance base during that summer. Scott told me if I got my CFII and took Flight Safety's Turbine Transition Course, they would give me an interview. He was anxious to have people who work well within the company."

Beckmeyer stayed in regular touch with Hall after that summer internship, the first critical step in the networking game. "I would send him an e-mail every so often. I also took Flight Safety's Master Pilot Program. That was very operationally oriented," to prepare for that first job. Unfortunately, when Beckmeyer called Hall in late 1996, things at the airline did not look good. "They'd just furloughed some pilots." Beckmeyer called Hall again in early 1997. When Hall said he was leaving the carrier, it looked to Beckmeyer as if his first intense networking sessions might be for naught (Hall was gone only a month before returning to Chicago Express). But all this time, he'd been looking on his own for flight instructor jobs. He had, in fact interviewed with Sporty's Flight Academy, when the chief pilot from Chicago Express called and offered him a job based upon Hall's recommendation. He quickly accepted. Perhaps Beckmeyer was in the right place at the right time, he thought.

Beckmeyer believes—now that he's an experienced first officer with 720 hours total time under his belt, half of that turbine—that "the people at Chicago Express give others the chance to learn a ton of great stuff from the captains." He recalls though that "training was the most intense thing I've ever been through. In the beginning you are just so overwhelmed you hardly have time to think. At the end of my first day on the line I was exhausted, but it was a good exhausted kind of feeling."

Beckmeyer knows that without that push from his internship with Chicago Express, despite having to pay for his training, "this kind of job would have been years down the line for me." He'd interviewed with

American Airlines for an internship during that summer of 1995, but realizes now that not getting that job was probably the best thing that could have happened to him since, "American Airlines would never have hired me at this point in my career."

He's now based in Dayton—an hour's drive from home—which gives him more time to spend with his wife and 11 month-old daughter. He also gets to say hi now and again to those gate agents at CVG who remember him from the days when he worked on the gates too. "They told me they knew they'd see me again someday, but in a pilot's uniform."

Regional Airline Training Grows Up

Pilot training for regional airlines is becoming more sophisticated. That's the good news. . . .

The thunder's crash was not overpowering, but it was loud enough to be heard over the whine of the large turbines that hung on each wing. The flashes of lightning around the departure corridor prompted the captain: "Tell tower we want to line up on the active to check the radar before take-off." The first officer complied. The captain knew well the telltale signs of a passing cold front but decided the risks were acceptable for the takeoff. The gusty wind outside varied the intensity of the rain against the cockpit glass.

The captain told the first officer he was ready, and after receiving clearance, he advanced the power levers. As the heavy aircraft picked up speed, the wind rocked the aircraft about. At the first officer's VR callout, the captain pulled back on the wheel and called for gear up just as the fire warning light and bell on the left engine sparked to life. Outside, the wind seemed to blow even harder as the crew struggled to keep the machine in the air using the drills they'd been taught over and over.

But the lack of full climb power and near max takeoff weight, combined with the high-density altitude this night, quickly showed their effects. As the airspeed began to decay, the captain instinctively lowered the nose of the aircraft, but in vain. The airplane's left wingtip was first to make contact with the ground, lurching the aircraft violently left so quickly that the two pilots saw only a blur of lights outside before the red explosions were added to the violent rocking. In the final seconds, the crew knew that nature had won. Then everything stopped.

What Happened

"Let's talk about what just happened here," the instructor said as he turned up the lights in the cockpit. The crew was still recovering from the savage, though nonfatal, crash they had just experienced. Many of you might have realized this story began in a Level C simulator. But most of you didn't know that this night, the crew training here was from a regional airline, testing the crew's skills against one of Flight Safety's EMB-120 Brasilias.

The level of sophistication in regional airline training has changed drastically (Fig. 5-14). Checking out in a regional airliner used to include performing a stall series in the aircraft in the middle of the night, often in actual IFR conditions. Another favorite was pulling a power lever back with the gear still in transit to simulate a V1 power failure.

For a number of reasons today, however, training at the regionals is catching up to the majors with state-of-the art full-motion simulators that allow an instructor to demonstrate a complete range of emergencies and unusual situations with no risk to the crew or aircraft. Just how dangerous training in the aircraft can be was demonstrated yet again in December, 1991, when a Business Express Beech 1900 crashed during a middle-of-the-night training flight, killing the three pilots on board. A BAe 3101 Jetstream belonging to CCAir also crashed during a training accident.

What made the Business Express crash doubly ironic, though, was that the carrier had just signed an agreement with Flight Safety International.

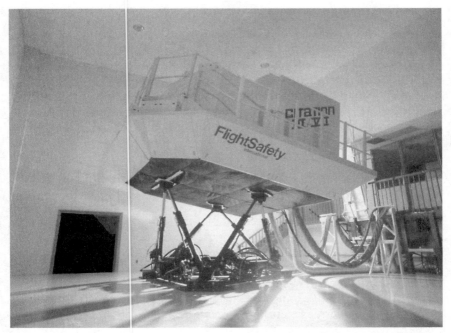

Figure 5-14. More and more regional airline training is conducted in simulators. *Courtesy Flight Safety International*

The training organization, based at LaGuardia Airport in New York, will perform all of Business Express's turboprop crew training in simulators. Business Express recently began acquiring British Aerospace BAe-146s, training for which will be conducted at the British Aerospace Washington Dulles training center.

Another reason for a more rapid conversion to training programs like those of major airlines is a certain amount of what Flight Safety Manager of Product Marketing Bruce Landsberg calls "societal pressure." Airline passengers who fly a regional-size aircraft painted in a United, USAir, or American paint scheme usually believe that the training standards demanded of these crews are the same as those of the pilots flying the MD-80 or B-737 they just connected from. Until recently, nothing was further from the truth, because regional training standards were approached with a more hurried pace—"a rush-them-in and rush-them-out kind of training," as one pilot put it.

Previously, most regional carriers, often still called commuters, were operated under Federal Aviation Regulations (FAR) Part 135 rules that grew into "scheduled air carrier" regulations from the ranks of the on-demand charter services. As the size and capacity of aircraft grew, though, they often passed the 30-seat or 7,500-pound useful load cutoff that transformed those carriers into Part 121 operators.

Two Training Curriculums

This caused a great deal of turmoil because it required some airlines to provide training departments and records that could cope with the differences between the two training curriculums. Some that still flew both sizes of aircraft had to operate both programs simultaneously as a money-saving move. The FAA doesn't currently allow a Part 135 carrier to train to Part 121 credit standards because the agency believes this policy might be in conflict with the FARs themselves.

In December, 1991, the Regional Airline Association (RAA) petitioned the FAA, requesting an exemption to Part 135 to allow regional air carriers to substitute pertinent sections of Part 121 to improve quality control and cut down on duplication. In their exemption request, RAA cites the FAA's own words: "FAA recognizes that the airman and crew-member training, checking, and qualification requirements of Part 121 will always meet or exceed the requirements of Part 135. This is consistent with the recognition that Part 121 affords the highest standards of safety in civil flight operations."

The FAA eventually changed the regs to remove most regional carriers from Part 135 governance and place them under the stricter guidelines of Part 121.

Elsewhere in the petition, RAA makes another significant point when it "recognizes the growth and maturing of the regional segment of the scheduled air carrier industry . . . because carriers operating under Part 135 are acquiring airplanes of increasing sophistication and are upgrading their training programs to take advantage of improvements in flight simulator capabilities and training techniques . . ." FAA is currently reviewing RAA's petition, which ALPA views as a major step forward for this segment of the industry.

Why have regional carriers waited so long to embrace simulators as the means of training their crews, even though majors have used simulators for decades? Most regional airline managers give the reason quite matter-of-factly . . . cost!

But training costs aren't always black-and-white issues. When Navajos, Metros, and Bandeirantes made up most fleets, taking the airplane offline at night and sending an instructor and a couple of trainee pilots out to fly when the airplane would otherwise have been sitting on the ground made good dollars and sense.

Training in Aircraft

One of the problems inherent in late-night training is that the aircraft to be used for training would often end up at an outstation late at night, requiring the training crews to position themselves away from base awaiting the airplane's arrival, often a large time waster.

The training sessions were seldom good learning situations because they were conducted in the middle of the night, when most people's brains are in the sleep mode. If the training crews broke something on the airplane during the night, that aircraft would often be unavailable for the first flight in the morning from that station, causing untold conflicts back at the hub. Then, too, maintenance would sometimes take a back seat to the need to upgrade a first officer or two.

Consider the overall quality of training. Simulating a good engine failure is pretty tough when the student sees the instructor reaching up to pull a power lever back, even if the movement is covered up with a piece of cardboard. When the instructor, trying to simulate a fire about to eat a wing, reaches over to hit the fire warning test circuit, the student knows deep inside that it's only pretend.

In fact, most aviation training experts believe that only about 25 percent of the emergencies and unusual situations can be simulated in the aircraft itself. Before the simulator, many other kinds of aircraft problems were only talked about.

Finally, as if all these other items were not enough, the strength of the cost-effectiveness argument truly loses its impact in the safety aspect. How does an airline determine the cost of the lives of an instructor and two or three pilots, not to mention the loss of just one aircraft while training?

Some airlines saw simulator training as relatively inefficient, though. One carrier was giving its new first officers 10 hours in the simulator and another 5 to 10 hours in the aircraft before they took their Part 135 second-in-command (SIC) ride. The company subsequently learned it could train new first officers to take the ride in the same 5 to 10 hours in the aircraft and save the cost of the simulator entirely.

To bean counters, the elimination of the simulator would seem to be an easy place to cut costs, but doing so certainly raises the question of quality. A pilot simply will not emerge as proficient after 10 hours of training as he or she will after 20 hours. The question then becomes, "Are companies trying to install competent first officers in the right seat, or merely training pilots to pass a check ride?"

Today, more and more airlines are training their crews to proficiency—until their knowledge of the concept has set in—whatever reasonable amount of time that takes. Another problem with the old-style cram method is that even if you can force-feed the pilots' brains to pass the test, their grasp of the material two weeks after the ride is minimal. This is fine when you're a cook and can open a book if you need to, but deadly when the right prop overspeeds in an airplane.

Simulators' Cost Effectiveness

British Aerospace Manager of Marketing and Business Development William Grayson outlines one of the best economic as well as safety reasons for carriers to train in simulators. "When training in an aircraft, only one pilot at a time receives credit. In a simulator, both pilots train as an effective team and receive credit for the same flight. So really, the aircraft would have to be twice as cheap to operate as a simulator" to be truly cost-effective.

Although some carriers are still sending first officers to one class and captains to another, the ability to train in a true crew concept is a benefit that shouldn't be minimized. With a simulator, regional crews can now learn flying and nonflying pilot duties the way they would actually happen in the aircraft on the line.

While the cost of a regional airline simulator is about $10 million, a few of the regional's major airline code-sharing partners view the cost as worthwhile. USAir purchased a Dash 8 simulator now installed at its Charlotte, N.C., training facility for use in USAir Express training. AMR Corporation operates an ATR-42 simulator at American's Dallas/Fort Worth training center.

In 1978, a United Airlines DC-8 with a landing gear problem ran out of fuel and crashed near Portland, Oregon. United management decided this would never happen again, "at any cost." The cockpit resource management (CRM) program was the result. While this program has been a part of major airline training for years, it's only just beginning to make its way into the regional airline system. Both Comair, a Delta Connection carrier, and Piedmont, a USAirways Express carrier, use CRM as a part of their regular training program.

The regionals seem to be looking at CRM and installing the program on their own timetables instead of merely reacting to an FAA mandate. Because carriers with an active CRM program have lower overall accident rates, instituting such a program should reflect positively on a carrier's insurance rates, too.

For the pilots, CRM is a win/win situation because they get specific human-factors training to help them cope more efficiently with problems on the flight deck. ALPA's chief accident investigator at Comair, Captain Mitch Serber, says, "The airline's CRM program will soon be linked with a line-oriented flight training (LOFT) program to add additional feedback to the cockpit crew training loop."

One regional airline pilot, however, calls the CRM program in the regional system merely a buzzword to keep the FAA off the carrier's back. He says the reason his carrier instituted the program was because management believed "the system is safe and the equipment is reliable, so most accidents must be caused by pilots." Other pilots said their CRM programs were sometimes no more than a short video or an even shorter speech from the chief pilot.

Outside Training

Tough economic times often bring innovative new programs to an industry. One controversial program that has gained momentum recently is the management of an entire regional airline's training program by a professional training organization and not by the airline itself. A case in point is the December, 1991, Business Express agreement with FlightSafety International. Any pilot now interested in employment with Business Express is automatically referred to FlightSafety, which conducts all the initial screening of new hires. FlightSafety then determines which applicants will be referred to Business Express.

FlightSafety conducts all new-hire aircraft-specific training for Business Express, too. But the real sting in this program comes from the $8,500 bill that the applicant receives for that training. Opinions differ as to how or why a young, low-time pilot applicant would want to pay for their own training, but a leading reason seems to be that with the current supply of pilots far exceeds the demand, so an airline can pretty much call the shots: Want to fly? Pay the bill!

The strongest motivator to the airlines, though, is the cost savings, because an airline no longer needs to fuss with the paperwork or cost of running a training department. This kind of "shoe-on-the-other-foot" program also allows the airlines to recoup some of the money they had been losing over the past five to seven years when pilot after pilot left the regionals for the majors, taking their valuable training with them after a year on the line.

Another thorn in the side of experienced pilots is that even those type-rated in a particular airline's equipment are tossed into the same pool with the inexperienced pilots. Experienced pilots must pay to get hired. All the airline does is sit back and look at the fully trained applicants that FlightSafety sends them. (This training procedure is sufficiently unorthodox that ALPA's president, Captain Randolph Babitt, has instructed the Association's Collective Bargaining Committee to review the pros and cons of such training and to make suitable recommendations for dealing with the issue during work-agreement bargaining sessions.)

FlightSafety conducts full-service training for four other airlines in the United States and occasional initial screening service for Atlantic Southeast Airlines (ASA).

While this program might make company accountants smile each time they look at the money they save on a class of new hires, many of the pilots we interviewed believe that the airlines that use this pay-for-training system are courting disaster. While no instances have been recorded where the quality of pilots flying the line has significantly diminished because of this kind of program, some ALPA pilots believe that day might be just around the corner. Again, this speculation is just that—a prediction of what might come to be.

"I think (the Flight Safety Program) looks good on paper," one pilot said. "These crews coming from the FlightSafety Initial Training program might really know the checklist and what to do if the antiskid fails, but these new training programs don't historically address the weakness of an aircraft. It takes an airline ground school, run by airline people, to address these kinds of things, the day-to-day problems you run into on the line—that's situational awareness. An instructor who might have come from teaching the Mooney program last week is not going to be able to teach a new-hire how to cope with an aircraft's drawbacks and how to be an airline pilot."

While many international airlines do put low-time first officers into the right seat as the FlightSafety program is doing, "Those pilots sit in a classroom with, say, Lufthansa for two years before they jump into the airplane," one pilot said. "Then they sit in the right seat for years before they can upgrade." He emphasizes what he thinks is a potential for tragedy: "Some of these new first officers could be moving over to the left seat with some very low total times as well as experience levels." ASA and WestAir are two carriers that eventually decided against turning their entire training departments over to FlightSafety after initial discussions. We shouldn't let the opinions of some cast aspersions on all, but we should also not let these predictions pass unconsidered.

One regional pilot sees the following scenario possibly unfolding because of what he believes is a potential conflict of interest: "Imagine an airline telling a pilot that even though that pilot paid for and passed the training, performance was marginal, but that the airline is going to give him or her a chance anyway. If the airline tells the pilots it expects them to really stay in line, this could really set the tone for how those pilots will react to a great many things in his or her future flying."

"Some examples might be the inability to say, 'No, I won't go out and fly around those areas of thunderstorms,' or 'No, I won't fly an aircraft that has not had proper maintenance.' Some of these kinds of pilots just won't have the good sense and experience to make good, sound decisions even though they technically meet the requirements. The company owns their soul, but I think having pilots that won't give them any resistance is just what these airlines want."

Because the airlines can save vast amounts of money with this kind of program, pilots are likely to see more, rather than fewer, of them. Whether the preceding predictions come to pass, however, only time will tell.

The Schedules

Flying schedules at the regionals tend to be similar to the majors, yet different. At a regional airline (Fig. 5-15), you could find yourself with only ten days off per month if you're the low person on the seniority poll. That means you're flying basically five days a week. But, I know a new Boeing 737 captain at a major airline who is only receiving 12 days off per month. What you'll find, however, is that it's not how much time off you receive, necessarily, but rather how that time is organized. If you have split days off or your trip ends late one day and begins early on another, your time off can seem even less than it really is. Asking to see a copy of a monthly bid line would not be out of line during a final interview. Depending on the carrier, and the scheduler, I have also seen regional airline schedules with lots of flexibility, sometimes offering many lines of 14, 15, and 16 days off each month.

Figure 5-15. AMR Eagle ATR 42. *Courtesy Robert Mark*

Seniority

A constant subject at any flight department, be it corporate or airline, is seniority. Basically, seniority is based on who showed up first at a corporation because the number of pilots they often hire are small at any one time. At an airline, your seniority is not just an important thing. As one pilot said recently, "Seniority is the only thing." Your seniority at a carrier will determine which aircraft you fly, what schedules you can obtain, when you can upgrade to captain, even when you can take your vacation.

Seniority numbers at an airline are usually established during initial training. The oldest member of the class is normally assigned the lowest seniority number. Some airlines today are choosing seniority numbers by a lottery system within the class, too. The seniority system is quite easy to understand. When two pilots bid the same schedule, the pilot with the lowest seniority number is the winner. If a new hire class is assigned different aircraft, the more sophisticated will usually be assigned to new hires with the lowest class seniority. If a pilot wants to live in a particular domicile city and he or she has a lower seniority number than you, you'll be stuck in a city that's not your first choice until your seniority number is low enough to be able to hold that city. Just in case you were wondering, your seniority number does change with your company longevity. Each time a senior pilot leaves the company, a new pilot beneath him on the seniority list moves ahead to take over the old number. If you're given the option in a hiring situation of a

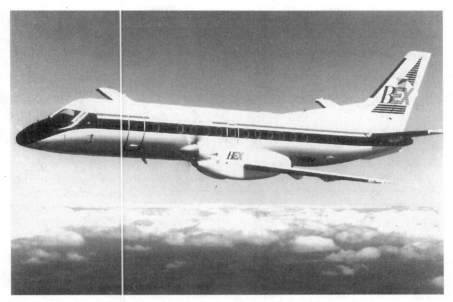

Figure 5-16. Saab 340B Business Express. *Courtesy AI Stats*

later class date, but the aircraft you want, I'd take the earlier class date. Remember, seniority is everything.

I spent three years of my flying life at a regional carrier (Fig. 5-16), and, for the most part, I enjoyed it. There's no doubt that the days are sometimes long and the pay could certainly be better, but the people were great. At an airline like United, where seniority numbers are currently near 10,000, knowing the people—other than the immediate ones you work with—is often impossible.

CommutAir--Keeping Its Pilots Busy

Pilots relentlessly pursue flying jobs for just one reason—because they want to fly. That seems simple enough, but in the real world of today's airline industry, yield, load factor, and scheduling concerns often play havoc with a pilot's desire to stay aloft for as much of their workday as possible. At some regional carriers, it is not uncommon for pilots to be scheduled for a 12-hour duty day and spend much of that day wandering around an airport terminal waiting for the next leg to begin. This just drives some pilots nuts. But one carrier based in the northeastern United States, may just be the answer to some pilot's prayers.

One of the things Bill Rabozzi really enjoys about his new position as a first officer with Plattsburgh, N.Y.-based CommutAir, is that the carrier flies its pilots a lot. But few complain. "I hate sitting around," Rabozzi remarked. "But I don't do a lot of that with CommutAir." The downside, if you're the type who wants to read the newspaper or grab a cup of coffee between legs, is, "there are not a lot of breaks. It really is intense. CommutAir is run very precisely and very efficiently," Rabozzi said.

In the Regional Airline Association's 1997 statistics, CommutAir, a privately owned airline flying as a US Airways Express carrier, is ranked 23d in size having carried 677,325 passengers to 23 cities in eight states. The company is owned by three airline veterans—Tony von Elbe, John Sullivan, and Jim Drollette. This nonunion carrier reported 1996 revenues at $80 million and revenue passenger miles for the same period as 118.6 million. Almost unheard of in the regional airline industry, however, is that a significant portion—30 percent of CommutAir—is owned by the company's 325 employees through an ESOP in company stock, giving pilots and other employees a voice in their own destiny.

The employee-stock ownership plan is in lieu of a traditional retirement scheme that, of course, is essentially unknown in the regional airline industry, too. "The owners felt they wanted to get employees involved in ownership," chief pilot Ray Baker said. "A captain with us for seven years might have enough to retire on." CommutAir also offers a quarterly profit-sharing plan for employees—that could total 30 to 40 percent of their base pay—after they've completed four consecutive quarters of employment with the airline. Rabozzi reports that "I get a passenger load factor and revenue update from the company each month, so I know the company is profitable. They keep you very informed." Pilots enjoy one week of vacation after the first year and two weeks after two years of service.

The company has capitalized on the single-aircraft concept pioneered by Southwest Airlines, significantly reducing training time and costs. CommutAir's 30-aircraft fleet consists only of the D model of the 19-seat Raytheon 1900. When the company was born in 1989, Baker remembers the two Beech 1900 Cs the carrier flew. They employed just 13 pilots then—10 of whom are still employed with the carrier.

10 Pilot Bases

Currently, CommutAir employs 232 pilots spread among 10 crew bases in the northeast. Three new crews were added to the roster as a result of the airline's recent change to a pure Part 121 operation. Bases include Plattsburgh, Albany, White Plains, Rochester and Utica/Syracuse, N.Y.; Boston and Worcester, Mass.; Portland, Maine; and Washington, Dulles Airport.

Plattsburgh, located on the northwestern shores of Lake Champlain is about 50 miles south of Montreal. The city, one of the carrier's most junior crew bases, was also home to the 1980 Winter Olympics. That means CommutAir crews understand the true meaning of the word snow.

Capt. Mike Cooke—a seven-year CommutAir veteran, said, "in the northeast, we see thunderstorms in the summer, wind in the fall, snow and ice in the winter, and fog in the spring." That means CommutAir pilots see plenty of opportunities to remain IFR current too, especially since many of their schedules include 10 takeoffs and landings per day, with a few encompassing 12 legs. Although the carrier does not fly a traditional hub-and-spoke type schedule, they regularly fly into major hubs such as Dulles, Newark, LaGuardia, Philadelphia, and Boston.

The average stage length at CommutAir is about 40 minutes. Some legs are as long as one hour and 15 minutes (Boston-Elmira) and as little as eight minutes (Plattsburgh-Burlington, Vt.). Cooke reports that the 10-minute turnarounds in a place like EWR can be tough on even an experienced first officer, but keeps the waiting-around time to a minimum. He added that a trip he flew recently was scheduled for seven hours of duty time and included landing and takeoffs at 10 airports—some twice in the same day. Most trips are out and backs that return pilots to their domiciles at night. There are some legs where crews—mostly those from Plattsburgh—encounter some dead-heading before the shift begins, simply out of the necessity to reposition aircraft; PLB is the company's primary maintenance facility.

Pilots with no previous turbine time like flying the Beech 1900. Ruben Gomez, a six-month first officer, said, "I like the 1900 because it has no autopilot, it's very fast, and it's a lot of fun to fly." Gomez also believes that the lack of an autopilot "improves your scan." CommutAir's aircraft are not minimalist versions of real airliners either and include a radar altimeter, EFIS systems, as well as TCAS 2, a step up from the less sophisticated TCAS 1 in many regional aircraft. Baker said the TCAS 2 was installed because it "provides pilots with better information and makes for a safer operation."

Everything moves quickly in the regional industry, so although no new aircraft are on order, it is relatively easy for Raytheon to turn over the keys of a D within as little as one month to six weeks should the carrier require it. The company has investigated larger aircraft for the future, too. Baker said, "We've looked at Saabs, Brasilias, Dornier 328s, and Jetstreams (41s), but in our market size the 19 seater makes money.

Right now, we break even with a 30 percent load factor." The company reports their average load factor as 43 percent, while on the Portland to Boston leg, "loads typically run at 90 percent."

Attrition to the Majors

Baker admits to seeing a "steady stream of people leaving CommutAir for the majors, right up to the day they passed the pilot records act. That slowed everything down." Thirty-six pilots—mostly captains—have left the company in the past 12 months. "When the paperwork problems get straightened out, that rate is going to accelerate." For now, however, the majors cannot hire a pilot until the background records check is complete.

Continued Hiring

CommutAir added 78 pilots in 1998. The airline will interview eight to 10 pilots per month in preparation for the eventual departure of a number of line pilots who have already been advised they have a major-airline job waiting. This should translate into classes of four to six pilots per month.

Not surprisingly, Baker reports that "scheduling is everyone's highest concern." CommutAir pilots bid daily trips in 28-day cycles, a considerably different program from the monthly lines of flying typical at most carriers. All pilots receive the same number of days off each bid period—12—so the value of understanding the intricacies of the daily bidding process can really pay off with large groups of days off sandwiched together, especially at the end of one bidding period and the beginning of another.

Cooke said he often flies "four days on and three off" and can have "four days on and seven off by creative bidding a day at a time." This gives pilots the flexibility to bid not only the days off they want, but also the types of flying they enjoy—if their seniority number will allow. First officer Karen Anderson said she "likes the day trips that start about 10 or 11 in the morning and run until 7 to 9 P.M. Some overnights, like Buffalo, begin in midafternoon and have you back in Plattsburgh before 11 A.M. the next day. On others, you could start early one day, be gone overnight and not get back until late the second day." The latter trip would add about 10 hours of flying time to a pilot's logbook.

Pay Policy

CommutAir pilots are only paid when they fly, hence their aversion to deadheading or long waits between return trips. A new captain is paid $30 per flight hour, while a first officer's hourly rate is $16. That's about

$28,800 for captains and $15,360 annually for first officers, based on 80 monthly flight hours.

A recent pay rule gave crewmembers a $1-per-hour raise annually. First officers are regularly evaluated by the captains they fly with to learn how well they are progressing toward upgrading, currently running at about two-and-one-half years. Captains share the results of those appraisals with their copilots after the flight. Cockpit crewmembers must complete a one-year company probation both as a first officer and again when they upgrade to captain.

Although the carrier was formerly a pay-for-training company, that policy was dropped in the spring of 1997, much to the chagrin of pilots who were in the last few classes to pay their way. Currently, FlightSafety still trains all CommutAir pilots up through the simulator ride, but the company picks up the tab. First officers receive one week of indoctrination, a week of aircraft systems and one week of simulator instruction. Each pilot spends approximately 14 to 16 hours of time in the simulator to prepare for his or her checkride, which is performed in the Level-D 1900 simulator with either a FlightSafety or CommutAir instructor at LaGuardia. First officers also receive a line-oriented flight-training scenario in the simulator to give them some real-world experience before their first revenue trip. The company uses Y cording after training that allows new first officers to listen in to all radio communications through their headsets, while sitting in the front row of the 1900 cabin on a regular trip.

A valuable aspect of CommutAir's training routine is that they don't force new first officers to sit on reserve for long periods of time just after they've finished training. The carrier has five lines of reserve, but breaks it up so that each pilot stands the watch for no more than two weeks each month. This guards against newly trained FOs forgetting a portion of what they learned before their first revenue flight.

Gomez remembers his first day on the job like it was yesterday. "It was overwhelming. We had a "nor'easter" off the coast of Boston and were in the clouds and snow all day long. Our final approach to Boston, the airport was down to minimums too. We landed about 30 minutes before the airport closed." He added, "Flying around the Northeast is very valuable to me."

Minimum Requirements

While the requirements to apply to CommutAir are still the regular FlightSafety standards of 1,500 total time and 200 multiengine, Baker said "we've seen some pilots with 6,000 hours. But mostly, our pilots are hiring on with 2,500 to 3,000 hours of total time." He reports new hires often arrive with previous Part 135 freight or charter flying experience as well. A current first-class physical is also required to apply

The process of hiring on with CommutAir begins by sending a resumé to Ray Baker, Chief Pilot, CommutAir, 518 Rugar St., Plattsburgh, N.Y., 12091. The company does not accept job-related telephone inquiries. If a pilot meets minimum requirements, their resumé is passed on to FlightSafety to begin the evaluation process. Many pilots interviewed for this article reported hearing about the company from someone already employed by CommutAir. Some reported being offered a job within a day of their final CommutAir interview. The current FAA paperwork snafu, however, is slowing that process down for CommutAir new hires, so three to four weeks is a more typical time to hear back.

What does CommutAir look for in the pilots it hires? "We look at the person," Baker said. "We don't board interview. It's very much an individual thing." The selection process must be working well too. Anderson said that "the people are great. Everyone is really cool to fly with." Gomez added, "I enjoy an atmosphere in the cockpit where I can ask questions freely and learn. I can go into the company and talk to anyone about anything. CommutAir is a good bunch of people to work for. It's a place to mature in aviation."

Added another pilot, "They're looking for down-to-earth people. Be yourself in the interview." Gomez added that at this airline, "it's not a typical captain–first officer arrangement in the cockpit. I know who is responsible, but here it is flying-pilot and nonflying-pilot atmosphere. That's important because I believe that everyone in the right seat is a captain in training."

Some new hires believe that since the company is populated by so many people focused on the individual, the CommutAir portion of the interview might begin the moment a CommutAir employee picks you up at the Plattsburgh airport.

Hindsight is always 20-20, but Anderson said, "I know I made a good choice. I know people who plan on staying at CommutAir for life, too." Rabozzi told applicants "never give up. You might have to struggle some to get hired, but don't ever give up." Gomez added that "a good attitude is extremely important." Mike Cooke explained that at CommutAir, "if you do the job, they treat you with respect." The distances pilots are willing to travel is also a solid barometer of a carrier's value to crewmembers. One CommutAir pilot flies in for work from Oklahoma.

CommutAir owners are currently looking for an IPO for the company, but Baker said "we're not large enough right now to go public." The company has been looking at extending their reach to cities further south from their current position as well as north into Canada. "We will probably expand into more smaller markets," Baker said. "That may mean we'd have to add another maintenance base as well." CommutAir also flies about 10 charters per year and hopes to do more.

Regional Pilots: Lifestyles of the Not so Rich and Famous

A few years ago, when I was still flying an EMB-120 Brasilia for Midway Commuter, I was at a party that was fun and uncomplicated. Because some of us did not know each other very well, we started talking about our jobs. When my turn came, I said I was a pilot. The faces of the people around me brightened noticeably—to most people, no matter how you cut it, hauling people around in airplanes is still a pretty glamorous line of work. Very quickly the air was filled with other people's lively stories of being trapped for hours on a runway somewhere waiting their turn in line for takeoff or of a trip filled with violent "air pockets."

As the conversation progressed, one of the men asked me, "What do you fly, a B-737 or one of those DC-9 jets?" My reply was, "A Brasilia." "What's that?" another asked. "It's a 30-seat turboprop aircraft. Very fast. Very nice machine," I answered him. "Oh, it's one of those little puddle jumpers, huh? You wouldn't get me on that thing. I didn't know real airlines flew those little airplanes anymore."

I found myself becoming defensive as I related how efficient the airplane was, and how much money our commuter subsidiary had made for Midway Airlines, not to mention how similar flying a large turboprop was to flying a jet. "Are you trying to get on with one of the regular airlines?" the man asked. "I don't know. Maybe I should," I said. This was not the last time I would find myself questioning my job as a regional airline pilot.

Two months later, I rode jump seat on an ALPA-crewed B-767 to Los Angeles. As I stood in the cockpit doorway and introduced myself, ALPA card in hand, the two pilots welcomed me aboard. The conversation prior to the "before starts" was animated and lively. These two pilots obviously knew each other well and enjoyed flying together. Passing through 25,000 feet on the climb, the conversations began about the state of the industry; and at one point, the captain asked what aircraft I flew. "A Brasilia," I told him. "Oh," was his only response, but very quickly I noticed the atmosphere in the cockpit change. The rest of the trip, I was almost completely excluded from the cockpit chatter. At the end of the flight I said goodbye and thanks, but the two pilots never even looked back at me. They only waved and said, "See ya."

My sense of alienation sometimes arose from contact with regular people and sometimes from other pilots. Once, a Midway DC-9 captain tried to keep me from using the crew lounge at Midway Airport in Chicago by saying it was only for the regular Midway pilots. I showed him my ALPA card, and he moved aside.

Am I What I Fly?

I began trying to make sense of what I was seeing. I had worked hard to reach 4,000 logged hours with an ATP. Yet some people seemed determined to judge me by the kind of airplane I flew, not the job I did—kind of like saying you must be someone special if you drive a BMW.

Worst of all, I began to second-guess myself, especially because my paycheck showed what value the airline placed on my service. When I started flying for Midway Commuter as a first officer, my annual salary was $11,700—that qualified me for food stamps! Because my wife had a good job, I was one of the lucky ones economically. But now, to add insult to injury, the people I had always thought of as peers seemed to look down on me.

I took the whole thing rather personally until I began talking to other Brasilia crews and, later, crews at other regional airlines. Pretty much everywhere, I learned, the crews who fly for regionals are regarded by many as the poor little brothers and sisters of the airline industry: by management, the flying public, and sometimes by other pilots. I started asking more questions.

Dash 8 First Officer Steve Varinsky summed it up: "Flying for the regionals is sort of like being in the minor leagues. Everyone looks at us that way—from the flying public to other pilots and even to other airline employees. Even when you travel nonrevenue, they look at you sometimes and say, 'Oh, you're only a commuter pilot.' I think this whole thing started a long time ago with people's perceptions of air taxis and commuters as companies that flew little 6- and 8-seat airplanes."

Today, though, the differences between the service a regional carrier provides and that of a major are disappearing. Many regionals are now operated under Federal Aviation Regulations Part 121, just like the majors. In fact, services are purposely being blended together to give the public, in former AMR Chairman Robert Crandall's words, "a seamless flight." In this way, passengers will believe they are flying on one airline's airplanes from start to finish, even if they begin on a United B-767 and complete their trip on a WestAir Brasilia painted in United Express colors.

Today, too, shooting an ILS approach down to minimums in a Dash 8-300 with 55 passengers on board really takes no more or no less effort than it takes for the crew of the DC-9-10 with 63 people on board that follows right behind. The newest regional aircraft are large, sophisticated, and quick. Some regionals today are even operating pure jet aircraft. But how regional crews are treated at the company level is where the major differences appear, on every subject from pay and benefits to scheduling.

Many regional airlines are reaping huge profits at a time when their major carrier partners are losing their shirts. One of the most successful, Atlantic Southeast Airlines, gleaned a $32 million profit last year. Comair, a Delta

Connection partner, soon to introduce jet service with the Canadair RJ, put a $2.9 million profit in the bank for just one quarter. Airline managers sometimes possess a bargaining chip because they have a captive audience in their cockpits. Where are their pilots going to go during tough economic times if there are a glut of unemployed pilots still looking for work?

Low Wages

According to recent ALPA studies, the average annual wage for a first officer at the lowest paid of all the ALPA-represented regional carriers, AMR Eagle Simmons Airlines, is $16,104. Even if you use the kind of simple logic most pilots love and say that this job is only half as tough as flying a 75-seat airplane, you'd find the Simmons pilot's wages to be about 80 percent less than the right-seat driver in that 75-seat jet. Imagine raising a family on $335 per week.

Jetstream Capt. Bob Phelan says, "We're looked upon more as slaves by the company: Just get out there and do your job. They think we should be paying them instead of them paying us."

Shorts 360 First Officer Pete Trimarche says, "On a scale of 1 to 10, I'd rate my personal satisfaction with my job as a 9. But I'd give my quality of life a three. I graduated from college four years ago, and I still really can't support myself on my income." Dash 8 First Officer Evelyn Tinkl says, "That professional pilots should not be put in situations where they qualify for government-assisted food programs."

What Is Fair?

A number of regional pilots were asked what they considered fair money for their work. Brasilia Capt. Ken Cooksey says, "No one expects a Brasilia pilot to be paid MD-80 pilot wages, but pay should have a realistic floor based on the number of seats in the aircraft and proportional to what other pilots are paid."

To some major airline pilots, the pay issue may seem to be only an irritant. But beneath the surface may actually lie a dragon soon to be faced. As airlines look for new, innovative ways to cut costs, operations that violate current scope language may emerge.

Northwest, USAir, and Delta plan to continue removing jet aircraft from unprofitable routes and replacing them with high-performance, high-capacity turboprops operated by regional airline crews working for C-scale wages. In fact, using new Canadair RJ jets, Cincinnati-based Comair plans to begin searching out new low-cost routes on its own that don't even involve a feed with its partner Delta Air Lines.

Of course, regional pay can be corrected, but whether it will be is another matter. One Comair pilot says, "Management's always yelling about how ALPA contract proposals will put the company out of business, but Comair has been profitable for 16 straight quarters." Comair's CEO David R. Mueller earns more than $500,000 annually in wages, with stock options bringing his yearly compensation to more than $1 million.

A 25 percent pay raise for each Atlantic Southeast pilot, right now, would cost the company an extra $3.5 million per year. Obviously, that is a cost Atlantic Southeast, with its $32 million profit last year, could afford. In Europe, Sabena Airlines regional DAT hired ex-Midway Airline's Brasilia captains in 1992. The airline started these U.S. pilots as captains at a beginning wage of $55,000 (U.S.) per year, plus cost-of-living adjustments.

Obviously, even though regional pilots love their work, their quality of life is often questionable at best. The hours are long, and the lack of duty rigs makes a pilot's time very cheap, leaving companies little motivation to build more productive schedules. When regional airlines experience crew shortages, they just lengthen the duty days for the pilots currently employed. ALPA's president, Capt. Randolph Babbitt, and Regional Airline Committee chairman, Capt. Stephen Ormsby, testified vigorously in 1992 on Capitol Hill on behalf of flight and duty time rules for regional pilots.

Although many of the pilots interviewed said they used ALPA benefits, most say money is a major limitation. A Business Express pilot says, "I'd like some of the insurance programs; but as a first officer, I just can't afford them." But many pilots believe the ALPA programs already in place are extremely valuable. Brasilia Capt. Dan Ford says, "I consider my ALPA dues as a 'free lawyer' card. Just one use of an ALPA attorney, and I'll have more than paid back the dues I paid over the years."

Most regional airlines have minimal retirement programs, if any at all, leaving regional airline crews and their families to their own fate at age 60. A Simmons Shorts 360 pilot has seen the handwriting on the wall with retirement programs at other airlines. He has "talked to too many guys who have lost out with their airline retirement plans, those A and B plans we all seem to dream about. I want an account with my name on it and my money in it."

Changed Domiciles

Express Airlines 1 pilots recently arrived at work to learn their airline's two main domiciles had suddenly increased to no fewer than 40 minidomiciles. With just a few days' notice, hundreds of pilots and their families were told to uproot themselves in a gypsylike move or find a new job. Because Express Airlines deemed the move necessary to improve its financial health, it did not consult the union.

Often, regional managers, at carriers where total pilot numbers and master executive council (MEC) strength is reduced, use a common technique to deal with problems. "If you don't like the procedure, there's the door." Other airlines use the "go ahead and grieve it" stance, knowing full well that while the grievance wheels are in motion, managers may still call the shots the way they choose.

Overall, regional pilots want to be thought of as just airline pilots. But a complaint sometimes voiced is the lack of respect they feel from their jet-pilot brothers. Saab Capt. Mike Sigman says, "The rudeness of some major airline pilots is what really burns me. I used to fly a DC-9, but I don't anymore. Just because we are flying something smaller and slower shouldn't make other pilots look down their noses at us. We are all professional pilot members of the same union. ALPA National treats us as equals."

But fault for regional carriers' troubles cannot all be placed on the backs of management. Years ago, the regionals were staffed with pilots who were often unfamiliar with the day-to-day operations of working with a union. Today, though, more and more regional pilots are recruited from former airlines like Eastern, Midway, and Pan Am. While many of these more experienced ALPA pilots understand the union, they report that many of the younger pilots still do not take an interest, or at times do not seem to have any idea of what ALPA is even on the property for, other than a pay raise.

Help from the Majors

Indeed, the education process for regional MECs is a tough one. But here is where the resources of the major airline MECs could be of great value to the regionals—through the experience these senior ALPA pilots could pass on to members flying at the regionals. The Delta MEC Central Air Safety Committee has greatly assisted the MECs of its Delta Connection carriers like Comair, for instance.

Can regional pilots see light at the end of the tunnel? Yes, but change will be slow in coming, something that may be tough for some pilots to swallow. Atlantic Southeast MEC's former chairman, Capt. Cooksey, says that while the pilots asked for a great many items they didn't receive in their recent contract, "ALPA National felt we did pretty well for only our second contract. The fact that in August 1992 the president and vice president of our airline asked to meet face to face with our MEC and

Negotiating Committee was a major step forward in labor relations at our company."

At WestAir—now out of business—then MEC chairman, Capt. Walt Blore, said, "If I accept what the company says at face value, WestAir has a very promising outlook, although the company is already asking for some concessions in our upcoming contract." And, contrary to what some regional pilots believe, many major airline pilots do know and understand the problems their regional airline comrades face.

B-727 Capt. Ken Adams says, "I have a great deal of respect for a regional pilot's job. We all share the same sky and we have to stick together." He believes that smaller aircraft need as much protection as the larger ones do, whether it's in equipment or duty rigs. He'd like to see all regionals operated under Part 121, which they now do.

Flow-through Agreements

One Delta pilot says that regional pilots are "doing one hell of a job; and in the future, all the majors will need some sort of agreement with the regionals if they are to continually replace their retirees with experienced crews. The military is just not a good source anymore." Prior to their bankruptcies, both Midway and Pan Am had flow-through agreements with their regional partner airlines that not only allowed pilots to move up, but also gave those pilots a reason to remain at the regional level until their number came up, reducing retraining costs for those airlines.

The MECs of USAir and its wholly owned regionals (Henson, Pennsylvania, and Jetstream) have formed a joint committee to try to formulate a flow-through agreement. Regional pilots do not seem to be asking for special treatment, just equal treatment and a fair day's wages for a fair day's work, but the regional pilots can't bring about all the necessary changes alone. They need the help of their major airline ALPA brothers and sisters. If regional pilots don't receive that help, all of us in the airline industry may be doomed to repeat many of the hard lessons that ALPA learned over the last quarter century.

6

The Majors

It seems that, in the world of aviation, if you tell people you fly for a living, they always assume that you fly for a major airline. Let's take a look at this segment of the flying world and see just what it contains that seems to make so many people want to work there. First of all, realize that no matter what other pilots might say, the size of the aircraft they fly is important to them. All of us pilots have big egos. It just seems to come with the territory. But, what you'll find is that flying the biggest aircraft might not always be the best for you.

Earlier, I spoke of the volatile state of the airline industry in this country. While airline bankruptcies seem to have stopped, for a while at least, we've lost some major players like Midway, Pan Am, Braniff, Eastern, and others. When I wrote about the industry in 1993, Delta and American planned to lay off pilots. Northwest Airlines was on the verge of bankruptcy, while America West and TWA were already in bankruptcy. Continental Airlines had just recently emerged from chapter 11.

So why all the blue news? Not to make you decide not to fly for the majors, but to make you aware that flying for the airlines, any airlines, is a game whose rules have changed over the years, as have the rules of corporate America in general. Or have they really? Twenty years ago, you considered yourself fairly safe if you managed employment with a United, an American, or a Delta Airlines, but, actually, there were lay-offs years ago. A friend who is now a captain for one of the top three major carriers was furloughed two different times in his first eight years with the airline; the first happened the day he left B-727 class.

Who's Getting Hired at the Airlines?

Data courtesy Air Inc., Atlanta, GA—800-AIR-APPS

Information gathered between 6/97 and 6/98

Despite the variety of experience levels that many airlines call their minimums, there are still some rather experienced people being hired. Let's take a look at some of the statistics. This doesn't mean you won't be hired if you don't meet these criteria, but this will certainly give you a snapshot of the newest airline cockpit members. Note: These numbers represent the averages of civilian pilots hired.

At the Majors, 98 percent of pilots hired had an ATP, while at the nationals that number was 68 percent, 88 percent at the Jet Regionals, and 56 percent at the prop regionals.

Sixty-one percent of pilots held at least one type rating, while at the nationals only 18 percent did. At the jet regionals the figure is 25 percent and 16 percent at the prop regionals.

Average flight hours at the majors was 6013. At the nationals, pilots averaged 3246 hours and 3833 at the jet regionals. The prop regionals pilots averaged about 3000 hours.

PIC time was also all over the board, with the majors seeing about 3567 in new pilots, nationals 2426, jet regionals 2732, and prop regionals 2183.

The average multiengine time was 4333 for pilots hired at the majors, while the nationals saw just 1734. At the jet regionals, that figure is 2187 and 1535 at the prop regionals.

At the majors, 87 percent of pilot hired had a four-year degree, while 78 percent at the nationals did. Sixty-two percent at the regionals had a four-year degree, while 67 percent at the prop regionals did. Not surprisingly, only 1 percent of pilots hired at the majors and the nationals had between two and three years of college. At the jet regionals that number was 3 percent and none were hired at the prop regionals with this amount of college.

Age was also an interesting factor in pilots hired. At the majors, the average was 34 years old and only 15 percent of pilots hired were over 39. At the nationals the average age was 32, with 12 percent of pilots hired over the age of 39. At the jet regionals, the average was 33.5 years with 18 percent over 39, while at the prop regionals, the figures were 31.6 with only 5 percent being over 39.

But an interesting fact that Air Inc.'s Kit Darby mentioned is that the airlines hire fewer pilots over the age of 40 not because they can't make the grade at that age, but "because only a small number apply."

A Perspective on Airline Hiring in 1999

Here are a few forecasts from some of the larger airlines for 1999. Airborne Express: 180; Alaska Airlines: 145; America West Airlines: 200; American: 800; Continental: 500–600; Delta Airlines: 500–600; DHL Airways: 75–100, Federal Express: 250–275; Northwest Airlines: 380–450; Southwest Airlines: 260–300.

Interesting note regarding SWA. There is a possibility that Southwest will relax its requirement regarding mandatory 737 type ratings. The latest rumor is that if an applicant does not have a 737 type rating, the applicant may be offered employment contingent on having the type rating prior to new hire indoctrination. This is only a rumor, but our sources tell us that it may become reality as early as next summer. We'll keep you posted.

Trans World Airlines: 275–350; Pilots ratified new four-year contract, bringing pilot salaries within 90% of industry standard. United Airlines: 950–1000; The floodgates are open, however; first-time applicants still face an 80% chance of rejection on their first interview. United Parcel Service: 220–250; US Airways: 100–180. AirTran: 320–380; American Eagle: 475–550; American Trans Air: 380–400; Atlantic Coast Airlines: 220–275; Atlantic Southeast: 250–275; Atlas Air: 120–150; Business Express: 100–220, Comair: 330–375; Continental Express: 400–500; Emery Worldwide: 75–80; Express One Int'l: 225–300; Horizon: 180–225; Kitty Hawk Air Cargo: 400–425; Mesa: 100–120; Midwest Express: 75; Polar Air Cargo: 75–100; Reno Air: 75–150; SkyWest: 380–445; Sun Country: 75–100; Piedmont: 125–200; Trans States: 300–375; Vanguard: 75–125.

As you can see, we expect 1999 to be another blockbuster year. In our estimation, an additional 12–13 thousand new pilot jobs will be created in 1999.

We've had a lot of calls in recent weeks asking us if earning a Flight Engineer rating is worth the money. In our opinion the answer in no. In most circumstances earning a Flight Engineer ticket will not substantially increase your odds of getting an airline job. There are only a handful of airlines that hire "Professional Flight Engineers," and the numbers are decreasing rapidly.

Not only that, even at the major airlines, more and more new hires are bypassing the engineer seat and moving directly into a First Officer position. We believe a much better investment would be in earning a type rating in a Citation or 737 or just spending the money building multiengine PIC. Remember that most employers do not consider a type rating of substantial value unless the pilot has 500 or more hours of pilot-in-command time on that aircraft. The exception, of course, is Southwest.

Courtesy Berliner/Schafer Aviation Consulting Group
888-745-6899, www.pilotswanted.com

PILOT HIRING PROJECTIONS
1993-2007

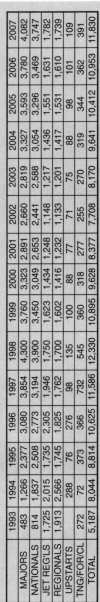

	1993	1994	1995	1996	1997	1998	1999	2000	2001	2002	2003	2004	2005	2006	2007
MAJORS	483	1,266	2,377	3,080	3,854	4,300	3,760	3,323	2,891	2,660	2,819	3,327	3,593	3,780	4,082
NATIONALS	814	1,837	2,508	2,773	3,194	3,900	3,450	3,049	2,653	2,441	2,588	3,054	3,296	3,469	3,747
JET REG'LS	1,725	2,015	1,735	2,305	1,946	1,750	1,623	1,434	1,248	1,148	1,217	1,436	1,551	1,631	1,782
REGIONALS	1,913	2,566	1,745	1,825	1,762	1,700	1,602	1,416	1,232	1,133	1,201	1,417	1,531	1,610	1,739
UPSTARTS		288	76	276	98	135	100	88	77	71	75	88	98	101	109
TNG/FOR/CL	272	72	373	366	732	545	360	318	277	255	270	319	344	362	391
TOTAL	5,187	8,044	8,814	10,625	11,586	12,330	10,895	9,628	8,377	7,708	8,170	9,641	10,412	10,953	11,830

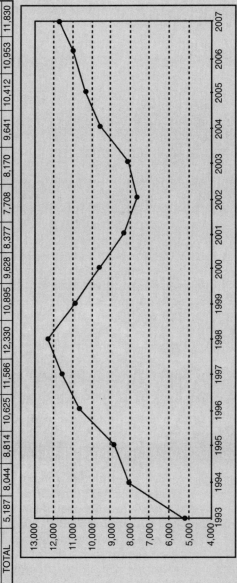

Figure 6-1a. Jet airline hiring by year. *FAPA.*

Major Airline Pilot Retirements by Year

Airline	Total Pilots	1998	1999	2000	2001	2002	2003	2004	2005	2006	2007	2008	2009	2010
ABX	766	3	2	4	4	13	12	19	24	29	29	30	25	32
AMR	9,488	216	226	164	121	153	160	208	207	262	315	403	459	370
AWA	1,407	15	18	17	15	39	48	48	39	34	48	50	52	49
CAL	4,245	97	102	116	113	115	144	115	160	258	317	246	219	209
DAL	9,667	360	374	364	368	391	263	247	231	219	290	326	280	270
FedEx	3,481	32	38	47	46	64	85	75	120	147	140	127	115	129
NWA	6,246	108	168	202	204	179	174	158	149	161	239	210	210	205
SWA	2,379	10	17	17	18	35	63	83	107	147	142	129	97	119
TWA	2,986	182	204	212	188	150	124	75	72	62	70	59	53	41
UAL	9,742	256	330	368	323	340	267	221	260	278	266	231	208	184
UPS	2,178	18	15	10	30	31	40	48	84	93	63	55	38	44
USA	4,979	65	97	94	113	127	149	146	161	229	273	197	232	195
	57,584	1,382	1,591	1,615	1,545	1,637	1,529	1,443	1,614	1,919	2,192	2,063	1,988	1,847

Copyright 1998 Air, Inc. - Atlanta, GA - 1-800-AIR-APPS

Figure 6-1b. Pilot hiring.

The airline business is full of ups and downs. (See Fig. 6-1a and 6-1b.) The pilot who is successful will be the one who learns how to roll with the punches. Sure, hiring is tough sometimes. It will get better, but you might have to wait a while as more pilots retire. Where do you think the new pilots are going to come from? So be patient in your job search. Don't become frustrated if the best airline job in the world doesn't hit you right away, because I'm telling you right now it most likely will not.

Getting Organized

Back to some of the same methods we used earlier for locating the airlines that are hiring. An important point is to be certain that you meet the minimum requirements for the airline you approach. It's a waste of everyone's time to shuffle the paperwork off to Southwest Airlines if you don't hold a type rating in the Boeing 737 (Fig. 6-2) because, no matter how great a pilot you are, you aren't going to get the interview. Southwest also requires at least 1,000 hours TURBINE PIC time. If you don't have that, don't chase them. Try something else. At United, for instance, the total time requirements are considerably less, about 350 hours total time. At United, they just don't happen to believe in the total-time god that so many other airlines have faith in. United happens to believe that there are other means available for the airline to determine whether or not you're the best possible candidate for their cockpit.

Figure 6-2. Boeing 737. *Boeing*

The Hiring Process at United Airlines—An Interview with Bill Traub, VP of Flight Standards and Training, United Airlines, (Retired)

"The personnel interview consists of two, or sometimes three, people meeting with the candidate. I must say that we would like every candidate invited to interview to processing to be successful. Candidates need to do their part by displaying a positive attitude and enthusiasm.

At United, flying the simulator is an extremely important part of the process. You can't be rusty on instruments and pass. The ride is all computer scored. We've also correlated that performance with a pilot's achievement level later on the job and that correlation is very high.

Essentially, the sim check measures instrument skills and eye-hand coordination. The candidate should look at this as an instrument flight test and understand they must prove themselves. The test may use a Frasca simulator or perhaps a DC-10 simulator. Regardless of the simulator, the test is the same. Believe it or not, they both fly the same.

You can make a mistake on the simulator check and still pass. If you make an error, we look at how you recover. If the rest of your ride goes to pot because of the error, you're probably not the person we want. It's like golf. Having played the course before doesn't make you a good golfer if you don't have the basic skills.

You'll shoot several ILS approaches that get progressively harder. Getting off altitude is not a killer, but the longer you stay off, the more points you lose. You begin with 100 points, too. 90 to 100 is 'A' work. 80 to 90 is 'B.' We've never hired anyone who had less than a certain score on the simulator ride.

We begin the interview with a little chitchat to break the ice and get to know you a little better. But our people will spend up to an hour, before they meet the candidate, going over the paperwork you've already sent us.

The interview is structured with six or eight areas of concentration. For example, we want to know about your commitment to the profession, how you've prepared for achieving your career goals, and how well you work in a team environment. As a clue about these, we might look at participation in sports or other extracurricular activities in high school or college. We believe that past performance is the best predictor of future performance.

We look at a lot of the nonverbal cues as well. Do you have a firm handshake, your posture, your neatness? If you don't look us in the eye very often, well--that doesn't project much confidence. We also try and learn how well you accept critique. We want to know what you've done well and what you have not. We want to be assured that you will grow and not continue to make the same mistakes. Nervousness is a judgment call on our part. If a candidate's nervousness interferes with performance in the interview, it is unfortunate, but you must find a way to overcome it enough to present a positive and professional image.

Candidates must also pass a first-class physical, but one that United doctors would give. There, we're trying to predict who these people are. Being overweight probably does not make a good first impression."

Traub also talked about some of the common "showstoppers" that candidates will want to avoid. "People who don't listen during the interview, who don't give us the answer to the question we asked. If there is a hint of dishonesty, we can tell. A history of checkride failures can be difficult to overcome, although one failure can be turned into a positive, by telling us what you learned, why you're now a better pilot. Logbooks must be accurate. Computerized versions are OK."

Finally, a poor driving record is a tough one to beat, as well as not having a license in the state in which you reside. If you don't follow the rules while driving, or with respect to licensing, how do we know that you will follow the rules when flying? Respect for authority is of the utmost importance.

After the simulator ride and the interview, all information is sent to a board for evaluation. Pending a successful medical exam, we'll make the candidate a job offer if in the judgment of the Board of Review, the simulator evaluation and the personnel interview meet United Airlines standards. Remember, approach the process with commitment to be successful. United's goal is the same as yours; we want you to be successful. Make the most of this wonderful opportunity.

While a subscription to *Air, Inc.* or *Flying Careers* or any of the other publications can be a valuable asset, these services aren't necessary to learn the basic information about an airline, although they can provide that to you at a cost we've already discussed. But considering how much information is out there, I'd sign up with one of these companies.

A few simple steps could save you a great deal of work and frustration later. First decide, by whatever means you choose, which airlines you intend to concentrate on. You might be tempted to say "any" at this point, but I caution you against that. Take the time to read about the different airlines before you apply. Perhaps you should set your sights somewhere other than the majors, or perhaps on some airline in addition to the big three. But without some organized plan to approach these carriers, or any other flying job, the work will become three times as tough.

First, pick up a box of manila pocket folders, the ones that are open at the top, with closed sides. Label one folder for each of the airlines where you intend to apply. Find a location in your home where the dog won't chew on them or the kids won't try to use them for coloring books. From this moment on, every piece of information you collect about this airline should be stored in this folder. If it fills up, that's a good sign. Start another folder, a part two, to continue the process. Record the date of every letter or update you send out, as well as the response, or lack of it. Look through all the folders at least once a month to make sure you haven't missed an update that was required or to learn whether or not a follow-up letter was ever answered. Then, if you're offered an interview, you'll have a veritable wealth of information on the airline to look over before you go in. Regularly scanning your files is going to be work, but the organization will give you a monthly update on just where some of the opportunities might be appearing.

You should consider adding a piece of useful software to your computer—a scheduler and database—such as ACT or Maximizer. Originally designed for salespeople, either of these two programs will help you more easily track all of the contact information you'll be sending out. Maximizer, for example, allows you to develop an individual file for each company you apply to. In that file, you can keep records of all contacts—when you last updated, who you sent it to, and what your totals were at the time. It also allows you to schedule the next event, such as when to update. If United wants an update in six months, the scheduler can be set to alert you when the time is right. If you're applying to many companies at the same time, you'll welcome a reminder of exactly when to make the next contact.

The Psychological Exams

If you haven't heard of the MMPI, the Minnesota Multiphasic Personality Inventory, hold on to your hats. This is one of the more common evaluations given to pilot applicants during some airline interviews. I took one at a major carrier a few years back. This test contains about 500 multiple-choice and true-false questions. It's designed to tap into many different aspects of your personality as well as determine how truthful a person you are. With this exam, there are so many questions that it's virtually impossible to beat it. Some people try to outsmart the exam and make themselves look worse in the end. The exam also revisits the same question more than once to learn whether you're consistent with your answers. An example of a question is "I never lie." True or False. If you answer true, most people would find that pretty hard to believe. If that's true, then say true, but realize that the question will appear again worded slightly differently. Conflicts in these answers help give an overall picture of your personality to the airline to assist them in their evaluation of you as a pilot candidate. Seldom would any airline make a hire or not-to-hire decision based solely on the MMPI profile results. This section is designed only to tell you that tests like the MMPI exist, and you can't study for them. But the good news is you can't fail the computer-scored test either. More good news is that not every airline requires these exams.

The Interview

There are so many variables in the interview process, depending on which airline you interview with, that trying to handle each on a case-by-case basis would be inefficient as well as possibly old news by the time you read this. One of the generalities that applies to most all the carriers, however, is that virtually none of them will hire you with simply one interview. One midwestern airline brings you back for as many as four interviews, while United, for instance, now believes that they have candidates pretty much figured out in two. There's just no hard-and-fast rule here. Some airlines treat you as total professionals and won't even ask you to take a written exam, while some others, I've learned, practically make you crawl on your belly like a snake from one phase to the next. One airline is known for asking extremely personal questions about you and your wife or girlfriend. These questions are designed not just to gain information about you, but also to see how you react in stressful situations. One comrade of mine was asked about how long he thought it might be before his wife became pregnant, as well as whether or not they were trying to have children in earnest or not. He told me it took every ounce of his strength not to get up and walk out of the interview. The point here is be ready for anything, but try to find someone who has been through the interview process. Personally, if I didn't know anyone, I'd hang out at the terminal and ask questions of the pilots I met who worked for the carrier I was interviewing with. Air Inc.'s preinterview counseling could also help.

Let's get to the main point of any airline interview. I tell you this based on the ones I've been involved with as well as those that pilots I've interviewed have related. The most important part of the interview is your attitude. One midwest carrier says there's not much an applicant can do to study for the interviews, so, "We just tell the applicants to be themselves." The most important part (other than doing some homework on the airline itself, either through recent research or by using the information you've gained and stored in your folder on the carrier) is to convey the idea that you're committed to working at that airline. No one is telling you to bounce off the walls when you arrive for the interview, but you must convince the airline that you want the job.

Interview Preparation: A Multistep Process

by Karen Kahn

Proper preparation for an upcoming interview can be one of the most important factors in determining whether or not you get the job. There are many things you can do to enhance your chances of a successful interview, even before you get the call or letter to schedule the actual interview date.

You can begin your preparation several months in advance by becoming informed about the industry in general and your target company in particular. If you don't subscribe to an aviation news publication such as *Aviation Week & Space Technology, Air Transport World, Business and Commercial Aviation,* or *Professional Pilot,* find someone who does and ask them to save you their old issues. You can also check into any of several online services (AW&ST maintains their own) and scan for articles that pertain to current events in the aviation field. Try to expand your horizons by reading in an ever-widening circle to include topics just slightly beyond your present professional level. Know what's happening in the world you wish to enter. You may be asked to discuss it in your upcoming interview.

Keep a file of articles on your target airline(s) and seek out their pilots and public; contact employees whenever possible to gain further insight into the workings of the company. Start networking to determine what you'll encounter in an interview and how to conduct yourself as a new employee. Ideally, you'd like to talk to someone who's been interviewed recently and can give you current information. Remember that senior employees are good sources of inside information but are rarely familiar with recent interview practices (unless they actively participate in the process). So ask the employees you meet if they know of anyone who's been hired within the last three months who might be willing to talk with you.

If you don't know anyone at "your" airline, join an online service and investigate some of the aviation forums like AVSIG, AOPA, or the ALPA (Air Line Pilots Association if you have access to it through a union member) pilot forums on CompuServe to seek out current pilots or employees who can give you some detailed "insider's" information. Remember, however, that "talk" is just that. Don't expect or ask for any favors such as requesting a recommendation from someone who hardly knows you. This puts the employee in an awkward position and could possibly boomerang later.

Check your local library or online news service for additional information (new routes, airplane acquisitions, catchy marketing campaigns, etc.) and follow your company's stock prices so you'll know what's happening on their financial scene in particular and in the airline industry in general. If "your" company is a public corporation, call your local stock broker and request a copy of the company's annual report. Obtain a copy of the airline's in-flight magazine and pay particular attention to their route map showing cities served and any code-sharing or alliances with other airlines. Save any advertisements you may find in national or regional publications. Your files will be brimming with information before you know it!

Now that you've researched their history, it's time to work on yours. Be sure your resume is current and concise—one page maximum. Gather your own records, including driving, credit, and educational. Make sure they are correct and speak well of you. If not, be prepared to explain the circumstances surrounding the questionable event. If you're not sure of how to handle a tricky situation, like an FAA violation or employment termination, get some professional advice to make sure you handle a potential minefield with care and confidence.

Gather your letters of recommendation and make sure they're current and pertain to the interview situation at hand. Producing a former letter of recommendation that describes your virtues for another purpose— perhaps a scholarship application—only confuses the interviewer and makes them wonder why you gave them one that's inapplicable to the task at hand. (Kind of like planning a cross-country flight with outdated weather information.)

Get to work on your reading list. Begin with a good text on interviewing such as H. Anthony Medley's *Sweaty Palms* or *Interview for Success* by Ron and Caryl Krannich and include airline specific titles such as Irv Jasinski's *Airline Pilot* interviews, Norris and Mortenson's *Airline Pilot Career and Interview Manual*, and Cheryl Cage's *Checklist for Success*. Each of these has a good list of questions you'll be expected to answer, so begin to compile your own responses that detail your personal, professional, and education accomplishments.

Once you're familiar with interview techniques, move on to the psychological testing section. Most every airline administers these tests and the more familiar you are with them the better. Most bookstores carry a generic text by ACCO entitled Officer Candidate Tests and you can order specific books from FAPA, Airline Ground Schools, and AIR, Inc. containing the types of tests that airlines are known to use.

The other important area to study is...aviation! You're expected to demonstrate a knowledge level commensurate with your flight time. Review the FARs, Parts 61, 91, and 135, the AIM, Aviation Weather, Jeppesen introductory pages, and your airplane FOM with special attention to limitations and emergency procedures. An ATP prep manual will also give you a good overview of what you're expected to know...and don't forget basic math and vocabulary skills!

Whew...seems like a lot of material to cover! It is, indeed, and you can see why it's important to start early and be thorough in your preparation. If you'd like help with your brush-up program, look into some professional interview counseling to polish your presentation and help you through any rough spots that may lurk in your background. This can be especially important if you have any "skeletons in your closet" or have had several unsuccessful interviews in the past.

Now's the time to plan for the future and do everything in your power to ensure your success. You wouldn't take a checkride without hours of preparation, would you? So plan to give that upcoming job interview your very best effort and you'll find the results well worth the time and effort you've devoted to it.

Karen Kahn is a captain for a major U.S. airline. Type rated in the MD-80 and the Lockheed JetStar, she holds an ATP, Gold Seal CFI: AIM, and is rated in gliders, seaplanes, and helicopters. In addition to being an FAA aviation safety counselor, she runs Aviation Career Counseling, a pilot career guidance and airline interview counseling firm based in Santa Barbara, CA. Contact her at 805-687-9493.

Just in case you were thinking that interviews only ask relatively easy personal questions, read on. Here are the first 25 of 201 question possibilities. The best part, though, is that I don't have the answers listed here. You need to think about them yourself.

The 20-Minute Interview

Well, friends, it's that time again. Whip out that ol' video camera, put on your best duds, and strap yourself in. You might be in for a bumpy ride. Try answering some of these questions gathered from recent interviews at United, American, Delta, and UPS. Read the following questions, then give them to a spouse or co-worker to complete your interview. Try videotaping the interview for later review.

- When do you have to declare an alternate?
- What are alternate minimums?

- Name all of the elements for a holding clearance.
- If you do not receive an EFC time, is your holding clearance valid?
- What are lower than standard minimums for 1600 RVR, 1200 RVR, and 600 RVR?
- Can you file an IFR flight plan under Part 121 without listing an alternate? If so, when?
- What is a hydraulic accumulator?
- Name and explain the three types of hydroplaning.
- What is the maximum crosswind limitation for the aircraft you are flying now?
- Have you ever bent or broken a company policy?
- What was the last book you read? Why did you decide to read it?
- An airplane flying from Denver to Cheyenne has a crew of three: a pilot, a first officer, and a flight attendant. Their names are Brown, Black, and White, but not necessarily in that order. There are also three passengers on the plane, named Mr. Brown, Mr. Black, and Mr. White. Mr. White's income is $9,999.99. Mr. Black lives in Cheyenne. The flight attendant lives just halfway between Denver and Cheyenne. His nearest neighbor, who is one of the passengers, has an income exactly twice his own. The flight attendant's namesake lives in Denver. Brown is a better chess player than the first officer. Who is the Captain?
- Who is the Speaker of the House?
- Who is the Senate Majority Leader?
- How many members are there in the House of Representatives?
- On an ILS approach, what is considered the final approach fix?

Courtesy Berliner/Schafer Aviation Consulting Group
888-745-6899, www.pilotswanted.com
For a list of a few hundred more possible interview questions, see Appendix B

American Airlines Interview Debrief

by David Manning

I was notified of my American Airlines interview on Monday, September 14, 1998. I selected Wednesday, September 23rd as my day. I was offered either the 23rd, 24th, or some time later in October or November. I chose the earliest date, partly because of my excitement, partly out of a sense of reluctance to get any lower on the seniority list.

I was Federal Expressed my additional paperwork two days after my notification. It detailed all the required documents. Most difficult for me were copies of all college transcripts, not just the university conferring the degree. After a flurry of faxes and credit card authorizations, I had my transcripts from three colleges and driving records from two states. I was missing one transcript on interview day, and they told me I'd need it before they could schedule a class date.

The Big Day

Arrive a bit early for the 0815 show time. Give all required documentation and logbooks to Barbara, the front office staff member. She's very friendly and professional, so don't be afraid to ask questions; however, she can add comments to the process, so treat her like gold.

The Applicant Profiles

1. Cont. Express EMB-145 RJ First Officer (former AA Flight Attendant), female

2. Mesa Airlines Beech 1900D Captain, female

3. Airborne Express DC-8 FE, female

4. Continental Express EMB-145 RJ Captain (Dad, two brothers, pilots at AA), male

5. Chrysler Corp. Hawker 800/Gulfsteam G-IIIIV Captain, male

6. Former U.S. Coast Guard Falcon Jet Aircraft Commander/civilian Learjet 60 Captain (Dad retired AA Captain/wife AA Flight Attendant), male (me)

Group welcome aboard and brief. Discussion of the state of affairs at AA hiring. One at a time, introduced ourselves with quick rundown on past flying experience. Now is not the time to read your doctoral dissertation— just a quick, "this is what I did." Leave to take Psychological Inventory, it takes about 45 minutes. It asks questions like, "Sometimes I feel so overwhelmed, I just want to hide" and you mark a scale of 1–5, Strongly Agree to Strongly disagree. No advice on how to take it, except to be honest. I don't know if it'll say that I'm deranged or whether it's just for a personality profile for your future personnel file.

Next, one at a time you'll be taken back into the depths of the office to review paperwork and make sure everything is there, certified, etc. I had to add the other two colleges I attended to my original application since I only put down my final university at graduation.

Time to go watch a quick ten-minute video presentation on the simulator profile. A quick break, then your group of six is broken down into three pairs. Your partner will assist you during your simulator. Important: you and your partner aren't competing! They want to see you both work together! My partner was the ABEX DC-8 FE. We were allowed 30 minutes to prebrief. We flipped a coin; she got the left seat. Sim profile was very basic. In fact, more basic than I expected or was briefed on. My partner's profile: one leg hop from CLE to ORD, fly the airway, shoot a Localizer approach to a freebie landing. Switch seats, and mine: I flew from ORD to CLE, although for me weather went below minimums in Cleveland and I returned to 01W to shoot the full NDB to mins (exactly what I would've done—yeah, right!). While you're the Pilot Not Flying (PNF), expect a question about holding, i.e., if given this clearance, how would you enter, what airspeed for a particular altitude. Whole session lasted from 1100 to 1240 or so. Upon return to the recruitment office, we were told no time for lunch because our interview was at 1300. Quick cup of coffee, then into the interview. At the same time, the other pairs were either coming out of their interviews or coming back from lunch. The good thing about going into the sim first and missing lunch was that we were done for the day first.

Interview

Two captains—a B-767 check-airman, and a Fokker F100 line-captain. First and foremost, they were the consummate professionals, but they did their best to make me feel comfortable. Offering a drink, asking if I wanted to take off my suit jacket, etc. They seemed very relaxed and wanted me to be relaxed, too.

When they interviewed me, they didn't just "ask questions" a la United Airlines. They actually crafted a set of questions into their own version of a short situation and asked me how I'd react or what I'd take into consideration when making my decision. It immediately put me at ease and let me speak "airplane," letting them know who I was as a person and an aviator. This is exactly what they want—to get to know you, not hear any gouge regurgitated. Questions/situations I remember (by the way, these are all on the gouge available somewhere on the 'net):

Initial—tell us about your flying experiences until this point.

If it snowed a foot overnight, how would this impact your preparation for a flight?

You're an FE. Your captain arrives at the jet looking ill and smelling of alcohol. What do you do? If he's insistent on flying?

What if the captain is low and slow on an approach?

How would my First Officers rate me as a captain on a scale of 1–10? Why?

What was your most pressing emergency? How did you handle it?

What considerations would there be departing from or operating out of high-altitude airports (used my experience with Telluride, Aspen, Eagle Co.)?

What makes a good captain?

Have you ever had a conflict with a captain? How did you resolve it?

What is the worst mistake you've ever made in an airplane? (Then they said, "It's okay, all mikes are off and the FAA isn't here.") Be careful...

Do you have a mentor? (I understood whom do you know at the company?)

What would you do if the captain was using nonstandard procedures but not unsafe?

Did I have any questions for them?

No systems or regulatory-related questions. In fact, when I got home and my wife asked me what technical questions they asked, my first reply was "none," but after thinking about it I remembered that I detailed a few of the situational questions in a technical manner as well as a procedural manner. Overall, very relaxed. They just wanted to see if they could stand flying with me for hours on end. It took about an hour and fifteen minutes. Remember to smile and be enthusiastic!

After waiting for my logbooks to be returned, I was released to leave. I was told I'd hear something in ten days to two weeks.

I was called back for the physical in 7 days. During the physical, they gave me a one-hour cognitive test that I had never heard anything about. It used a light pen and a computer and seemed to be testing hand-eye coordination. I used the pen to touch a number—five, for instance. Then I'd move on to the next page and it would show me a series of new numbers and ask me to touch the number from the last screen. I had only a few seconds to decide. There were no results of that test given to me before I left. They made it clear that you were graded for speed and accuracy. The attendant took the computer grade before I left, but I didn't see it.

My networking: I had plenty of friends who were pilots and took a business card and kept a phone number from everyone I ever met. It really was easy to meet people. Some companies want character references; others simply want people who have flown with you.

Thanks to David Manning, who is currently awaiting a class date with American Airlines.

The Top 10 Mistakes Made by Job-Seeking Pilots—and How to Avoid Them!

1. Misrepresentation

Not all the skeletons in your personal closet can be overcome. But lying—about anything—on an application or during an interview can be a career killer.

Practice answering sensitive questions—especially if you have a DUI, incident, etc.—with an interview coach who specializes in airline interviews or another pilot who has recently been through the hiring process. The better prepared you are for the question, the less temptation there will be to misrepresent yourself.

2. Showing up for an Interview Unprepared

Some airlines interview as many as five qualified applicants for each one they hire. Preparing for an interview—regardless of whether it is a major or regional airline or even a corporation—is not something that can be successfully accomplished in a few minutes. Time spent honing interviewing, nonverbal, and other skills will be well spent. While most pilots are accepted or rejected for employment based upon several criteria, the interview is often the first step, and the first point, of elimination.

Prepare, prepare, prepare. Keep detailed, well-organized files on each company to which you've applied. Know everything about the company there is to know—not so you can dazzle them during an interview, but so you can speak with confidence and answer any questions intelligently. Taping your practice sessions and reviewing them with a good coach is a great idea.

3. Being Too Revealing

Just as it is important to never misrepresent yourself on an application or during a human resources interview, it is equally important to answer only the questions being asked, or you risk sabotaging yourself.

Answer completely, but keep your answers short and don't ramble. If the interviewer wants to dig a little deeper in some area, let that be their decision. Again, the more practice interviewing you do, the more confident you'll feel when the big day arrives.

4. Downplaying the Importance of Interpersonal Skills

Interviewers—both from human resources and CRM/line representatives—want pilots they can see themselves working with for several days at a time. Communicating effectively in the recruitment process is imperative to getting the job; communicating effectively in the cockpit is pivotal thereafter.

One of the most effective ways of honing your interpersonal communications skills is by practicing the toughest of all questions—conflict resolution. Tackle each question individually, making plenty of eye contact while watching your nonverbal cues. Remember, too, an airline interviewer may ask a question with as much interest in the demeanor with which you answer as what you say.

5. Showing up for the Simulator Evaluation Unprepared

In most cases, employers don't expect you to fly the simulator as if you regularly flew that particular aircraft. However, buying a little sim time—especially if you can learn what type you'll fly during the evaluation process—to sharpen your instrument scan and get a feel for the sensitivity of the simulator will be beneficial. Simulators don't fly precisely like an airplane, only close. The time to refresh your memory about that fact is before your evaluation, not during the session.

6. Being Unwilling to Relocate

Some jobs require pilots to live in locations they might find unattractive, at first. But applicants should remember that such moves can be temporary and that jobs that help build quality time are often difficult to locate.

Look at each job opportunity carefully before rejecting any. Remember, reaching your career goal is a process of small steps—none of them permanent—until you reach your career position.

7. Believing Your Qualifications Will Speak for Your Experience

Regardless of how many thousands of hours you have logged, airlines will ask you about your flying experiences. A common mistake is thinking that you won't need to talk much if you already have 10,000 hours.

Be assertive during an interview, but never cocky. Remember, despite your significant experience, the resumés of the next three pilots interviewing—as well as those of the last three—are probably similar to yours. You'll need to make an effort to convince a potential employer that you are the best pilot for the job. Have at least one story in the back of your mind ready about how you successfully coped with some conflict in the cockpit or dealt with an inflight emergency.

8. Not Understanding the Cyclical Nature of the Industry

Bailing out of the aviation industry is a personal decision. Many pilots have quit their flying jobs and ceased to log hours during downtimes, only to regret their decision once hiring resumed. Succeeding in this industry usually requires constant focus and unwavering dedication.

Be persistent and patient. If a civilian pilot can't dedicate a significant period of time to building a career—time during which he or she may be poorly compensated—it may be time to look at another career.

9. Lack of Correct or Current Career Information

Using gouge as a supplement is OK, but don't rely on hearsay as your only source of career information. The aviation industry is unique. Knowing when application windows are open, where to send paperwork, how much of an application fee the company charges, minimum qualifications, etc., are an important part of a successful job search.

Purchase some form of career information service. But evaluate the service before you buy, to make certain they're not simply regurgitating old, incorrect, or bad information.

10. Being Unprepared for a Written Exam During the Interview

Most companies will ask pilot applicants to complete some form of written knowledge exam during the interview process. These run the range from a simple 10-question exam about regulations, ATC procedures, and weather to a complete 70-question Instrument, ATP, or Commercial knowledge exam

A number of books are available to help you review for the knowledge exams—if you know which one the company uses. There are also the standards that every pilot should study before any interview: the FARs, AIM, Aviation Weather, and both NOS and Jepp IFR charts.

Honorable Mentions

1. Not Holding a Four-Year College Degree

Just looking at the statistics on the education level of pilots who get hired will tell you not to shortchange yourself here. Correspondence (external degrees) programs now make getting a legitimate degree easier for pilots of all ages and work schedules.

2. Being Unaware of the Competitive Qualifications of Pilots Who Are Getting Hired

Building on the education issue, pilots should know which certificates, ratings, and kinds of experience are required and which ones are preferred. Then, make sure you have both.

©1998 Bossard Publications Inc. d/b/a *Flying Careers* magazine. Used with permission.

Simulator Checks

When you arrive for the interview at some airlines, you'll be expected to fly a simulator and demonstrate your skills as a pilot. All airlines, however, don't require a simulator check, although many regional carriers will. Corporations will not normally have a simulator on premises to be able to test your skills. Never arrive at a simulator check cold, with no recent IFR training. Unless you're a real ace, the chances of getting through the test are slim. Here are some other facts to consider because good instrument skills are what produce a good simulator session (Fig. 6-3).

Figure 6-3. Flight Safety F-1000 simulator cockpit.

How to Prepare for a Prehire Simulator Check

There are different opinions about whether or not pilots should buy simulator time before going to an airline interview. While there is no single answer to the question "Should I buy some time?" some pilots have found they needed it. Others have been successful landing a job without buying it. The bad news is, if you wrongly evaluate your need to buy simulator time before a prehire checkride, it could cost you the job.

In the grand scheme of things, the prehire simulator check most airlines put pilots through is only a portion of what these aviators are evaluated on. Also considered is a pilot's aeronautical knowledge and overall physical and mental fitness for the position. All too often, unfortunately, pilot applicants treat the prehire simulator check as an afterthought to the interview process, not worthy of nearly as much planning as other portions. This can be an dangerous mistake, according to sources.

When the airline I was working for went bankrupt a few years back, the closure threw hundreds of well-qualified and current pilots on the street in search of new opportunities to ply their skills. Tom (not his real name) was hired almost immediately by a USAir Express carrier on the East Coast flying an SD3-360, notorious for a high cockpit work level in bad weather because the aircraft has no autopilot.

To Buy or Not to Buy

After a year of this kind of hard IFR flying, Tom received a letter from United Airlines for the big interview. During a number of telephone conversations, I asked him whether or not he planned on getting some simulator time under his belt before he flew the prehire simulator check at United. He said 'no," believing the many hours of hand-flown IFR during the past year would suffice. Besides, he said, "the simulator at United was a Frasca. How tough could that be?"

A few weeks later Tom and I spoke after his simulator ride and interview. Apparently the Frasca that United runs people through at Denver was a bit more slippery than that lumbering Shorts he flew. His scan in the unfamiliar Frasca was a little slow, as were some of his course corrections while he tried to get used to the simulator's sensitivity. When the letter arrived a few weeks later, Tom was surprised that the nation's largest airline had decided "not to continue processing you at this time."

So what went wrong? After all, IFR currency should be enough to pass a checkride for a job, right? And isn't IFR proficiency what the airline is really looking for anyway? Well yes and no. While the need to demonstrate your ability to fly an aircraft under IFR conditions is certainly the intent of the checkride, the spirit of the ride encompasses much more.

The ride will also—in most cases—measure your judgment, your knowledge of ATC procedures, as well as your overall use of cockpit resources, such as a copilot in many cases, according to airline sources. No airline is hiring you simply to manipulate the controls. They want a pilot who matches a set of their own special criteria that tells them you are the person who will operate their aircraft safely and efficiently in all flight conditions.

Simulator Ride's Importance

Shane Losasso, president of Jet Tech, said, "The simulator ride is very important. Look at the number of [pilot applicants] who are turned away after a simulator ride. It is substantial." The strategy then for applicants is simple. Treat every part of the interview process—including the prehire simulator check—as if it were a 'make or break' part of landing the job, because it is.

When John Shogren, now a first officer on a Boeing 747 for United Parcel Service, arrived in Pittsburgh for a prehire simulator check with USAir, he said, "I did no training before the ride. In retrospect, however, if I'd taken some simulator time before the USAir ride, I might be flying for them now. I did that poorly. I had been hand flying a Brasilia and a Shorts 360 for nearly 1,000 hours and I thought my skills would transfer easily, but I was wrong." Shogren's logbook showed about 2,700 hours total time when he took the USAir ride.

Bill Mayhew, former vice president of flight operations at Wheeling, Ill.-based Airway Flight Service, said, "A lot of people who come to us for prehire simulator checks have flown aircraft, but [they] are not used to flying simulators. They are not used to the sensitivity a simulator might offer. Proficiency in an airplane does not necessarily equal proficiency in a simulator."

Process Viewed Differently

What makes simulator checks confounding to applicants is that pilots of various skill levels often view the process so differently. Todd Carpenter, a former C-141 instructor pilot, did not purchase any training either, although he had flown only about 500 hours during the previous 12 months prior to his prehire ride. "I felt pretty current before the rides with FedEx and Airborne," he said, "since I hand-flew the approaches and departures in the 141 most of the time."

Not All New Hires Buy Time

Carpenter believed the training was unnecessary before the Boeing 747 ride with FedEx, in particular, because "I did not believe they wanted me to fly a 747 perfectly. I think they wanted to see I was someone who could learn, as they went along trying to get a good feel for the airplane."

Jenny Beatty, an MD-80 first officer with Reno Air, added another spin to the subject by explaining that "most airlines do not expect someone with 1,000 hours to fly like someone who has 5,000." Beatty also did not purchase any simulator time prior to her prehire simulator checks because, "I try really hard to maintain my skills...I felt I was prepared. But I also do some armchair flying by sitting in a chair and mentally going through both aircraft and IFR procedures. A large part of training your body is training your mind." But many pilots believe they need every point possible during a prehire ride and refuse to give up a single one easily.

"Worth the Money"

Tim Doreen, a Jetstream 32 first officer for Atlantic Coast Airlines said, "The entire simulator check and interview is a stressful time anyway, and you want to feel as comfortable as possible. If you don't pass the sim, you don't get to talk to anyone (at some airlines). If you don't get to talk to anyone, you don't get the job. I flew a Frasca in Denver before my ride with United, and it was definitely worth the money."

Citation pilot David Hilsdon said "the main reason I took some simulator training before my prehire check ride is that there is always a difference between the way simulators fly. Since I considered my checkride with FedEx to be a career-changing one, I wanted to put myself in the best light. I figured even an hour in the 747 would give me a leg up."

And what does an airline really want anyway? According to Losasso, "Each company tends to look for basically the same thing, but goes about the process just a little differently. The end result is that the company can probably correlate how current someone is on instruments or whether they seem to understand instrument procedures well with their initial training washout rate."

Reno Air did not always use a prehire simulator ride for its applicants, according to Jim Hubbard, the carrier's director of training. "We were hiring people with lots of experience in our aircraft. Now we're hiring less-experienced pilots. We're really looking for basic instrument flying skills and airmanship...anticipating power changes, that sort of thing."

Good Instrument Skills Crucial

Hubbard also said that Reno tries to keep the ride as objective as possible so "a person's lack of experience in the MD-80 does not count against them. We have a series of standard profiles and choose them at random. In a one-hour flight they are given a briefing ahead of time, as well as written instructions without having to know anything about the MD-80. It wouldn't hurt someone to have some MD-80 simulator experience, but it's designed so that any ATP-rated pilot with good instrument skills should perform very well."

The other side of the prehire coin is Southwest Airlines, which requires no simulator check before hiring a pilot. Paul Sterbenz, vice president of flight operations at Southwest, said, "We do not use prehire simulator checks because of the time and energy it takes to do them in the first place. But also, the objectivity of what you find out is somewhat suspect. It tells you whether or not a person can fly a simulator, but it doesn't tell you that much about what their general airmanship is in terms of situational awareness and getting around in the real world. Secondly, our requirements are pretty high. We require an applicant to possess a type rating in the B-737 before we will interview them. With those considerations, we are having pretty good luck with a very low failure rate in training. We don't have any indications that another screening device like a prehire simulator check is really necessary."

What can you expect to see on a prehire simulator ride? Nothing outrageous, really—nothing more than a profile to perform some specific flight maneuvers. Here's a look at a sample profile from one airline's prehire simulator check. Flights often begin with the aircraft already airborne, as this one does, flying straight and level at 250 knots. First, decelerate to 200 knots and accelerate to 250 knots again in level flight. Next, climb at 250 knots, descend at 200 knots and accelerate to 250 in the descent, and then, a turn. Next comes steep turns in both directions and then an entry to a hold. In this profile, flown in a Boeing 747, no instrument approaches were made.

If you've decided to make the extra effort (and incur the expense) to buy some training time before your simulator ride, some problems you'll face are what simulator to use as well as how to locate the company that owns a simulator you can rent. Opinions about which simulator to use vary.

What Type of Simulator?

Don Buchanan, president of Allied Services, a simulator training company, said, "Fly the heaviest simulator and the one that is the toughest to fly." Todd Carpenter advised that "If you had no heavy time at all, I would buy some time in a simulator like the [B-]747." Mayhew added, "Try and find out what simulator will be used on your ride and try and get something as close as possible."

Local networking with other pilots, as well as reading aviation publications, seem to be the way most pilots learn about the right simulator company. But one relatively new method is with a personal computer, by searching any of the dozens of World Wide Web aviation sites available on the Internet. Other sources of information are the online services' aviation forums where hundreds of airline and corporate pilots—as well as the want-to-bes—hang out.

A word of caution if you decide to seek advice through your computer: Take the information you hear with a grain of salt because you never know for certain who you are corresponding with unless you verify it in another way.

Can you afford simulator time before a prehire ride? A better question is whether you can afford not to buy simulator time before your ride. How important is the job you're interviewing for? How will you feel in five years if you don't get that job at United or FedEx or Northwest that could eventually pay you more than $175,000 per year because you did not want to spend $500 to $1,000 for some additional training? Hilsdon said, "If you have to beg or borrow the money, do it."

One caveat, however: Not all pilots hired by airlines, of course, buy prehire simulator time. Some pilots "ace" the simulator evaluation portion of the interview process without the expense of time in a full-motion-or other-simulator. Some pilots who are flying often for an airline decide that the expense is worth the peace of mind it gives them going in to the interview. It's an individual decision and not something that should be taken lightly.

At Jet Tech, the prehire simulator package costs $375 and includes an hour of prebriefing about the profile, an hour in the Boeing 737 simulator, and a half-hour of debriefing. At Airway Flight Service, $225 buys two and a half hours in a Frasca 142P, very similar to the one used at United Airlines. "Tom" spent nearly $1,200 for the Boeing 747 training time he purchased. And also, don't let too much time elapse between your training and the ride. More than two weeks and you're probably wasting your money because you'll have forgotten so much of what you'd learned. Remember too, prehire simulator training is not the place to relearn IFR procedures and regain your IFR currency. You can do that for less at your local fixed-base operator in a smaller, less expensive simulator, sources said.

Figure 6-4. Airbus A-300. *Airbus Industrie*

Tom received his interview letter from a major airline a few weeks back and called me to talk about it. The first thing he mentioned was the need to buy some time in a simulator before he headed out to Los Angeles for the ride. From his contacts in the industry, he learned that the simulator this airline uses is also for rent and he bought three hours. He wanted every advantage he could muster during the interview process this time. He told me he was ready this time for the ride like he'd never been for the one at United.

There was one more call from Tom last week, although it was hard to understand what he was saying because he was yelling and screaming so loudly. He had just been awarded a class date with that major airline. I asked him whether the money he spent on the training was worth it. "Best money I ever spent," he replied. (Fig. 6-4.)

The Schedules

Major airline schedules, as well as the pay, are the reasons why most people are interested in these jobs. Airline schedules can offer as little as 8 days off per month or as many as 20, depending again on the routes and the aircraft flown. Most major airlines operate under FAR Part 121, which prescribes a maximum of 100 flying hours per calendar month for a pilot. But before you run off to tell your friends how little time you'll have to work each month, if your salary guarantee is only based on 80 hours or so each month, realize that 100 flying hours means just that, the flying portion. Another important consideration is how many hours of duty, which includes preflight and postflight chores and sitting around between legs of a flight or around a hotel on an overnight.

In general, a pilot who flies ORD-HNL (Chicago, Honolulu, about a 16 hour round-trip flight) might only make the trip about 5 times a month before he or she will have flown up to that 80-hour guarantee (Fig. 6-5). The flight probably required a minimal amount of sitting time because, once off the ground, you fly until you land in HNL and go to the hotel for your rest period. On the return, it's basically the same thing back to ORD. Pretty cushy job, most likely flown in a Boeing 747 or DC-10, so the pay is good. Obviously, these routes are highly sought after because the time off is tops. This is where a high seniority number will be worth its weight in gold, for without it, anyone who's more senior will outbid you.

A more average route might be ORD-HOU-MEM-DCA (Chicago, Houston, Memphis, Washington, D.C.) and then overnight and possibly DCA-ORD the next day. In a Boeing 737, for instance, the pilot will most likely have flown about seven hours in two days. What you might not

Figure 6-5. Boeing 767-300. *Boeing*

notice at first glance, though, is that in Houston, the crew sat for two hours before the flight to MEM, where they might sit for another hour and a half before heading to DCA. It can make for a pretty long day. You have various ways to choose your schedules each month, based on days off, maximum flying time, minimum flying, certain days off, certain bases, and so on. Today, the work of choosing your schedule with an airline, called bidding, has been significantly reduced because much of the tedious sorting work required years ago can now be accomplished with a personal computer (more in chapter 8).

TWA recently initiated a new bidding system that has taken a major leap forward in its use of computers and database technology. Rather than having the company develop and print lines of flying that pilots must choose from each month, TWA pilots are now invited to bid their monthly flying considering all of the variables to the schedules that reflect their own values—and there are dozens and dozens.

For example, if pilots want to fly weekdays and have weekends off, they can make that a high priority. If they want trips that begin midday— great for commuting pilots—they can make that a priority as well. If they want maximum takeoffs and landings each month or minimum approaches, such as they might pick up on international trips, they can make their bid reflect this.

There are over 100 variables that TWA pilots can enter into the computer software from their home PC to the company each month. The airlines then build the lines of flying for each pilot based upon their desires and their seniority. The line the pilot eventually receives is as close as the computer can make it. This does change one very traditional aspect of airline flying, however. It practically eliminates pilots flying with the same crewmembers for a month at a time.

Duty and Trip Rigs

The major airlines, unlike the regionals, provide their pilots with pay from the moment they leave their base until the moment they return. Some regionals only pay their pilots when they fly. If the aircrafts break down or gets weathered in somewhere, they don't get paid. Major airlines use two systems to assure proper pay to the crews, the *duty rig* and the *trip rig*.

The duty rig assures a pilot of a specific amount of pay for a certain amount of time on duty, regardless of how much actual flying was involved. For example, a 1:2 duty rig means: one hour of pay for each two hours on duty. If a pilot signs in for duty at 11 A.M. and flies one hour to the destination and sits for two hours before flying back home, he or she only flew two hours total. But, if the pilot signed off duty at 5 P.M., he or she was on duty a total of six hours. But the pilot will be paid on a two-hours-duty-equals-one-hour-pay duty rig, so he or she will actually be paid for three hours.

A trip rig is similar in that it pays pilots a minimum amount for a given time away from their home base, regardless of how much flying was involved. If a trip rig were, say, 1:3, a pilot would be paid 1 hour of pay for every 3 hours away from base, regardless of the time flown.

Additionally, pilots are paid an hourly per diem allowance when they're away from base and are also eligible for various free-space-available passes on their own or other airlines.

The Big Bucks

This is what everyone calls it. An airline job is where pilots believe they'll find their fortune, so let's take a quick look at some of the money involved in this end of the industry, with figures from the *Air, Inc.* pilot salary survey.

I've picked three airlines at random: Continental, an airline that recently emerged from bankruptcy; United, the nation's largest carrier; and Southwest, an airline that runs against the traditional hub-and-spoke concept of connecting passengers. Continental is IACP; United is represented by ALPA; and Southwest is represented by the in-house Southwest Pilots Association (SWPA). (See Table 6-1.)

Table 6-1. *Pilot salaries at selected airlines (1998)*

Starting pay	Continental	United	Southwest
First year (per month)	$2,500	$2,336	$3,011
Second year	$3,451	$3,479–$4,892	$5,080
First Officer Fifth year	$6,844	$7,096–$9,003	$6,839
Captain Tenth year	$12,007	$10,951–$15,785	$11,701
Max pay as Captain—Annual	$146,460	$191,232	$143,508

A few things to keep in mind about the preceding scale. A B-747 first officer is paid the same as a B-737 first officer. At Southwest, the airline only operates one type of aircraft, the B-737. At United, the scales vary after year one, depending on which seat you sit in and in which type of aircraft. In year five, for instance, a United B-727 first officer is paid $7,096 per month, while the high end is limited to the B-747 first officers at $9,967 per month.

At the regionals, retirement plans, other than a 401K, are virtually unknown. At most of the majors, retirement is a very serious affair. The two major plans are the A-Fund, a defined benefit retirement plan and the B-Fund, a defined contribution retirement plan. The A-Fund pays a defined amount each year to the employee beginning at retirement based on the pilot's earnings. The B-Fund gives the pilot a monthly retirement sum based on money added to the account over the years by the company, the employee, and accrued interest (Fig. 6-6).

Figure 6-6. Airbus A330. *Airbus Industrie*

The choice of airlines is still substantial, although many of the old players of the industry—like Midway, Braniff, Eastern and Pan Am—have departed. While all of the jobs those carriers sucked from the aviation industry have not returned, there's hope on the horizon with the new start-up carriers. (Fig. 6-7.) Where some people were astonished at the thought of regional pilots paying for their own training, some start-up airlines may also require various sorts of financial concessions from pilots just to pick up the job too. Pilot pay at most of the start-ups is also considerably less than most major airlines.

I spoke with Peter Larratt, a Boeing 737-200 and 737-400 training captain for British Airways. I asked if he would recommend this career. "Yes," he said. "There are really several aspects of flying professionally that appeal to me. One is the initial challenge of flying large aircraft. Then, the challenge of managing the entire operation. Each day, a new job is started and you see the job through to a finish. Generally, too, you don't take the work home with you. You're essentially your own boss, even when you're a junior member of the crew." When I asked if he had any tips to pass on to future generations of pilots, Larratt said, "For the first few years, maintain a low profile and learn your trade well. You might be the best aircraft handler in the world, but that's a very small part of the job of a professional pilot. You can't teach experience."

Figure 6-7. Boeing MD-11.

Labor Organizations

Finally, let's talk about an organization that you'll most likely encounter during your aviation career . . . unions. The airline industry is probably one of the most unionized of all industries. Corporations and their pilots tend not to be union, however, leaving negotiations for pay and benefits up to the individual pilots themselves, just as you'd expect in a regular job. Flight instructors also are seldom unionized, nor will you find unions in pipeline patrol, banner towing, or many other flying jobs. These tend not to have unions because the operations are just too small, often just a few pilots and aircraft. Regional airlines are mostly represented by unions, although there's more than one. One regional carrier is represented by the Air Line Pilots Association (ALPA), while another is represented by the Allied Pilots Association (APA). At the majors, most pilots will be represented by ALPA, while some carriers like Southwest Airlines or United Parcel Service pilots have their own in-house unions. (Fig. 6-8.) One noteworthy exception to ALPA representation is at American Airlines, where pilots are represented by APA. However you might feel about unions, if you remain in the aviation industry long enough, you're probably going to have to deal with one. That can bring on some interesting choices.

A list of most of the airline industry union's Web sites can be found in Chapter 8.

Figure 6-8. Northwest Airlines pilots are represented by ALPA.

Unions: Can Pilots Exist Without Them?

Gregg Watts developed his love for flying a little later than some—age 23, actually—when his older brother Glenn, an agriculture pilot, started letting his younger sibling hang out with him at the airport. He took Gregg along once when he sprayed for mosquitoes in an ancient Aztec. One ride at 200 feet and 160 mph and the younger Watts was hooked. Gregg won his private pilot's certificate in six months and worked two jobs to pay for the rest of his ratings. It was no surprise to the rest of the family when his first paying job turned out to be as a spray plane pilot, too.

A few years later, his career took him to the left seat of a pewter gray Beech 18 hauling night freight along the Gulf Coast for $1,200 a month—before taxes. He upgraded to a Cessna Caravan for turbine time and then to Lockheed Electras for Part 121 experience, each step a part of his plan to someday command a jet aircraft under Part 121 himself. In January 1991, Watts began his current professional sojourn with Southern Air Transport, a worldwide freight carrier, where he now flies in the left seat of a DC-8. It's been a dream come true—almost.

"When we decide to become pilots, we know up front that we are going to miss some things at home, our children's birthdays, our anniversaries, graduations, and even weddings sometimes," Watts said. "It's just one of those realities of the job. But there also has to be some balance between giving your company what is fair and seeing your family." An aviator's lifestyle can be rugged and often takes its toll on even the strongest families. "On one trip, I was told I would be gone 30 days straight. I knew that, and I was prepared for it. When the first 30 days ended, the company told me I'd be out another 30 days because there was no one to replace me. When I protested, the chief pilot told me he had 400 resumés on his desk and asked me what I wanted to do. I flew the trip."

Strained Relationships

The aviation business is tough—on pilots and on managers. Both are often forced to push the other in directions that wreak havoc on professional and personal relationships. But when one group possesses authoritarian powers over the other, the system can swing out of balance quickly.

Pilots often feel there is only one way to maintain their equilibrium. What Watts termed a "change in management's approach to getting the job done," brought Southern Air Transport pilots to the doorstep of the International Brotherhood of Teamsters. The IBT, now the bargaining agent for Southern pilots, is currently negotiating that group's first labor agreement.

Trade unionism has been a part of the American work culture since the late nineteenth century when Samuel Gompers, an early president of the American Federation of Labor and the Council of Industrial Organizations fought the early oratory battles with the titans of the new industrial revolution over higher wages, shorter hours, and more freedom for workers. Much of what Gompers accomplished has been carried down through the years by thousands of other workers building the union movement piece by piece.

Union Numbers Declining

At its peak in 1945, approximately 35.5 percent of the American workforce was unionized. Today, union membership—still locked in a two-decade tailslide—has declined to 14.5 percent of working men and women. In 1995 alone, unions lost 400,000 members. Despite the total decline in union membership nationally, the U.S. airline industry remains a stronghold, one of the most unionized of all professions.

According to Air Line Pilots Association spokesman John Mazor, "at least 80 percent or more" of eligible pilots join that union. And in the grand tradition of past union members, the aviation industry is still a platform from which thousands of pilots speak out each year about issues that help create today's pilot working environment such as, safety, duty schedules, and merger considerations. The recent standoff between American Airlines' pilots represented by the Allied Pilots Association and the management of AMR Corp., the parent company of American, clearly demonstrates the passion of the movement.

Frustrated by stagnant wages in recent years, as well as conflicts over who will fly the new regional jets—mainline or commuter pilots at American's four regional carriers—American Airlines' pilots walked off the job at midnight on Feb. 14, 1997. The highly-charged job action lasted only a few moments, however, until President Clinton—citing that significant economic loss would reign—stepped in and ordered pilots back to work. He called for a presidential commission to study the problem and asked both sides to return to the bargaining table. The two sides eventually signed a new contract.

But American Airlines is not the only airline mired in labor troubles. United pilots were initially unhappy about the pay raises they were recently offered. This despite the fact that they, along with the airline's mechanics, possess a controlling interest in the carrier. In early March, however, pilots agreed to a new pay structure and turned down the pressure cooker that had begun boiling in January.

Additionally, a strike by one union can create difficulties for another union as the American strike did. Since the American pilots want to fly all jets and protect their own jobs, they could be putting the jobs of regional pilots at risk—pilots who are also union members, albeit a different union.

Also, these commuter pilots seemingly stand to earn higher wages if American Eagle—like many other large regional or national carriers—gets the nod to fly the jets.

Continental Airlines' pilots are also beginning contract talks, only their second in 12 years. Continental's pilots got their first contract in 1995. Northwest pilots, recent contract negotiations ended in a three-week strike. At USAirways (formerly USAir) intense negotiations continue to gain pilot concessions that will reduce that carrier's costs, currently the highest in the industry, according to industry statistics.

Unions Offer "Protection"

One USAirways first officer said that, "Union membership is the only protection we have against the [Frank] Lorenzos, [Carl] Icahns, [Robert] Crandalls, and [Stephen] Wolfs of the airline industry," naming some of the more well-known airline chief executives of past and present. In a defeat for ALPA last fall, that union was ousted as the bargaining agent for the pilots of all-cargo FedEx, to be replaced by the in-house union called FedEx Pilots' Association, led by newly appointed president Capt. Mike Akin.

In the aviation industry, unions are pretty much a fact of life. But Bret Henry, a pilot for Horizon Airlines believes, "Unions help at some companies that have poor management, but when you have a leadership that values safety as a No. 1 priority, you don't need the outside help."

Becky Howell, a Boeing 737 captain at Southwest Airlines and a member of Southwest Airlines Pilots Association, an in-house union, said that even though that airline is unionized, "We are fortunate enough to have a union that is very close to management. We still realize that we're in this together. Cooperation with one another gets the most benefits."

Continental Airlines Boeing 747 first officer, Ed Neffinger, said, "Unions are not always necessary, but management generally forces them on to the property by their actions. Once a union is in place, the relationship is necessarily adversarial, but need not be confrontational."

If you join a union airline, you may hear the term agency shop agreement, where pilots are not forced to join the union, but must pay for the maintenance of the labor agreement in place at a particular carrier. This occurs because even though a union's membership might not include all pilots, the union is still the sole bargaining unit for all the pilots.

Nonmember pilots have no vote and are not required to take part in union activities. At American Airlines, APA spokesman Tom Kashmar said, "There is no agency shop arrangement [present]. Pilots may choose not to be members of the union and pay no maintenance fee. However, only about 50 pilots—[about one-half of 1 percent]—of American's list of about 9,100 have chosen not to join."

What happens when you don't join the union at a union carrier? Legally, there should be no pressure upon nonmembers. But sometimes, that union member, nonmember relationship can become strained.

Many months after the 1985 United Airlines' strike, nonmember, as well as union member pilots who crossed the picket line to fly aircraft reportedly were ostracized in the crew room at most United bases. Small clicker devices were handed out to union pilots to make their feelings known. Anytime a pilot who crossed the 1985 picket line entered the lounge, dozens of clickers went off until all heads turned to acknowledge the offender. Ten years later there reportedly remains some animosity between both groups of pilots. Some captains simply limit conversations in the cockpit to essential business when flying with pilots who crossed the picket lines in 1985.

The probationary period of one year—even at an airline—is based on the Railway Labor Act. During this time at American Airlines, pilots do not pay dues and do not enjoy the protection of the contract in force at the time. A pilot may be represented by a union officer—usually another regular line pilot—in meetings with the carrier's management upon request, however. At the end of one year and provided the pilots pass their 12-month checkride, they become full dues-paying members of APA.

Dues at carriers vary considerably too, but ALPA carriers charge about 2 percent of a pilot's base pay per year, while at other newer union carriers like Southern Air Transport, dues are expected to cost 1.25 percent of base pay per year. Dues at FedEx's in-house union, for example, are one flight hour's pay per month.

Many pilots believe the primary benefit of union representation is better wages. It would be misleading, however, to categorically state that unions always mean higher wages. Pilots at American and United are concerned that, despite representation, their salaries have not kept pace with the changes in cost of living.

Role in Safety

Most unions also display a formidable role in aviation safety. ALPA's Safety Department numbers 15 full-time staff people, while ALPA field workers—line pilots volunteering their time—add almost 600 more workers to the cause.

Unions offer pilots protection in disputes between a cockpit crewmember and management. The union representative stands in on such meetings as a witness to what was said and also to ensure that all provisions of the current labor agreement are followed closely. Should there be an FAR violation, union pilots enjoy the availability of legal assistance to mediate between themselves and the FAA. Unions usually offer pilots access to various insurance programs as well, such as loss-of-license insurance.

But one of the major advantages to flying for a unionized carrier is the labor agreement that outlines exactly what a pilot can expect from management and what kinds of behavior management may expect from pilots.

An example might be the limiting of a duty day to 14 hours. The pilots will not remain on duty longer than that, and management should not ask them to do so.

An advantage of most contracts is that they are hammered out locally, by locally appointed negotiator pilots working with management, not members of some union hierarchy simply flown in for the occasion. This also aids in contract enforcement later, since many of the same parties will come together during this phase as well. "First contracts are often modest and can take a year or more to negotiate," said ALPA spokesman Mazor. "But they become something to build on as both pilots and management get more comfortable with the process."

If there is a downside to pilot unions, it is the fact that they must represent all members of the bargaining group even those that might not be superb pilots. One Midwestern regional carrier captain's poor flying habits were so well known among the pilot group that many first officers preferred not to fly with him. When his employment was terminated for another allegedly near FAR violation, the union was forced to negotiate the rehiring of this pilot, much to the dismay of his coworkers.

Another potentially unpleasant aspect of unionization is "that some pilots tend to want someone else (the union) to manage their career," said one pilot at a major carrier. One Airborne Express DC-8 pilot said that "Nonunion pilots with little knowledge of unions and their role in the industry (are the biggest threat)."

There are other threats to unionized pilots as well, according to Shem Malmquist, a Boeing 727 captain at FedEx. These include code-sharing passengers and freight on foreign airlines. "It's incredible to me that a passenger can buy a ticket on most any major U.S. airline, climb aboard a jet operated by a foreign flag carrier, and go someplace and return without ever setting foot on an aircraft operated by the airline named on the passenger's ticket. It's a very ominous trend."

Another practical example of a union's power can be demonstrated over the issue of pilot pushing, like Watts being forced to fly 60 days being a good example. In another, a Part 135 nonunion carrier allegedly attempted to force one pilot, currently a first officer with USAirways, to fly a Cessna 402 with an engine that would not develop full power. When the pilot complained, the chief pilot allegedly told him to either fly the airplane or find new employment.

Unions Can Change

Sometimes, one union will be ousted as the bargaining agent in place of another, such as was the case at FedEx late last year. ALPA was out and the FedEx Pilots Association was in. Some of the union animosity began at FedEx when they merged with Flying Tigers in 1988. Tigers was represented by ALPA, and FedEx was nonunion. Eventually, however, ALPA did win at FedEx, much to the dismay of many pilots there who were staunchly nonunion. As with many merged carriers, the split between the two pilot groups actually widened as ALPA attempted to negotiate its first contract with FedEx management. Eventually the pilots began arguing more among themselves than with management putting FedEx management in an enviable position.

ALPA membership never climbed above 65 percent during its tenure at FedEx. In late summer 1995, the organization of the FedEx Pilots Association began as pilots decided they wanted representation at FedEx, just not ALPA. ALPA eventually won the election-certified on Oct. 29, 1996, but much animosity reportedly remains in FedEx cockpits today over the issues stirred up during ALPA's tenure.

Despite a show of unity during contract negotiations in 1998—98 percent of FedEx pilots joined the union—when both sides arrived at the stare-down point in those negotiations, the pilots backed down under FedEx CEO Fred Smith's threat to restructure the entire company and leave the pilots out in the cold.

Logically, career decisions about whether or not to join a union shouldn't be made with only the heart, but with the head as well. An essential part of being well informed is reading publications that are not prounion that will offer distinctly different perspectives of both management and labor. In a recent issue of *Air Transport World*, a magazine aimed primarily at airline executives, the publication's editor, Perry Flint, called United Airlines' pilots greedy, believing their request for hefty pay raises abrogated their agreement signed during the leveraged buyout of the airline in 1994.

Is there a union in your future? If you remain in the aviation industry, that answer is probably yes. But whether you will take an active role in a union can only be answered with a maybe. A union's strength depends on the active help of many pilot volunteers at the local level, many who may not be used to directing their own destiny to the degree that a union membership can thrust upon them. But without that help, without that involvement on an individual level, there is no union.

The Upgrade Decision

Captain, oh my captain! Crew scheduling says the airline will choose 10 first officers for upgrade to captain in the next few weeks and your seniority number puts you in the middle of the potential group of candidates. You'll finally have the opportunity to make your own decisions using the knowledge you've gained over the years of sitting in the right seat. You're going to be in charge.

"The ego can be a big deal for some people (in upgrading), also the pleasure of being the boss and running the cockpit the way you like it," said Federal Express B-727 first officer Shem Malmquist. The decision to upgrade is an individual one, a decision made by most pilots based upon a variety of factors. It also turns out that upgrading to the left seat, as well as that fourth stripe, is an opportunity some pilots choose not to accept if their airline employer allows them that choice.

It was concluded years ago that when the upgrade opportunity first appeared, a first officer who had been sitting in the right seat for 10 or 15 years would jump at the chance. United Airline's vice president of flight standards and training, Bill Traub (now retired) reports, "Since the average upgrade time is about seven years here, it is not very common for a pilot to turn an upgrade down. In fact, we delve into this issue during the initial interview. Those who think of their profession first are the kind of person we look for."

"We've not seen a time when someone would not upgrade when offered the opportunity," said Ken Krueger, director of flight operations at Midwest Express, a single-hub airline. "But a new hub (the company recently added Omaha, NE) may change all that here," he added.

Today, however, the upgrading pilot faces a "Pandora's box" of questions and decisions that could force that pilot to choose to remain in the right seat. If money were the only issue, one B-747 pilot summed it up very nicely. "The object of the game (whether to upgrade or not) is to maximize pay without working. Period! If you can make more money sitting shotgun, then it's a better deal."

The chance to command an aircraft is expected to be a part of most first and second officer's career plans. Indeed, most first officer's job descriptions portray them as a captain in training. Traub said, "We hire pilots to be captains at United and we like to see them upgrade at the earliest possibility. But, we won't argue if a pilot's family or possible commuting problems would prevent them from upgrading." American Airlines also hires its pilots to be captains, not permanent first or second officers.

Although airlines like American, United, and Southwest hire only potential captains, each airline expects its pilots to upgrade at a different time. Some companies also are willing to allow a first or second officer to remain permanently in a noncommand position, although somewhat reluctantly. At United, for instance, if a pilot chooses to remain in the right seat of a B-767 for the remainder of his or her career, the airline allows that as an option. Some pilots change their mind about upgrading later, too. Traub said a problem with pilots who remain too long as a first or second officer is "people that stay an extended period in one seat often find the upgrade process difficult. They tend to stagnate. Their command skills just don't develop as a first officer. Some second officers have almost forgotten how to fly as well as how to command."

At American Airlines, a pilot's right-seat time is limited. "You must upgrade within one year of the time when a pilot junior to you chooses the left seat," said American B-767 first officer Gary Mendenhall. Southwest Airlines' Chicago chief pilot Lou Freeman said a pilot cannot choose to remain indefinitely in the right seat. "Everyone here is captain-qualified when they walk in the door. We hire you to be a captain." At Southwest, a pilot can delay the upgrade process if, as a junior captain, he would have to hold a reserve line. But once a Southwest pilot can hold a regular line schedule, he or she must upgrade. The only other exception: Southwest will not force a pilot out of their domicile to accept upgrade in another city.

Continental Airlines does not force its pilots to upgrade. Many of the airline's Los Angeles-based DC-10 first officers could hold a captain's schedule in another city, but choose to remain in Los Angeles because they enjoy the southern California lifestyle.

USAir pilots also won't be pressured to move to the left seat until they choose to do so.

Many pilots believe the decision to upgrade is one they wanted to make on their own terms, or at least as much of their own as a seniority system would allow. Few said they did not wish to become a captain at some point in their career.

One of the first items most pilots consider in the upgrade decision is the salary increase a move to the left seat brings them. At Midwest Express, first officer's pay skyrockets nearly 70 percent after an upgrade to captain. The pay raise for a United pilot to upgrade to the left seat initially can range from about $35,000 for a senior first officer to as much as $75,000 for a junior one. Continental pilot Kaye Riggs said his position as a Los Angeles-based DC-10 second officer pays "about $50 per flight hour." A junior captain at Continental earns more than $80 per flight hour. At USAir, B-737 captain Bob Gaudioso said, "a senior B-767 FO could get paid more than an F-23 captain. The incentive to upgrade then is very small."

The high-end pay rates at the regional airline level are considerably less than the majors, making the difference in pay after upgrade considerably smaller as well. The importance of the upgrade at the regional level, then, might not seem as important. First officers flying a sophisticated aircraft like an ATR-72, a new generation glass cockpit turboprop, may qualify for food stamps. Turboprop pilots seldom pass an opportunity to upgrade because of financial considerations. Pilot Robert Lastrup confirmed: "The regionals are not known for paying their pilots very well, therefore without the money, there is no lifestyle choice." According to BE-1900 pilot Dave Oberlander, another factor in the upgrade decision for a regional pilot is the "need to build valuable [pilot-in-command] time. I also think that once a regional pilot makes over $40,000 per year, they will also choose lifestyle over money."

The pay disparity between left seat and right in corporate flight departments is not as vast as in the airline industry, since beginning pay rates in an aircraft, such as a Lear or Falcon 10, are substantially more than their first-year airline counterparts—in the upper-$30,000 range to the low-$50,000 range.

In the corporate world, however, the upgrade process operates somewhat differently. Many companies hire first officers who are either already are type-rated in the aircraft the corporation flies or pilots who they believe easily will achieve that goal. Since many corporations fly with a co-captain arrangement, achieving left-seat status in a corporate flight department also does not seem to carry the same sort of need to command as the airline industry, especially since most co-captains usually exchange seats on different legs or days of a trip—something unheard of in the airline industry.

A corporate flight department upgrades its right-seaters after they have gained a certain total time in the aircraft that varies greatly from company to company. But, a corporate flight department normally makes upgrade choices based more upon pure talent, with an eye toward the bottom line, and not so much on company seniority. When a pilot is deemed ready for upgrade training, they go. That plan does have its natural checks and balances too, however. Falcon 50 pilot Valerie Dunbar said, "I needed the complete concurrence of all eight of the other pilots in my flight department before upgrade training was approved."

At the major airlines, the upgrade decision usually is based on lifestyle. But lifestyle means different things to different pilots. Most want to earn the largest salary possible with the least amount of time away from home. And while more money can often buy a better lifestyle, if pilots must commute to another city to hold a captain line, the extra money may not be worth it. For instance, Riggs is based in Los Angeles by choice. A resident of San Luis Obispo, Calif., he commutes about an hour to Los Angeles for his trips. He could hold a DC-10 first officer position in Newark, NJ, or Cleveland, but chooses not to. "To me, the lifestyle of California is more important than the money," he said. Currently a reserve pilot in Los

Angeles a Newark line would offer more than the 12 days off a month he now has, but "If I were based in Newark, I could spend as much as five hours just getting to work. And commuting to Newark is difficult because the flights are always so full." He also said he would not be interested in living in either New Jersey or Ohio, so commuting would be a must. Riggs estimates 70 percent of Continental's pilots commute to work.

Los Angeles-based regional pilot Andrew G. Weingram said, "I already fly banker's hours, 110 hours per month, based in [Los Angeles] If I become a Las Vegas captain, I'll be getting up at 4 A.M. to work 16-hour days to fly 60 hours a month. No thank you!"

Besides the personal time lost, commuting to grab that left seat costs money. A junior captain and a reserve line holder usually is required to live within a two-hour call of the airport. If they happen to reside in another city, this means, "a pilot could possibly spend five nights a week in a hotel waiting for the phone to ring," Freeman said. That is not going to be a cheap. Some pilots choose to share the expenses of a small apartment in their domicile city with several other commuter pilots. These "crash pads" help defray the cost of living part-time in another city, but on a night when most of the pilots who share the apartment all need a place to stay, the quarters can become quite cramped. Sleeping bags and cots are common fixtures in crash pads.

Robert Lastrup said, "Commuting is a necessary evil. With so many changes in the airline industry...it's not really practical to move with every job or domicile change. Commuting is very stressful, too. For instance, you just get home from a trip and you're already planning your strategy for getting back to work. What about the weather? Should I leave a day early? You may even lose a whole day commuting on both sides of your trip. Commuting like I've explained it will never be worth the money."

Robyn Sclair, a Detroit-based DHC-8 captain said she passed on her first upgrade opportunity in the SA-227 Metroliner because "I just didn't want to be a Metroliner captain. In general, I was just tired of the airplane." Sclair waited an extra six months until her seniority allowed her to move from right seat on the Metroliner to left seat on the DHC-8. She saw her pay "just about double" when she passed her checkride. She also said, "moving to another domicile for the captain job would not have been a problem...but, maximum days off is also very important." At her Northwest Airlink airline, Mesaba, there is no policy to force a pilot from the right seat to the left. Dunbar flew as co-captain on a Falcon 50 and a Hawker 800 for Phillip Morris until some recent cutbacks in its flight operations.

"In our department, the upgrade process was not that competitive. Everyone upgrades eventually." Although the firm had bases in other cities, "each one was autonomous. Seniority at one base did not necessarily allow you to take a slot at another if you wanted to." She felt airline seniority is predictable while corporate seniority and its outcome on the upgrade process were just not as well defined. Most corporate pilots, like Dunbar, are salaried employees, so their pay is the same if they fly 10 hours per month or 110.

Southwest's Freeman said he could not recall a pilot who did not choose to upgrade when offered the opportunity. "A captain at Southwest is the greatest job in the world. We let you take an airplane out and just do your job. We pay you well, you make your own decisions, and there's no big brother looking over your shoulder." But the upgrade decision is not always so easy to see. At American, Mendenhall said, "As a first officer on an international widebody, I have a good schedule with 18 days off per month. If I accepted a position as a junior captain, I'd have 10 days off per month [flying] an F-100. I make as much money now as a junior captain. To me, the schedule is more important."

Lastly, some pilots won't upgrade because of a fear of failure during the training process itself. America West B-737 first officer Ray C. Phillips said, "We have first officers who won't even attempt upgrade. Some have tried and failed and simply won't try again." Freeman said Southwest had very few failures in the upgrade process. However, "If you fail the first upgrade, you'd go back on the line for 90 days. If you fail the second, you'd be terminated from the company." Dunbar also remembers corporate pilots whose attempted upgrades ended in failure and termination from their job. A United pilot who fails upgrade training returns to the right seat for 18 months before getting another chance to become a captain. Although Traub also reports second failures to be a rarity, "we'd also be much less inclined to offer a pilot a third opportunity to try."

When pilots get the opportunity to sew that fourth stripe on, they will have more decisions than just how to spend the money once the training is completed. The decision to upgrade or not involves many factors about lifestyle—money, days off, commuting, and others. A command position is the ultimate goal for most pilots, but pilots should be certain the first decision they make as a potential captain is whether or not becoming a captain is right for them (before they begin the process to upgrade). But, at some airlines like American, once pilots prove their ability to upgrade, they have the opportunity to bid back down to a first officer where schedules most probably will be better—if pilots' egos allow them to go back to being a first officer that is.

7
More Jobs

In case you were beginning to think that the airlines have the only flying jobs in the world, let's take a look at some other opportunities.

Charter Flying

There are people who believe that charter flying is only a time-builder until a real job appears. If you believe that, you could be missing out on a significant piece of the job market. A friend of mine has been flying for a charter operation at a fixed-base operator here in Chicago for 12 years, and he loves it. The operator has grown from a combined piston/turbine fleet 10 or 15 years ago to a pure turbine operator today. My buddy is typed in a Falcon 20, a Learjet, and is soon headed to Flight Safety for training in the G-III. The hours are varied, so while he knows just what days off he will have, there's always the possibility that he could be called in on his day off. But, the destinations are almost always different, as are the passengers.

While this pilot is flying jet equipment, there are hundreds of charter operators at smaller FBOs who are flying piston equipment (Navajos, Cessna 414s) and light-to-medium turbine equipment (Piper Cheyenne, Beech King Air). (See Figs. 7-1 and 7-2.) Those FBOs will also offer significant possibilities. A call to the National Air Transport Association, the FBO trade association in Washington (703-845-9000), could put you in touch with a listing of hundreds of fixed-base operators. Often, some of the more successful FBOs and their charter departments aren't necessarily located in major U.S. cities. Recently I visited an FBO in a small midwestern city and was honestly surprised to see the size of their flight department, some 23 pilots flying King Airs, Learjets, and Falcons on mostly a five-day-per-week schedule. Normally, however, charter flying tends to be an on-demand kind of service. The salary of charter pilots tends to run all over the scale, higher in the big cities and lower in smaller towns. The rate also varies with the type of aircraft you fly, but, as with life, everything is negotiable.

Figure 7-1. Charter flying could be in a Piper Seminole. *Comair Aviation Academy.*

Figure 7-2. A Beech King Air 200 is also a popular charter aircraft. *Beech Aircraft.*

Charter Flying: The Good, the Bad, and the Ugly

Historically, if you wanted a career as a professional pilot, there were only three routes — the airlines, corporate aviation, or the military. But the recent boom in airline hiring has thrown open alternatives some pilots may not have once considered. One such possibility is Part 135 charter flying. Thanks to a robust economy, the charter business is also expanding as never before. Andy Cebula, vice president of the National Air Transportation Association, the trade organization for charter companies and FBOs, reports that "we're seeing 18 to 20 percent increases in charter business over last year."

Unpredictable schedules have always made it tough for charter companies to hold on to pilots, but the shortage is becoming worse. "Attrition has been terrible," said Monty Lilley, President of Congressional Air, based in Gaithersburg, Maryland. To counter that, though, Cebula added that "In a tight employment market, you'll always do whatever you can to hold on to employees." In some parts of the country, this is already translating itself into limited improvements in wages and schedules, the two items that have traditionally sent pilots scurrying from charter as soon as any airline called.

Gil Wolin, President of Denver's Mayo Aviation added that "It's just the nature of the Part 135 charter business that to maintain your position in the marketplace, you need to control costs—so charter just can't pay what corporate flight departments or airlines can. We've just resigned ourselves to a high turnover rate." Fred Gevalt, publisher of the Air Charter Guide, admits operators "are having a tough time holding on to pilots because charter always has been just a stepping stone. But you can credit charter with providing a real-life training ground for pilots."

Part 135 companies typically fly unscheduled, on-demand charter flights carrying passengers, cargo, and sometimes both. The overnight delivery of checks is a typical 135 operation. One of the great business benefits of aircraft charter is they'll fly into airports not normally served by the airlines, Some charter departments even deliver enough service to a customer in a year that they begin to look like a corporate flight department to regular company passengers. The cost to charter an aircraft per mile is normally much higher than that of a corporate aircraft, but for a company that does not want the responsibility or the work of managing their own flight department, "Aircraft charter is the most productive use of air transportation," said Gevalt. "You simply buy it when you need it."

The key word in charter flying is on-demand, a clause that has proven a great boom to the pager industry, the typical way a charter operator communicates with cockpit crews. On-demand means flying on the weekends, in the middle of the night, or even on Christmas. An airline, for example, can't make money when its $100 million aircraft is sitting on the ground in Des Moines AOG—or on the ground—awaiting parts. They'll often charter an aircraft to carry mechanics and parts from a maintenance hub to the crippled machine. Hang out around any FBO and look for the people asleep in the snooze rooms. They're probably charter pilots who've been up since 2 A.M.

Gevalt said that in the United States, "508 charter operators are currently using some 1585 turbine-powered aircraft in Part 135 service (figures for piston-powered aircraft under 135 are not available)." The total number of aircraft operating under Part 135 was also considerably larger until just a few years ago when many that were operated as commuter carriers were mandated to upgrade to the higher training and operating standards of Part 121. Currently, Part 135 flying applies to nonscheduled operations in turboprop aircraft of less than 10 seats. On-demand service in jet aircraft of less than 30 seats, may also be operated Part 135, as well as single-engine aircraft that meet certain equipment regulations. Some scheduled 135 service—a tiny percentage—still exists, in small aircraft that will not meet Part 121 requirements.

One Part 91-135 regulation maverick is fractional ownership, an area of corporate flying—and pilot hiring—that has flourished under the leadership of Executive Jets' leader Rich Santulli and now being copied by other companies such as Raytheon and Bombardier. Many of the larger charter services actually provide backup lift to Santulli's NetJets on a regular basis. While operating under Part 91, fractionals provide on-demand service to their owners while following Part 135 standards, such as crew duty times. One of the main benefits to fractionals remaining Part 91, is that Part 135 requires weather reporting service or VFR conditions at the destination before landing. Part 91 pilot can shoot an approach to any airport, in any weather. But the FAA is currently taking a long hard look at fractionals to determine whether they should be brought under the Part 135 umbrella, remain Part 91, or possibly have a new set of regulations developed specifically for them.

One very necessary key to a successful charter operation is a good charter pilot, someone with a pleasant personality and the ability to easily converse with customers during the flight. Mayo's Wolin adds that "We establish a profile (for pilots). We don't just fly airplanes. We hire people to fly people. Until the customer is delivered to the other end, our crew's job is not done. Pilots have to be Boy Scouts—total technicians are not going to succeed in the long run. Pilots have to think about the little things, like whether a customer likes Pepsi or Coke. They have to listen to a customer's concerns. Charter pilots also develop a rapport with people that you'll never have in a 737 after you punch on the autopilot."

Although the range of aircraft flying Part 135 is vast—everything from a Cessna 172 to a Piper Navajo to a Citation to a Gulfstream G-5—most major charter flight departments are finding turbine aircraft—both turboprops and pure jets—to be their main sources of bread and butter. Charter companies also do not ordinarily own their aircraft. Most are leased back to the 135 operator by a local aircraft owner, typically a corporation. The benefit to the aircraft owner is additional income to offset fixed costs as well as a significant tax break.

Piloting charter aircraft is more demanding than many other facets of flying, simply because Part 135 crews don't have the benefit of a licensed dispatcher to help out with many of the chores such as checking weather and filing flight plans. Charter crews check their own weather, makes their own go, no-go decisions and normally load the aircraft with coffee, ice, snacks, catering, and even passenger bags when they arrive. A charter crew will also seldom see the same city twice in a week, or even a month. Brian Ward, a pilot for Mayo Aviation said, "We buy some food for early morning flights and tidy up the aircraft interior between legs. We also clean the aircraft when we get home at night," but adds that "We don't clean the exteriors."

Like their brethren at the Part 121 majors, the FARs limit the number of hours Part 135 pilots may fly. By contrast, Part 91 pilots have no duty or flight time restrictions to get in the way of an overzealous boss. In a one- or two-pilot cockpit, no Part 135 pilot may fly more than 1400 hours in any calendar year. By contrast, a Part 121 pilot is limited to 1000 hours (Most pilots fly considerably less.) In a quarter, the figures are 800 hours and 300 hours respectively. In a single day, the Part 135 pilot can fly a maximum of 10 hours if there are two pilots aboard, while under Part 121, the limit is essentially eight.

While Part 135 flight times limitations are vastly different from those of Part 121, they are a significantly better option than none at all, sometimes. Tom Deutsch, who now flies for a major US airline, remembers his recent job flying bank checks for Airpac in Seattle, WA. "I started in light singles like the Cessna 172RG and 177RG and worked up through the Senecas, Navajos, and Beech 99. I'd often average 14-hour days. Sometimes I'd only log two hours of flight time. The rest was sitting around." Part 135 also has duty limits. Part 91 has none. 135 pilots can be scheduled for no more than a 14-hour day, normally followed by a 10-hour rest period.

By contrast, some Part 121 operators still fly both continuous duty and reduced rest schedules. In the first, a crew takes the last flight to an out-station, often arriving about 10 P.M. They head for a hotel for a few hours and then take out the first flight in the morning, often garnering no more than three or four hours of sleep for the evening. Some airlines make crews fly four of these in four days. Reduced-rest schedules call for a crew to arrive at the outstation—often after a full day of flying—and return for duty exactly eight hours after they shut down the engines. The following day is often comprised of another full day of flying. Even charter flying schedules are not as tough on a person's body as some of these.

Part 135 operations require pilot training that on the surface appears similar to that of the major carriers. Training is where you first begin to discover the variations on Part 135 regulations, in this case, the content and the delivery of training material. For example, FAR135.245 says that a second-in-command must "hold at least a commercial pilot certificate with appropriate category and class ratings and an instrument rating. For flight under IFR, that person must meet the recent instrument experience requirements of Part 61 of this chapter." Essentially, anyone who has had three takeoffs and landings and some ground instruction could find themselves in the right seat of a turbine powered aircraft.

Some operators train in the aircraft at their home base due to cost of training issues and to fly a Piper Navajo, a few hours of in-flight training may make sense. But that kind of training in a fast turbine-powered aircraft flying at high altitude—while it may often meet the letter of the regulations—would hardly be safe. More and more operators are opting for a more traditional Flight Safety or Simuflite training environment when possible. For Kathy Julien, now a Falcon 2000 pilot, her new position as a pilot for ACM Aviation in San Jose, CA included a trip to Flight Safety as soon as she was hired into the right seat of the company Citation 1. Pilot training should definitely be a subject for discussion during the first interview.

With the numbers of pilots joining the major airlines on the increase—nearly 13,000 expected in 1998 alone—some Part 135 operators are investigating methods to assure themselves of a return on their investment in pilot training. Monty Lilley says, "There is no loyalty to any employer in this business." Lilley is strongly considering a training contract for pilots on any of the six piston and turbine aircraft his company flies. Mayo Aviation's chief pilot Cody DieKroeger said, "We have a training contract (at Flight Safety) that we value at $6000. If the pilot leaves prior to the end of the first year, they reimburse us on a prorated basis." Pilots at Air Castle, based in the Northeast, pay for their first type rating at Flight Safety before they begin flying. Stuart Peterson, an Air Castle pilot adds that "Once you pay for your first rating, the company pays for the rest."

On the other side of that argument, Jack Stockmann, director of operations at Wayfarer Aviation in White Plains, NY said "We believe that safety can only be obtained with thorough training, but we don't ask pilots for training contracts or agreements. We believe the fundamental of a good relationship is trust that works both ways." Wayfarer's fleet consists of 26 aircraft, the smallest being a Beech 400A and the largest a Gulfstream 4.

Schedules are a subject certain to raise the blood pressure of some would-be Part 135 pilots, as well as many who currently fly charter. Part and parcel to on-demand flying is that scheduling, the when and where of charter is often difficult, if not impossible to predict. Many married pilots avoid charter flying for this reason alone. For that same reason, many managers like hiring single pilots with few responsibilities. The Catch 22, however, is that single pilots tend to be young and focused on airline jobs down the road.

What might a Part 135 schedule look like? Wayfarer's Stockmann said, "The crews are off two hard days a month. They have to be available to fly the rest of the time. But since we do high-end charter, we have more notice and more time to plan. Pilots here feel we do recognize them and not treat them like a machine." At Air Castle, Peterson reports they work "20 days on and 10 days off and average about 500 flight hours per year." In the Denver operation of Mayo Aviation—they also have a few Arizona bases—Cody DieKroeger reports "pilots are on five days and off two. They fly about four to four and a half days a week, which translates into about 55 hours a month." Mayo tried a six days on and three days off schedule but found pilots were simply too exhausted. In their Arizona medical evacuation bases, Mayo's crews live in a trailer at the airport for a 12-hour shift. Kitty Hawk Charters' operations manager, Mike Joseph said from their Detroit base that "our pilots are on call 24-hours a day, seven days a week." Kathy Julien, has "no hard days off. In August I was flying almost 21 days. Then there is always pop-up stuff." The regs actually state that Part 135 pilots need have only 13 days off in a quarter.

In the Midwest, one operator—who declined to be named—felt they had a solution that kept pilots happy by providing them with "a life." The up side for this operator is that they'd not lost a single pilot since they instituted the new schedule. In their scenario, there are three dedicated crews—all salaried—for each of the company's jets. Each day, one crew is scheduled as the first to be called, the other the second, the third is off. When there is one trip scheduled, the first crew knows they'll get it and so on. While the crews don't know their destinations, they can plan their lives. Pilots can also trade trips with other equally qualified pilots But, as in all charter operations, crews can still be called when a pop-up trip occurs. One element of charter flying that bears repeating is that while pilots may be on call for many days during the month, they will not normally fly all of those days, resulting in a day off. It just isn't a scheduled day off for planning purposes. Most charter pilots carry their uniform and an overnight bag with them everywhere they go while on call.

Another overdiscussed element of charter flying is the need for pilots to carry a pager. Some hate it while others see it as a part of the territory, even when it goes off at 1 A.M. DieKroeger asks potential Mayo pilots how they feel about carrying one during the initial interview. But no matter what a pilot's feelings are about pagers, without them, a pilot on call would need to remain near the telephone—always!. The response time once paged varies widely and is dependent upon the type of operation. Mayo Aviation's air ambulance pilots report they can sometimes "be in the air within 10 minutes." They're required to be within 30 minutes for regular charter flights. Stuart Peterson said "we have a half-hour call out to be airborne within an hour at Air Castle." Kitty Hawk Charters requires their pilots to "be off the ground within 45 minutes," according to Mike Joseph. At ACM Aviation, Julien says they need to "be at the airport within two hours of the call."

The pay range for Part 135 pilots also varies significantly by the type of aircraft and the part of the country where they fly. Some companies offer pilots a weekly salary, others pay per hour, some a combination of both. Many pilots receive a per diem on the road, others simply cover actual costs. Rebecca Traudt, a charter pilot and flight instructor with Mount Hood Aviation in Troutdale, OR is paid per trip on some segments and per hour on others to pilot the C-172 the company flies under 135. On one regular 100-mile freight run, Traudt is paid $28 per trip. If she's flying on a per-hourly basis she's compensated at $14 per flight hour. At Congressional Air, Lilley said "some pilots get paid $150 per day and others receive $25 per hour."

Top-end salary at Mayo Aviation is $50,000 according to DieKroeger. Wayfarer's Stockmann reports "first officers begin at $50-60,000. Captains on small jets begin at about $68,000." The top pay at Wayfarer—typically in the largest aircraft they operate, such as the G4 and the Challengers—hovers at "about $112,000." At Kitty Hawk Charters, Joseph said "We pay $250 per week for a pilot to fly the turbine Beech(Beech 18 turboprop conversion)." Peterson reports that Air Castle pays $24,000 the first year in the Lear, with about a $6,000 raise the second. Lear captains earn about $40,000 and Challenger pilots about $48,000." Twenty full-time pilots work for Air Castle.

So, do you have the necessary experience to consider Part 135 flying? That depends upon the company and the aircraft you might fly. While some operators don't require experience in the aircraft they fly, previous Citation time would seem to be a step up during the initial interview for a company operating that aircraft. At Mayo Aviation, you'll need 3000 hours total time, 1000 multiengine and 2700 PIC to be considered. Wayfarer wants 2500 total time, an ATP, and a 4-year degree. Kitty Hawk wants to see a minimum of 1500 hours, of which 500 is turbine time.

And just when you thought there could not be too much more variation within this segment of flying, you'll find that hiring procedures are vastly different as well. Most charter operators reported receiving dozens of unsolicited resumés per week, many of which did not meet their minimum qualifications. DieKroeger said that "Because our hiring requirements are high, we conduct an initial phone interview with only about one in four applicants. We talk about their mountain and previous charter flying experience." Applicants selected pay their own way to visit Mayo in Denver for the one-day interview. This includes "a written exam about ATP procedures." This is followed by a one-on-one interview with the chief pilot, to see how conversant a person might be. "Then they'll fly an AST-300 simulator for about an hour." DieKroeger interviewed 30 pilots to fill their recent class of seven. The company currently employs 40 full-time pilots.

At Wayfarer, the process is even more intensive. But Stockmann believes that "if you put in the time to find a quality employee, you'll save much more time and possible problems later on. We have a responsibility to our clients." At Wayfarer, the assistant chief pilot conducts all the hiring and begins the selection process with a telephone call to assess the applicant's experience and goals. They even ask an applicant why they'd want to work for Wayfarer rather than an airline. "Some applicants will say they are looking (at the airlines), but we ask them for two years with us," said Stockmann. He said a successful candidate will visit the White Plains HQ at Wayfarer's expense for a more in-depth interview. "We listen to their view on things, how they think, and assess how they might fit into the company. We look for a good demeanor, a team attitude, good interpersonal and listening skills, and a strong knowledge of cockpit resource management (CRM). They'll also meet with the company president.

"If that is promising, we schedule a simulator ride in a Falcon at Flight Safety." Stockmann says it's actually more advantageous for a pilot not to have flown the aircraft before. "We're not looking so much at their ability to make the airplane fly, as to how they ask the questions about an aircraft they don't know. Do they use CRM skills? You can understand their knowledge base and skill level (that way). We ask about a lot of situational issues too." Next come background checks, from driver's records, to a security and even a credit check. All listed references are called as well. The final piece in the hiring pie at Wayfarer is to meet with the CEO of the company that owns the aircraft. If that is successful, an offer is made. The entire process often takes just a few weeks to complete. Wayfarer has 87 pilots on board.

At Congressional Air, applicant processing is rather simple since the President does all the hiring. "I like to see how someone dresses. I like to hear what they've been flying and what they think they can do for me," said Lilley. "If I like what I see, the applicant will take a checkride in one of our aircraft." Lilley adds that "the FAA checkride is not as hard as the one I give an applicant. We know them pretty well before they get hired." Congressional Air employs four full-time and four part-time pilots.

Since charter flying often appears to be a moving target, why would anyone want to live with a pager and never know when they were leaving home, nor when they were coming back? Tom Deutsch said, "When you fly for the airlines, it is a long time before you are a PIC. The nice thing about (some) 135 flying is that you're on your own." George Dunn said, after flying 135 part-time for three summers in Maine, "You really learn how to handle yourself in stressful situations. You may be sweating on the inside, but with passengers so close to you, you learn how to be professional. That's what these people are paying you for." He added that "A lot of people don't give a Part 135 operation credit for what they do. It's often just a few people with a few airplanes putting their heart into the business.

Brian Ward flies charter because "The people are really great. I enjoy that I get to see different places and meet different people. I also think this is a good stepping point. I'm just not a fan of commuters and the amount of work and responsibility for what they get paid." Oregon's Rebecca Traudt said "I don't think there is a downside (to 135 flying). I've heard other pilots say they would never fly a single engine aircraft at night in this rugged (Oregon)terrain. If the flight is planned well and the airplane is in good shape, you'll know how to handle emergencies. You also learn to keep your wits about you and keep your options open." Stockmann said "Pilots think that charter is only high demand, for low pay, with lots of standby, a minimal benefits package, and minimal security. At Wayfarer, compensation and quality of life are two major issues."

When asked what he liked about being on call almost constantly to fly a Lear 55 on Part 135, one disgruntled pilot had a simple answer, "Nothing!" Kathy Julien countered that." If you're a pilot who loves to fly this will be more satisfying. I fly really great equipment to some really great places. A great day is going somewhere fun and finding out you're staying for a few days. I also think Part 135 is a great training ground. If you work hard, you can work your way up to almost anything you want—if you can stick it out."

The differences between Part 91 (corporate) and Part 135 (charter) flying

Part 91	Part 135
	Duty day restrictions
none	14-hour maximum per day

Part 91	Part 135
Flight time restrictions	
none	10 hour maximum per crewmember
Rest period requirements	
none	10 consecutive hours in 24-hour period
Annual training	
not required	yes, in type
Management personnel required	
none	director of operations chief pilot & director of maintenance
Additional equipment requirements	
none	cockpit voice recorder, digital flight data recorder in aircraft of 10 seats or more. TCAS 1, ground proximity warning system
FAA approved operations manual required	
no	yes
Drug and alcohol testing	
no	yes

(Courtesy of Wayfarer Aviation)

Profile: Diane Powell, Charter Pilot, Lear 35

Diane Powell was luckier than some people. She had a great mentor for her adventure into a professional pilot career—her father, a man who just happened to also be a certified flight instructor. But while her father was a catalyst for her career, Powell still had to overcome the same cash-flow problems as other aspiring pilots while she learned to fly. She worked full time while learning to fly and got a break on her lessons from the FBO that employed her.

She also picked up a degree in business along the way to her other ratings. "Having a degree is more important today," she said. "But I wanted one for myself as a fallback, just in case I couldn't fly someday." She won her commercial and multiengine rating in 1989 and added an instrument rating in 1990.

Powell flew long, grueling days on traffic patrol around Fort Lauderdale in a Cessna 172 and logged 600 hours in one year. "I took off at 5:15 each morning and flew until about 9." She'd return to the airport in the afternoon and fly another few hours for the evening rush. There were also no breaks during the flying she said. "If you had to land for any reason, the bosses were not happy." Powell built time flying photographers around Florida too when she landed an opportunity to fly newspapers in an Aztec to Nassau from Opa Locka, just north of Miami. She also picked up her CFI rating along the way and even managed another full-time job with Miami's Metro Dade Port Authority as a dispatcher to pay the bills.

Now, with a little over 1500 hours total time, Powell says "I was up surfing around on an America OnLine (AOL) aviation message board one day and connected with a Lear operator who said he'd be in Palm Beach and asked if we could meet. They were concerned with my low time, but said they liked to give new pilots the chance that someone once gave them. Unfortunately, I didn't get the job. But they kept my resumé while I returned to flight instructing."

Then Powell had another opportunity to turn pro when she was hired by St. Louis-based Trans States Airlines. "The two-day interview for the job was pretty stressful," she said. "They really wanted to see that you already knew the airline-style procedures. I went through three simulator sessions and quit. I never did get the simulator thing down pat," she recalled. "Coming from general aviation, the transition from little airplanes to turbine flying was tough. They gave us no time at all to learn. They didn't seem to care how tough it was."

She recalled leaving St. Louis felling pretty despondent. Just after her return to Florida, the phone rang. "It was the Lear operator. He said he needed someone in the right seat—right now. I took the job. After two solid days of training, I was flying 135 trips in a Lear 35. This is something I always wanted to do and couldn't be happier—despite the fact that I must carry a pager. One of the best things here is that they let me fly left seat on the empty legs."

Powell—who says she'd eventually like to fly for United—remembers some of the words of wisdom her father passed along. "He said you should help someone else when you have the chance because it always comes back to you. Sometimes the pilots I've met always seem to be out for themselves and no one else. They're worried about only their job."

Diane has already begun the process of giving back to the industry that has given her so much. Her connections online are making that process even easier for her too. "I just helped a guy get a job in a Merlin because he was online. Networking really is the key to finding a good job. I find that online I can send an e-mail to people and ask lots of questions and they send me plenty of information back. I find the message boards there to be very helpful."

Corporate Flying

Many corporations have lost valued pilots to the airlines in recent years, but today many of these pilots are thinking twice about their jobs, as have many new pilots. Corporate flying can offer not only the fine personal rewards of flying professionally, but also financial rewards that are very similar to the airlines.

The corporate scheme of operation, however, can be quite different than the airline business, and often, corporate operators aren't terribly keen on hiring former airline pilots into their flight departments. Part of the reason is that a corporation doesn't want to spend the money training an airline pilot whom they believe might leave at any moment to return to an airline cockpit should a new opportunity present itself. One friend of mine flew corporate all his life and eventually moved to a chief pilot slot at a company flying a Boeing 737.

Corporate Flying

**An Interview with
Janice K. Barden, President
Aviation Personnel International,
New Orleans, LA**

Janice K. Barden has been telling aviation companies who to hire for 25 years. A veteran industrial psychologist, Barden has worked with many Fortune 100 corporations, in addition to airlines like PanAm, American, Braniff, United, and National. She remembers that "in the late '60s, you couldn't get on if you were over 32."

"Right now," she says, "airline hiring is stimulating everything. A good corporate flight department is still one of the best kept secrets, because pilots keep these jobs for a long time. While I think corporate hiring will remain close to what it is right now, we'll be retiring many of these pilots and the corporations don't know yet where the replacements will come from."

Barden's company, Aviation Personnel International, is a research organization dealing in personnel. There is no charge to the applicant for API's services. "At some companies, we handle everything for a flight department, from hiring of pilots to hiring of mechanics and dispatch personnel," Barden added.

Barden said that "the difference between corporate flying and other types is that these pilots are going so many different places each day and doing all the work with the people in the back as well. Today, a corporate pilot needs to think more than just flying. Remember that a pilot is in a key slot when they are responsible for the top management people at a company. They should pick a good school to attend to get their basic education. The corporate world recognizes a degree from a good university." Barden added that most corporations today are also looking for pilots who see themselves in a flight department management slot in the future, as well, so advanced degrees are also becoming more valuable.

API has an extensive screening program for applicants that they only offer to "about three people out of a hundred who walk through the door," said Barden. "To find a good pilot, I believe we must study people, their weaknesses and their strengths," she said. "For a corporate pilot, attitude is very important. In fact, it is the number one factor. Is the pilot good with people, a good team player? Do they have good values and are they willing to work? Corporate flight managers look for resumés that have some depth to them."

But Barden also knows physical fitness is very important. "We haven't placed a smoker, for instance, in many years. We'll ask, "Do you work out? What is your cholesterol level?"

What else does it take to begin a corporate piloting career? "For an entry-level position at the corporate level, a pilot should have a degree," Barden says. "A business degree will be an asset too, as well as good technical and interpersonal skills. We test them and learn about their leadership potential too. They have to believe in themselves and be self-directed. They need to be easy to get along with. As far as their flying skills, they need an ATP and should have been exposed to some great flying experience at this point too. That's where charter or regionals can really be some great training grounds. 2000 hours total time—at least—in turboprops and jets is the best too."

Many pilots may not be considering a corporate piloting career for a number of reasons that Barden believes she already understands. "I think that years ago there were war stories about flying in the middle of the night, but that seldom happens any more. Most of these companies operate under Part 91, but fly to Part 121 standards. This is a good businessman career. I think there is also a great pride of ownership in being a part of a corporate operation. They are a very close-knit group. They are only as strong as every one of the links."

But don't expect the road to hiring on at a corporation will be easy. "When I do career counseling, I tell pilots this is one of the hardest industries to break into," Barden said. "They've got to pound the pavement and make their own opportunities. They must develop their own network."

AvCrew Update

On AvCrew's Interactive message board, (www.avcrew.com), they asked the following questions: "How has airline hiring affected your flight department? How will it affect the department in the future?"

Here are some responses:

"Almost impossible to find crews for our aircraft. Salaries, though, have not risen to the expected levels."

"Consensus is [that] if the airlines would start very experienced corporate pilots at a higher starting salary, many would jump the fence. How many forty-year-old-plus pilots can live on $26,000 a year? I would have to liquidate everything."

Corporate International Captain

"It really hasn't affected our department. Most of our pilots have been here in excess of 12 years or more. The salary is excellent and the equipment is great."

"Airline hiring is reducing the total pool of qualified pilots available in the flying industry, especially in the 30–40 year old category with jet time. Overall, this is forcing corporate compensation rates up, which is good."

"Higher than expected turnover in the last six months."

"It currently has not affected our department (four pilots, "heavy" business jet); however, if the company cannot keep pace with other airlines in salary and retirement, we would be fools to believe people will stay for the 'good' of the company."

"The rollover rate has gone up at an astronomical rate."

"My co-pilot is interviewing with an airline and will probably get the job."

Corporate owned and operated aircraft come in all shapes and sizes, from the low end of a Piper Navajo or Beech Baron on up through King Airs, Cessna Citations, Falcons, Learjets, Gulfstreams, Challengers, to even airline equipment (Fig. 7-3). Schedules in corporate flying are sometimes a bit more unsettled than at the airlines, and somewhat similar to charter flying. Some large corporations employ a crew scheduler, and often the crews know a number of days in advance who is scheduled in what airplane to go where. Sometimes, though, the flight could be a last-minute trip scheduled because of some unforeseen business situation.

Figure 7-3. Learjet 45. *Bombadier Aerospace.*

Corporate flying can also involve a great deal of sitting around while you wait for the boss to complete his or her business. But because most corporate pilots are salaried, you're not losing money. The extra time could provide you the opportunity to catch up on your reading or even do some work for a business of your own. I never travel anywhere without my trusty laptop computer with its word processing software stored inside. Part of this book was written on a two-day layover I had during a Cessna Citation trip.

Salaries at the corporate level tend to be higher than charter flying, but there also tends to be fewer of these jobs around and, as at the airline level, the current competition is tough. The Business and Commercial Aviation Salary Survey shows that the range is wide and dependent on the type of aircraft and often the location of the base. Table 7-1 shows a few examples.

Table 7-1. Captain position average annual wage

BAe-1000	$ 66,385
Cessna Citation 3	$ 55,185
Gulfstream IV	$ 81,185
King Air 300	$ 49,968
Navajo	$ 34,279
(See Fig. 7-3a.)	

Figure 7-3a. Global Express. *Bombardier.*

Rates for first officer would be a percentage of the captain wages, some-times as little as 40 percent, often much higher, depending on whether or not the first officer is also type-rated in the aircraft. Often, many of the smaller turboprop and piston aircraft fly as single-pilot operations, which can keep any pilot busy in a complex terminal area or when the weather is bad.

As you begin the search for a new job, you'll want every conceivable extra point on your side, such as possibly a type rating. But buying a type rating has become quite a controversial subject when cost versus value received is considered. Some airlines, like Southwest, require a B-737 Type Rating to even apply. Others, like Midwest Express, basically say don't bother because they're going to put you through their own training pro-gram anyway. Then airlines like United say sure, a type rating is worth something, but only if you also have a fair amount (about 500 hours) of PIC time to go along with the rating. In the corporate world, a type rating can be an asset, but if you're typed in a Learjet and the job is for a first offi-cer on a Falcon 50, the type rating might not count for much, other than the fact that you're trainable. Let's take a closer look at type ratings.

Will a Type Rating Get You Hired?

In a profession where tens of thousands of dollars can easily be spent just to land a bottom-of-the-ladder job, is an additional $10,000 to $19,000 for a type rating worth the gamble?

If a pilot's goal in picking up the type rating is to hire on with an airline, he or she should look at how the potential employer evaluates a type rating in the interview. During an airline interview, a pilot applicant is evaluated in many different areas. How pilots present themselves and how they perform on airline written exams often can be the deciding factor. If the decision is between two equally qualified pilots, one with a type rating and one without, you'd imagine that the pilot with the type rating would win out. But, because most pilots never actually learn why they were or were not hired, that premise is pretty tough to verify.

The only major or national airline that requires a type-rating is Southwest. The company will not interview applicants who don't have a B-737 type rating. Southwest also operates a B-737 type-rating school, and about 35 percent of the school's students get hired at Southwest after the type rating is earned.

Viewpoints from other major airlines include that of USAir spokesman Jim Popp, who said, "Our minimum requirements don't ask for a type rating. A type rating is just reviewed along with all other qualifications a pilot might have."

America West, which operates a type-rating school, also doesn't require a type rating. At America West's Contract Pilot Training, Dee Rush said, "Of all the people who have gone through our type-rating school, only a small portion have been hired. There are many factors that are taken into consideration when a pilot is hired."

United Airlines spokesman Joseph Hopkins said, "United is seeing an overall increase in the qualifications of its applicants, but the type rating really is evaluated along with all of a pilot's other strengths."

At regional carriers, company officials are singing the same tune as the major airlines. Glen Bergman, chief pilot at Business Express Airlines in Windsor Lock, Conn., said, "To have a type-rating is important because it shows us that the pilot can pass a type rating course. We do, however, begin all new applicants now in the BE-1900, regardless of experience in other aircraft . . . I would much rather see previous turbine experience."

Drew Bedson, assistant chief pilot at PanAm Express (now out of business), said, "Possession of a type rating doesn't really affect anything in terms of being pulled for an interview. What's more important to those who do hold a type rating would be whether or not they really have used the rating as part of their job. If someone walked through the door with a type rating in a Jetstream he picked up at a school but had no practical experience flying the airplane, then he has no more experience in that machine than he would have on completion of our approved training course, so the rating wouldn't change anything."

Paul Rogers, formerly with FAPA's Aviation Job Bank, said this of his corporate clients looking for pilots: "The type rating at the corporate level makes a pilot applicant look a bit better than the competition, but there are plenty of people with lots of experience who just don't present themselves well during an interview. Corporations are more interested in a well-rounded professional applicant who works well with passengers and can be an asset to the company in other ways with their education or skills."

Airline statistics show that, in 1998, nearly 61 percent of the pilots hired by the majors were type rated (Fig. 7-4). In 1988, two out of ten pilots hired at the nationals already were typed. These numbers have been on a downward slide from past surveys.

It's not known how many of these type-rated pilots bought their rating and how many got the type rating from a previous employer. Also, it should be noted that all types of type ratings are included in these percentages. The type rating could be in any turbojet-powered aircraft or in any aircraft that has a maximum gross takeoff weight of more than 12,500 pounds.

Some pilot applicants believe that when their application is screened, the extra points from a type rating are what they need to get an interview or land a job. Pilot Mike Roebke said, "I bought the 737 type rating because most of my time in the last five years was in crop dusters and I didn't think anyone would look at me. I would definitely do it again if I had to." Roebke recently was hired as a B-737 first officer by America West.

Figure 7-4. A complete type rating can be obtained in a simulator.

Although pilot Henry Schettini was planning for the future when he paid for his type rating, he still can't find work. "I really felt my opportunities would be better if I had the type rating. I still think it's good that I have the rating, even if I don't have the job right now."

It could be just a hunch on the part of the applicant that says the type rating is worth the expense. For some it's not. As one regional airline Brasilia captain said, "I've spent enough money on this career. Another $10,000 to $15,000 just to possibly add one point in my interview score is just too much to ask."

The schools that sell air carrier type ratings believe the rating is the pragmatic way to approach the interview process. Nancy Wilson Smith, manager of customer service at Dalfort Aviation says, "Having the type rating makes a pilot more marketable. It proves you're trainable." America West's Rush said, "If you bought your own type rating, it shows you're pretty serious about your career." Aero Service's chief instructor, Steve Saunders, explained that, "Hiring on with a major is extremely competitive . . . A type rating will give you that competitive edge." Ray Brendle, owner of Kingwood, Texas-based Crew Pilot Training, said, "The airlines know that, with a type rating, the pilot has made the transition from civilian general aviation or military flying to the air carrier side."

If a pilot decides on a type rating, he should be prepared for a whopping bill, somewhere between $5,000 and $20,000. Rates vary by aircraft type and previous pilot experience. And a pilot shouldn't worry if he or she only has 2,000 hours total time. As Saunders said, "We're seeing considerably more lower-time pilots entering the type rating program . . . many from commuter operations." Pilots who never have been type rated in a turbojet aircraft will be considered an initial student by most schools. If the pilot were already typed in a turbojet, he or she would be a transition student. Pilots moving from the right seat of a B-737 to the left seat are upgrades.

An important decision for a pilot is in which aircraft to type. Currently, there are the B-707, B-727, B-737, B-747 (Fig. 7-5), DC-8, or DC-9 to choose from. A few of the schools soon will offer the B-757 too. The most prudent choice should be the aircraft that will do the pilot the most good all around. The B-707, B-727, and DC-8 were fine aircraft in their time, but they're part of an ever-decreasing portion of the aircraft fleet. Certainly, a DC-9 type rating is similar to the newer MD-80 series, but, for overall usefulness, most critics agree that the B-737 is the aircraft to fly. Crew Pilot Training's Brendle said, "The 737 is the most popular type rating in the industry because about 50 percent of all major and national airlines use the aircraft." Does this mean that a pilot's money is wasted if he or she types in a B-727? Perhaps. As Joe Marott, manager of Southwest Airlines Training Center, said, "It (a B-737 type rating) is still a requirement here at

Figure 7-5. Boeing 747-400. *Boeing.*

Southwest to apply for a position. Only the Boeing 737 type rating would meet our requirements for employment." So, a pilot should choose the aircraft carefully by first deciding just who it's he or she wants to work for. Many smaller aircraft type ratings also are offered, but they aren't nearly as useful or valuable.

Smart type rating students should choose a school like a good shopper buys a new car. Students should be wise, knowledgeable, and ask questions until they're satisfied. Price is only one aspect. Dalfort Aviation's manager of flight standards, Ben Williams, said, "A pilot should consider where the school does its training, how long the course will take to complete, and how long the school has been in business. Some schools will organize the ground school in one location, then ask the student to travel to a second for the simulator training and perhaps a third for the aircraft training. In this case, the cheaper school would not be a bargain."

Some apparent benefits might be intangibles too. Southwest's and America West's schools are part of those corporate structures, but both schools make it abundantly clear at the start that attending their school will not guarantee a pilot an interview with that company. On the other hand, a few private schools reportedly tell students indirectly that they have an affiliation with a particular airline. In most cases this is untrue. Keep in mind that all schools aren't created equal simply because they're FAA-approved. Consider the staff of that school too. Advanced Aviation

Training's president, Robert Mencel, said, "Each instructor here is currently serving in a training capacity with a U.S. airline." At other schools, this might not be the case. Currently, if a school is designed to accomplish at least 90 percent of the training in a simulator (most are), the school must be FAA-approved. The enormous cost of aircraft training virtually assures FAA approval too, but students should ask before signing up. In most cases, an initial type-rating student need only have a commercial, multiengine, and instrument rating to begin. In other phases of training such as transition, the requirements can vary, so check with the school prior to enrollment.

What can a pilot expect from a type-rating program? First of all, most schools will require some form of a deposit in advance to hold a position in the class, ranging from 20 percent to the entire cost of the course. None of the schools interviewed provided any financing, so be prepared to find a loan if necessary. Most schools agreed to a specific amount of training for a specific price, but no school offered a guarantee of the rating. Additional training required will involve additional funds, which the student should ask about first. At one school, the simulator costs $375 per extra hour, while time in the B-737 goes for $45 per minute.

One bright spot on the horizon for potential type-rating candidates is the new federal Veterans Training Act, the GI Bill of the 1990s. Pilots who served in the U.S. armed forces might qualify for this professional training cost assistance of up to 60 percent of the type-rating bill. Pilots should contact the local Veterans Administration office for details.

According to the U.S. Master Tax Guide, a type rating may be deductible on a pilot's personal income tax return, too. The guide says that "education expenses are generally deductible if the education undertaken maintains or improves a skill required by the individual in his employment . . ." To be certain, though, pilots should check with an accountant about their specific situation before signing up.

While the price of lodging was included in the package price at only one school, all had some sort of deal with a local hotel to provide accommodations and transportation to their facilities during the student's stay. The length of that stay for the course varied considerably from two and one-half weeks, to the longest at six weeks. How a pilot can carve a hole in his or her schedule for that kind of training is another problem a student must solve before committing to the training.

The schools run by Southwest and America West place their contract type-rating students into open slots along with their regular company pilot training classes, so students are able to talk and learn from the instructors and students already working for an airline. Southwest's Marott says, "The best thing in our school for type ratings is that we use the same manuals, the same procedures, the same simulators, airplanes, and instructors that are used to train our regular line pilots." Currently, Southwest's school teaches only the B-737-200 course to contract students. Marott said the school opened in the spring of 1987. "Since then we've graduated about 175 B-737 type-rated pilots. Of that number, approximately 60 were ultimately hired by Southwest Airlines," said Marott.

Saunders says Aero Service's B-727 type-rating course includes 120 hours of actual classroom ground school. The student then moves on for 16 to 18 hours of cockpit procedures training in a nonmotion aircraft simulator. Then they'll spend 18 to 20 hours in the full-motion simulator, usually in teams with about 10 in the left seat as pilot-in-command and 10 in the right seat performing first-officer duties and observing. Finally, the checkride is performed both in the simulator and in an aircraft the company leases. In most cases, an initial student can't complete all the training plus the ride in the simulator only. There can be some reductions in training time for the rating, based on previous experience. These changes are made on a case-by-case basis by the FAA's Principal Operations Inspector, who oversees the school.

America West's Rush said, "The ground school lasts 12 days. This is followed by 5 days of cockpit procedures training and then the FAA oral. The student next moves to about 12 hours in the simulator and about one hour of aircraft time to finish off the takeoff and landings. America West currently is under contract with the FAA to provide all that agency's initial type training to its air carrier as well as maintenance inspectors." Crew Pilot Training's Brendle says that, although the total amount of ground school there is similar to other schools, his "is approved for 80 hours of home study. The student must show by a test on arrival that he has completed the required work." Students receive an additional 40 hours of ground school when they reach the classroom. Brendle remarked that "some schools send you for the FAA oral right after ground school, without ever having seen the inside of the CPT or the simulator. When a student learns about an aircraft from a book only, without ever having had the chance to move a knob or switch, they give a weak oral."

Information such as ground-school time, simulator time, cost, reputation of the school, along with other considerations listed previously, should be kept in mind if the pilot feels he or she needs the type rating because of qualifications required by a specific airline.

In the end, a pilot must decide if the type rating is worth the money. That decision should be based on experience as a pilot and whether the pilot feels that his or her credentials will match up to those of other candidates applying for the job, as well as whether the company to which he or she is applying requires a type rating. It's a decision that must be made at the right time under the right circumstances.

Fly for the Federal Government

An often overlooked area of flying is with the federal or state government. Positions often exist for both fixed-wing and helicopter pilots, but the search can take time because there are so many agencies to check into. The FAA, for example, uses pilots to fly its fleet of flight check aircraft, while the U.S. Customs service monitors border traffic with their aircraft. Some of the publications we've already spoken about might carry ads for government pilots, and certainly a number of flying employment services will offer publications that list flying jobs for the government. An alternative to searching the various publications for government flying jobs would be to contact the various agencies directly. A call to the Federal Information Center at 800-366-2998 should get you started with phone numbers and addresses of federal agencies.

Some of the U.S. agencies that might need pilots are:

- Department of Defense
- U.S. Customs Service
- Federal Aviation Administration
- NOAA
- Department of Transportation
- NASA
- Defense Logistics Agency

In addition, state and local governments might use pilots. Your state's department of aviation would be a good place to begin the search (Fig. 7-6).

Finding a Flying Job after Age 50

Finding a flying job as you near, or pass, retirement age is more difficult than it would be if you were 25. But, as the United B-747 cabin decompression near Honolulu and the DC-10 crash at Sioux City both proved, there simply is no substitute for experience. The captains of both aircraft were near 60 years of age. Some companies recognize the experience factor in their hiring practices—but not all.

If you're an older pilot in search of a flying position, you may have to work some to locate them. According to some projections, 15 percent of new hires will be older than 40, some possibly as old as 57 or 58. Even if

Figure 7-6. Many agencies of the government fly civilian aircraft. *United Technologies.*

you retired from the airlines when FAR 121.383 suddenly made you totally ineligible to fly revenue trips anymore, there is hope in jobs other than the airlines. But the road to work contains a number of potential hurdles.

If you're still young enough to work for a Part 121 carrier, you'll compete with younger pilots, which a company looks at as long-term investments. Even though you already may be type rated in a B-737 with 10,000 hours total time, you may not get the job. Attitude plays a big role. For example, some carriers find that crews coming out of 10 or 15 years of flying with another airline "do not mold themselves very well into the ways of the new carrier," as retired DC-10 captain Jim Minning said. "These pilots have very strong ties to the old airline." Minning today ferries transport category aircraft for corporations.

Another problem facing second officer rehires is the cockpit intimidation factor, real or imagined. A man or woman who commanded a B-727 for 14 years may have a difficult time transitioning into a subservient role in the cockpit. A relatively new captain, on the other hand, may well have difficulty flying with a much more experienced pilot constantly looking over his or her shoulder.

Capt. H. McNicol of Flight Crews International, a crew-leasing company, said that, "It's the persistent pilot who finds a job...the man or woman who does not give up and is willing to be aggressive in his or her search."

As Minning said, "Sometimes, getting work is like being a racing car driver who needs a sponsor. You must go out and look for them." He also said that "a contact network is very important." In order to find a job after retirement, Minning said he "kept very involved in our industry and made a great many contacts that I kept over the years. Many of these people remembered me."

If attitude and aggression are important in a job search, so is a realistic view of the world, especially for the pilot who no longer can fly for a Part 121 carrier. In hundreds of other professions, thousands of good people are put out of work for one reason or another and must face the fact that the comfortable, well-paid job they once had may not be replaceable. So too with the airline industry. Sixty-five-year-old Hal Ross, currently flying a Citation III for a corporation in Southern California, believes, "You have to be pretty open about what you accept. I just don't think there are that many good jobs out there for pilots my age." Ross also alerted pilots to the fact that in many jobs today "you'll get paid what the market can bear, but they certainly will not be the airline salaries some of these people are used to."

What are some jobs for older pilots and how did pilots find them? What are the companies looking for and where are the majority of jobs coming from? For pilots who have reached 60 years of age with a Part 121 carrier, the easiest solution to keep flying, if the airline allows it (most do not), may be to accept a position as a second officer and move back to the flight engineer's panel where the regulations do not stipulate a maximum age. Lee Lipski, age 62, successfully moved from the left seat to the back seat at Continental Airlines. "I took a tremendous pay cut to move from captain to second officer...but a lot of us age 60 pilots are just not ready to retire." Certainly though, the future will see the eventual elimination of these jobs as more and more airlines receive aircraft designed for two-member crews. Also, some pilots may find it difficult to deal with the politics involved in such a move. Younger flight engineers don't like such reversals because those more senior engineers may block their own progress to larger aircraft with higher pay or may cause them to be the ones who get furloughed because they have less seniority than the older pilot, and of course, pilots on the job search don't like the idea because it reduces the number of positions available for new hires. For now, however, the option exists.

Another possibility is to work as a Professional Flight Engineer (PFE). These jobs are mostly at nonscheduled charter and cargo carriers since most of the major airlines use only second officers who are expected to upgrade. A PFE must hold a flight engineer's certificate in the particular aircraft the employer flies and often is required to be a licensed A&P mechanic. Most companies that use PFEs do not have a maintenance facility or a company representative in all the cities they serve. The PFE is the company mechanic there to make repairs and sign logbook entries and maintenance releases that otherwise would ground a valuable aircraft. While the number of PFEs in use is not staggering, companies that use them include Airborne Express on DC-8s, American Trans Air on B-727s and L-1011s, Evergreen on B-727s, B-747s and DC-8s, Key on B-727s, Reeve Aleutian on L-188s and B-727s, Southern Air Transport on L-388s and 747s, and Tower Air on the B-747.

Another option is crew leasing. You don't need to be a type-rated crewmember to take advantage of companies that lease crews to airlines around the world on short-term contracts. California-based Airmark Corp.'s President Ron Hansen remarked that he receives only about half of his resumés and applications from Americans. The remainder arrive from around the world. Hansen says he has "many leased crewmembers over the age of 50," such as a number of ex-Eastern Airlines pilots in Japan on the L-101 1. While Hansen admits "the pay is good," his comments on American crews is food for thought during your job search. "Americans are generally very difficult to work with on these foreign assignments. They want more money, they complain about the housing ,and in general consider many things inferior to the United States. I can't say I really blame them, but this is where some of the jobs are. The alternatives might be flying freight out of Detroit or mail for the U.S. Postal Service where you get paid less than a truck driver. Many of the choices are not very good for pilots today."

FAR Part 135 contains no age restrictions—many former Part 135 commuters now operate under Part 121—so it is a viable alternative for some pilots. Paul Cassel, age 65, flies a BE-1900 for a small commuter in California. Paul knew he wanted to continue flying and made his plans accordingly. "I just don't think that after flying as a captain for so many years I would have enjoyed being out of the actual flying end of things, so staying on as a second officer with my previous company wouldn't have worked." At age 65, landing any kind of job can be tough, but Cassel reports, "it was not difficult to find this job on the 1900 at all. This regional was glad to have me and my experience." Several regionals interviewed admitted that even though the law restricts them from discriminating on the basis of age anyway, they would welcome the high-time older pilots if they could get them to apply. After flying with a Part 121 carrier for many years, some pilots don't want to fly for a commuter because they are used to having someone else do so many jobs they must now do by themselves, such as manifests and flight plans. Also, the salary at a commuter is not as good as the majors, nor are the duty limits of Part 135. Cassel says though that "There are some really good things about flying a commuter. I'm usually home every night and I don't have the time zone changes anymore."

Another aspect of Part 135 flying could be with a charter company like Columbia, S.C.-based BankAir, although Chief Pilot Jeanne Cook admits she doesn't get many resumés from airline types because "they tend to look for more glamour than we can offer." BankAir is an all freight operator of approximately 28 aircraft such as Lears, Jet Commanders, and MU-2s. Former airline pilots "tend not to like this kind of flying, but it certainly isn't because we don't want them. Personally, I think some pilots like flying for a 135 company more than a 121, since the pilots seem to have more input here."

An older pilot also may want to consider Part 91 corporate aviation as a viable option to the airlines. In a recent poll conducted by the National Business Aircraft Association (NBAA) only 25 percent of their members have a mandatory retirement age of 60 for pilots. (Whether or not this is legal is another issue since the regulation that mandates age 60 retirement applies only to Part 121 carriers.)

Corporate flying is a new kind of flying for some. With virtually no duty restrictions, a corporation is free to fly a great deal in a short amount of time. Pilots may have regular schedules with assigned days off or possibly could be on call and carry a pager to call them for flight duties. Corporate Citation pilot Hal Ross had only good things to say about his employer whom he found through his own networking efforts. "I did not really meet any resistance to my age in my search for work. The most important thing to the companies I've spoken to is the state of my health. Once they find out it's OK, they listen to my qualifications. My experience and my safety record are what got me this corporate position. I'm on call seven days a week, but I probably only fly about two or three each week."

Government flying is yet another possibility. While FAA regulations prohibit pilots older than 60 from flying for Part 121 carriers, there is no such age restriction for pilots who want to fly for the FAA. Lee Lipski said, "It's very frustrating to be a check airman in a Boeing 747 at 59 years, 11 months and 29 days and then find yourself totally disqualified three days later (when nothing else has changed). I can't fly a revenue trip because I'm now 60, but I could hire on with the FAA itself as an air carrier inspector and even fly their B-727 after 60."

The same is true for most state governments. Richard Wray of the State of Illinois flight department said "there are no age restrictions here. One of our 13 pilots is 64, in fact. Age plays absolutely no factor in our deliberations on whether or not to hire someone. It's the person's qualifications we're most interested in." Wray went on to say that many state fleets are larger than Illinois's so more opportunity may be available in those other states. Most states give a preference to state residents and veterans. The largest aircraft in the Illinois fleet is a King Air BE-200.

A pilot looking for work must consider the possibility of a nonflying job too, and they are available if you hold the proper qualifications. A spokesman for FlightSafety International said they recruit their instructors from the ranks of retired military and airline pilots as well as those who have found themselves between airlines. FlightSafety wants its instructors experienced in the specific aircraft they will run the simulator for. FlightSafety emphasizes too, that it's a pilot's experience, not his age, that decides who gets the job. FlightSafety currently runs 36 training centers around the country and employs approximately 800 instructors. There are many other flight schools that may need instructors and it also may be possible to hire on as an instructor in a major airline's simulator training facility, such as America West's.

Space limitations prevent covering all the kinds of flying jobs around, so don't forget agricultural flying or forest fire fighting or even basic or advanced flight instruction at a local Part 141 school. No doubt finding work as you get older is tough, but the successful pilots seem to be those who use their networking skills, a good list of possible employers and plain persistence.

Air Inc.'s Kit Darby had this to say about pilots who believe they did not, or will not be hired because they are too old—past 45 years of age. "Get over it! Age is not a disadvantage. It used to be, but numerous lawsuits have made the airline very gun-shy because the federal law against age discrimination applies. Sure, fewer people are hired over age 45 or so, but few actually apply. These older pilots are their own worst enemies. It's now a much better environment to hire older pilots than in years past.

Landing a Flying Job
Outside the United States

If searching for a flying job leaves you somewhere between angry and distraught, take heart. There's another source of potential work waiting to be tapped . . . flying overseas. But don't expect flying work outside the United States to be easier to find than in the States. You might find it tougher to get hired by an international carrier. You also might find yourself traveling further to work each week, sometimes thousands of miles. Residency outside the United States also will accentuate the enormous differences in living conditions and customs from those you're accustomed to.

Before you begin shipping resumés to Bahrain and Taipei, check one very important item of your personality . . . your attitude. Carefully consider the changes you might put yourself through to keep flying, especially outside the United States. Some pilots look at the possibilities and turn in their Jepp bags forever rather than endure weeks or possibly months away from home. One former PanAm pilot, a type-rated captain, found initial employment in Alaska, but complained about having to commute from Fairbanks back to his New York home each week. Then, just after he completed training, his new airline furloughed the entire class and he found himself working in the Middle East on a short-term contract with no benefits. When he returns home now, he hopes for a jump seat ride or buys a ticket, and he said he thinks the Fairbanks-Kennedy Airport trip was not such a bad ride after all.

If you're thinking of commuting 10,000 miles to work, ask yourself if this routine is practical for you. It's not difficult to pack everything you own into a few suitcases and move to Europe or the Pacific Rim if you're single, but if you have a house and family here in the United States, a cockpit job based in Hong Kong could make commuting next to impossible. If you don't take your family with you, a consideration might be how long you're willing to be away from them. How long will your marriage survive with you out of the country? One pilot, who requested anonymity, left the United States to fly freight in the Middle East with his marriage intact. After four months away from his bride, the letter came to tell him she wanted a husband who resided at least in the same country. He left Bahrain hoping to rescue the relationship, but found it was too little, too late.

Maybe a short-term overseas contract might be a better idea. You'll stay current and probably keep your family intact while you wait for times to improve in North America. Dublin, Ireland-based Parc Aviation, the pilot leasing arm of Aer Lingus, regularly uses pilots on six to nine-month overseas contracts on aircraft as small as an EMB-120. Another option could be to take your family with you to Saudi Arabia or Indonesia. But know how your family feels about this before you accept a job. How will your teenagers enjoy living in a land that doesn't have MTV or a Blockbuster Video, or where it might be impossible for them to stop with their pals at the McDonalds down the street? Some overseas positions might not provide for nor encourage you to bring your spouse or family with you because, for example, some Middle Eastern countries don't allow their women the freedoms Western women have.

And then, there are the security considerations. Americans often are in great danger in other countries merely because they're Americans. Is it a good idea to expose yourself to such conditions? If you decide the job is worth such a risk, then the job search begins.

A recent aviation magazine editorial said the time to network is before a U.S. pilot needs to look for work. Many pilots found international work through a tip from someone else—usually another pilot—who saw an ad or heard about someone looking for crews. In 1992, a number of former Midway Commuter EMB-120 pilots found contract employment in Belgium from a tip passed on through the local Air Line Pilots Association (ALPA) office. You'll have to spend time on the phone calling airlines, leasing firms, and old pilot pals for leads. As Parc Aviation's, Tim Shattock said, "The more experience you have, though, the better your chances are."

Whether you get a tip from a friend or connect directly with a leasing firm or airline outside the United States, expect a market that doesn't give Americans preferential treatment. And you'll have to meet international requirements that vary widely among countries and employers. Cargolux's vice president of flight operations, Graham Hurst, said the quickest way to find employment with a European Economic Community (EC) airline (Fig. 7-7) today is to "get a European passport and a European license. We don't have any restrictions to hiring U.S. pilots except that we're supposed to try to find Europeans first." France and Germany, too, are notoriously tough places for U.S. pilots to find permanent work. British Airways is quite open about not hiring any pilot who's not either a British citizen or holder of an EC passport. U.S. pilot Larry Schweitz, now a Boeing 737 captain for an Egyptian charter airline said, "I think it's as difficult to find work overseas as it is in the United States, especially with the new EC. It gives European pilots a leg up on American pilots." The EC is in the process of uniting all of Europe into a single economic and monetary unit having: a powerful central bank and a single currency by 1999, common approaches to foreign policy and defense, and central-ized authority in such areas as the environment and labor relations. The stance with labor will be for European employers to give preference in hiring to Europeans.

Besides the roadblocks the EC might create for U.S. pilots, a potential stumbling block for some could be the language requirements. While English is still the international language of air traffic control, many for-eign airlines would like to hire pilots versed in another language. Former Northwest pilot Mike Henderson, now flying for KLM, said part of the requirement to fly as a permanent crew member at KLM is to "learn to speak Dutch to be able to make the normal and emergency cabin announcements." Henderson said he takes Dutch language classes at his Amsterdam crew base. Similarly, EVA Air encourages, although it doesn't require, pilot applicants to learn Mandarin Chinese.

Figure 7-7. An A320 is often flown on many international routes. *Airbus Industrie.*

Shattock said, "Language other than English is not normally a problem (for hiring by Parc Aviation)." Cargolux's Graham Hurst said, "If you live in Luxembourg, being able to speak French and German is an advantage." A former PanAm pilot said that, while flying for All Nippon, "I learned the Japanese numbers just so I could do the weight and balance their way. I think the other Japanese pilots really appreciated the efforts I made to learn their language too."

The pay and benefits on international jobs vary greatly. When the above-mentioned pilot began flying a B-747 on contract for All Nippon from Tokyo, the pay and benefits of more than $100,000 per year were a significant increase over what he earned at PanAm, where he was at the bottom of the B-747 pay scale. (The contract initially was between PanAm and All Nippon. When PanAm ceased operations, another outfit picked up the contract.) Henderson said his wages were about the same as what he'd been paid at Northwest.

Major medical care is available, although some contracts might not provide this benefit. Schweitz's individual short-term contract in Egypt (he heard about the job via word-of-mouth instead of a crew-leasing firm) keeps him current on the 737, but, he says, "There are no benefits. Either party can cancel the contract with 30 days notice. I also have to pay my own way up and back whenever I return to the United States because I receive no pass privileges." At All Nippon, one pilot said, "The pass policy was fairly restrictive . . . but they gave us two positive space international tickets each year." As part of the basic compensation package, some international companies (such as All Nippon) provide living quarters for Americans living abroad, but many will expect you to find your own lodging at your own expense.

Because all countries require some form of work permit for noncitizens (besides a U.S. passport) Americans find it helpful that many of the international airlines and corporations assist the employees in getting their paperwork in order. Most wait until the pilot arrives at the new-duty station to complete the paperwork, although Henderson remembered KLM sending him a complete packet of material before he ever left the United States. Another pilot, however, said that when he began flying for All Nippon, "We did all the work for visas and permits on our own."

How do pilots find overseas flying work if they don't hear about it from a friend? It can be difficult. Australia's Qantas said it "only advertise[s] for pilots in the Australian press." Some airlines, like Taiwan-based EVA Air, search for pilots with ads in "internationally distributed aviation magazines." Pilots should consider word-of-mouth as a supplemental source, too.

Search all the aviation publications, both U.S. and international. These are usually available at the library. And don't overlook the *World Aviation Directory*, also available at most libraries. Robert Orr, who flew freight for DHL from Bahrain, said, "I think the *World Aviation Directory* would really help because it lists addresses to international carriers and leasing agencies throughout the world." Finding out whether that company is hiring or even accepting resumés is where the pilot's work really begins, however. One pilot said he believes that a pilot seeking work outside the United States "must spend the money to call people directly who are working for an airline or company you're considering."

Parc Aviation's Shattock said, "Our (contract) placement of a pilot can depend on the season of the year, a pilot's qualifications, and their experience level. A 737-400 captain, for example, needs about 500 hours PIC to be hired. Although we hire mostly captains, the first officers we do use would also need at least 500 hours plus a type rating."For an overseas flying job, U.S. pilots usually need some type of flying certificate issued by the host country. However, it's common for the host employer to arrange and pay for the certificate. Henderson said, "In Holland, they have a B-3 license issued by the Dutch equivalent of the FAA that allows me to fly here. It requires me to maintain my FAA physical and U.S. flying currency, however." Canadian Airlines pilot Brian Rasmussen, also a former Eastern Airlines pilot, took no chances with his future. "I became dual rated with an ATP in both Canada and the United States while I was still flying for Eastern. When Canadian agreed to hire me, I was already current in Canada." To fly in Japan, one pilot said, "We had to start from scratch to qualify for a Japanese license, and the training was as tough as any I've ever been through." Because Luxembourg doesn't issue a pilot certificate higher than a private, potential Cargolux crews must obtain a Luxembourg validation to their U.S. license. Hurst says, "This means a U.S. pilot can fly a Luxembourg-registered aircraft as long as he maintains his American license and medical."

It's difficult to say whether the scale tips more toward a pilot finding international employment through a contractor or through the airline itself as a permanent employee. Cargolux's Hurst said, "We have about 30 nationalities (of pilots) working here, quite a lot of American pilots and flight engineers, about 20 out of 120 pilots." Most are permanent employees, but some are on short-term contracts. Hurst also said that, because of the large numbers of well-qualified pilots around, in 1993 "we might slow down on looking at potential pilots that don't have jet experience."

Cargolux hopefuls must have 1,500 hours minimum time, with at least 500 hours of pure turbine time. If you don't have 500 jet, 1,000 heavy turboprop time will do. The airline intends to lease a few short-term aircraft, but there's a chance it could take on more permanent crew members. Cargolux will be one of the first carriers to fly the freighter version of the Boeing 747-400. Parc Aviation or Sam Sita in Monte Carlo handles crew leasing for Cargolux. Those seeking permanent positions apply directly to Cargolux.

While most jobs outside the United States are for air-carrier-rated pilots, some agencies and companies hire corporate crews too. Houston, Texas-based Aramco, for example, hires fixed-wing and rotorcraft pilots to fly in Saudi Arabia. One crew-leasing firm manager, who requested anonymity, said he also recruited regularly for rotorcraft and fixed-wing pilots in all parts of the world. His last round of contracts were for Learjet, G-2, and G-3 rated pilots to fly in the Middle East. His firm advertised the openings in newspapers of major cities like Houston, Dallas, and Miami, but said they try to stay away from interviewing pilots from places like California because "those pilots just want more money than we can possibly offer." His last Lear pilot was paid approximately $50,000 per year, with a 30-day vacation and a one-bedroom apartment provided in Saudi Arabia. He warned pilots, however, that in places like Saudi Arabia, even though the U.S. pilots work for an American company through their contract, "over there, the Saudis call the shots."

Both you and your family must prepare for culture shock if you fly outside the United States. An All Nippon pilot said,"At least half the

American pilots over here (Tokyo) wouldn't touch Japanese food . . . it just tasted different. The Japanese are a very structured society too, so in the cockpit, there's always a boss, a worker, and an elder kind of attitude. These companies also have a very different attitude towards their employees and what their employees' responsibilities are to the company. They seem to expect the employee to be grateful for having a job, so they're really not very interested in providing great travel benefits or things like that. Even the scheduling is hard set by the company. The pilots have no input."

Schweitz said, "There's a tremendous culture difference (in Egypt), and if you're not prepared to keep an open mind it can become a very difficult situation. There's a big difference in the way they treat foreigners in Egypt, too. It's a lot more than just going up in an airplane. Cairo is not Phoenix."

Certainly one of the brightest highlights of flying overseas is the tax break available to a U.S. pilot. If you maintain a permanent residence outside of the United States for more than 330 days in a year, the IRS might allow you to exclude your first $70,000 a year from U.S. taxes. That tax break can afford you a significant pay raise for a one-year contract, depending on your salary. Any money you earn will be free of U.S. taxes. However, Mike Henderson said that, while his money was free of U.S. taxes, "We'll soon be paying Dutch taxes of about 12 percent for a U.S. pilot." Pilots who are not sure if their pay is subject to taxation by the host country can get that information through the host country's U.S. embassy in Washington D.C. or through the U.S. Embassy in the host country.

A flying job outside the United States is not easy to find or sometimes maintain, but if you're the adventurous type, it can be quite profitable. If there's any advice for a U.S. pilot looking for work outside of North America, Shattock probably said it best from Dublin. "Aviation is a very cyclical business with many good and bad times. I hope we're starting to climb out of this aviation recession and that will give us a need for more pilots. Don't ever give up . . . keep on looking."

Surviving the Loss of a Job

While it may at first sound a bit contradictory to offer an opinion on how to survive being put out of work in a guide designed primarily to help you find a job, it isn't really. Although job loss is a subject many pilots have never dealt with, losing a job in this business can be just as much a part of the industry as finding a new one.

As you'll read in the accompanying piece, most companies seldom disintegrate in a heartbeat. Most start with a slow, downward spiral that they just never pull out of. The question is, how do you prepare yourself for a job loss? Easy. By always taking the pulse of the company you work for. That means learning all you can about the finances of the company, their customers, when they add a base or new city, and when they lose one. Read the *Wall Street Journal* and the *New York Times* online to hear what the analysts are saying.

If you fly for XYZ Corp. and are planning to upgrade in their new aircraft a few months after it arrives and read that the company has had its second worst quarter in history, a bell should go off—one that says you may be lucky to keep your job let alone worry about upgrading. If your airline begins retreating from cities, you should start asking questions about how that might affect your career. There are plenty of other methods available to learn what's going on within your own company, like talking to the chief pilot surely, but why not talk to people in finance? The flight department often hears about things after the decisions are made—not before.

But if the inevitable occurs and you find yourself either furloughed or simply canned because the flight department or the airline shuts down, first of all, take the time to grieve. No rational person wants to be out of work and no pilot I know feels good about hearing that his or her services are no longer needed. (One caveat is that I'm assuming you did not do something to cause the company to let you go. If that's the case you don't need to grieve; you need a new line of work.)

Assuming again that the company fell out from under you, the most important thing you can do is to believe that there is hope, that although things did not work out quite the way you'd expected this time, they will improve. You will find a new job. But no matter what happens, don't ever give up. Take it from someone (me) who did just that after he had three jobs pulled out from under him. I left flying for a while and was totally miserable as I searched for work in a new field, looking for that real job that never appeared because my heart, my passion, has always been and always will be in aviation.

There's more to the loss of a job, too. When I say don't give up, that is only the beginning. That allows you to go out and have a drink to your dead job, but also to begin focusing your energies, not on a job, but the next job, because there will be one, if you don't get lazy. It's easy to find a job after you lose one, but will it be a job that pays the rent, or one that takes you up a step from where you were? That's important.

Go back to the plan that you've hopefully put together and take a look at the kind of position you've been working in. Is it really where you want to be? If you've been flying freight and you hate not being around people, perhaps in addition to the job loss, this is the boot you need to send you in a totally new direction. Why NOT apply to a regional? Or if you really are a people person, why NOT begin looking in the corporate sector?

Surviving the loss of a job really boils down to planning ahead, just as you would when planning a flight. Many pilots I've interviewed recently explained why they had a degree in business or some fairly lucrative subject—"Just in case I couldn't fly." I'll grant you that never flying again is an extreme case and something that hopefully few of you reading this book will ever face, but it highlights the strategy again—plan ahead!

So practically, this means that you never toss away all those business cards from corporate or airline pilots you've met. You never know when you may need a reference or may want to pick up the latest details on the hiring market for corporations in your area.

Midway Airlines Revisited

When the phone rang about 11:45 P.M., I was already asleep and now quite annoyed at being disturbed. The past few weeks had not been good at Midway Airlines or its subsidiary, Midway Commuter, who I flew a Brasilia for, as employees began to wonder when they would hear more official details about the buyout from Northwest Airlines. Nearly a month had passed since the buyout plans were announced, but with each passing day the anxiety levels rose among the employees. Many never said anything specifically, but you could see it in their eyes. They were worried. They had good reason. We'd been sitting in Chapter 11 bankruptcy status since March. We knew Midway could not last forever. The word had just come down that Northwest was pulling out of the buyout deal for reasons unknown. The date was November 13, 1991.

The voice on the end of the phone was our MEC Chairman, Ty Hackney. "Rob! Wake up!...It's all over buddy! Midway ceased operation effective at midnight tonight." So there it was. The final struggle was over. The work that ALPA had initiated many months before and taken to the MEC at Northwest that began the buyout talks in the first place had all come to nothing. I thanked Ty for the call, hung up and went back to bed. It was not a sound sleep however. I was unemployed.

In 1979, Midway Airlines was the brainchild of Irving Tague and David Hinson (former FAA Administrator) and the darling of the deregulation set. A new airline for a new age of airlines.

But Midway Airlines actually filed with the Civil Aeronautics Board (CAB, since disbanded), on October 13, 1976, nearly three years before the dam of airline deregulation broke open, the last airline, in fact, to file before deregulation became the law of the land.

The two entrepreneurs believed the concept of a no frills, Southwest Airlines-type carrier with a peak and off-peak fare structure to be a sound one. The first Midway Penny Fares promotion in 1979 would offer passengers a standard $30 one way trip to Detroit for only 30 cents. But first, this new airline needed airplanes, money and an airport. Initial capitalization of Midway Airlines was a mere $5.7 million.

Finding an airport was the easiest of obstacles to overcome as Chicago's Midway Airport was then a virtual ghost town since the airlines and their runway-hungry Boeing 707s and DC-8s had pulled out in the mid-60s. Initially, employees numbered less than 200, but using three DC-9-14s purchased from TWA, Midway took the name of its home airport for its own and in 1979 began serving Cleveland, Kansas City, and Detroit. Frank Hicks, Midway's first director of flight control knew the airline was catching on when he relinquished his near terminal parking slot for one four or five rows further away to make room for more passengers

The airline continued to grow. In 1980 it gained five more DC-9s to the fleet, as well as four more destinations, LaGuardia, Washington National, Omaha, and St. Louis. To mark the airline's first full year in service, Midway made 850,000 shares of common stock available in the airline's first public offering. The stock sold in just two hours.

Because of the 1981 PATCO strike, a number of Midway reversed a number of route decisions and dropped service to Boston and Orlando. Many people believed this retrenchment displayed the kind of rapid fire decision making necessary to run an efficient airline. Capt. Jerry Mugerditchian, who became the Midway pilot's Master Executive Council (MEC) Chairman in 1985 and who is now ALPA's vice president-administration and a United Airlines first officer, says, "Midway had experienced excellent growth and expansion in 1981, but things were changing rapidly. There just was little consistency to pay, work rules, and minimal training. Airline management was very short-staffed." Without clear-cut policy manuals, the airline changed the rules whenever they thought necessary. Pilots don't like lots of changes.

The first two MD-80s appeared in 1983 as Midway changed its marketing strategy to reflect its new niche—MetroLink, all business service at coach fares.

In mid-1984, Midway management agreed to purchase the assets of bankrupt Air Florida. While the airline, according to then vice president of flight Dick Pfennig, was not quite ready for an expansion yet, "Air Florida was one of those opportunities that was there..." This move produced a new north-south presence for the airline through the new Midway Express and the fleet of Boeing 737s, which the airline gained with its new Miami base. Midway airlines' reach now increased to the Virgin Islands through Miami. The airline now employed 2000 people.

By the end of 1984, the airline carried 1.3 million passengers to 13 cities through its Midway airport hub. Also in 1984, difficulties with management arose when it began integrating the Air Florida pilots into the Midway Airlines seniority list. "That was the straw that broke the camel's back," Mugerditchian said. By the middle of 1985, ALPA was representing Midway pilots.

Again in 1985, the company began to reconsider its operating niche as the search for profits continued. Midway had suffered losses in both 1983 and 1984. Midway Metrolink was a failure and Midway Express was out because passengers and travel agents found them to be too confusing. John Tague, then Midway's director of airline planning said "It (1985) was a very difficult time for the company. But we truly believed that what may have been painful for a few would be beneficial for many."

David R. Hinson, spearheaded the recovery plan. He had recently been named chairman and chief executive officer. The plan was a no-holds-barred attack on excessive costs including a 10 percent reduction in personnel, a suspension of all capital expenditures, and the elimination of Newark and Topeka from the route structure. Midway decided two-class seating was the direction to take and returned their MD-80s to the lessors.

By mid-1985 however, the airline began turning a profit and added three new cities. By December, Midway carried two million passengers in one year for the first time.

By 1986, improved yields became the buzz word. Midway reported net income of nearly $7 million and annual passenger loads inched closer to three million. The carrier also established its first links with local commuters to feed the jets.

As 1987 opened, the airline's management believed Midway rated national-carrier status as they placed an order for 8 MD-87s and options on 28 more. Midway President Jeffrey Erickson announced, Project New Attitude. Midway's vice president of customer service Lois Gallo explained, "The program recognizes that Midway employees are the experts at handling customers and gives them greater latitude to do just that. We regard our employees as responsible, capable experts and trust them to make proper decisions." Midway employees were overjoyed.

Midway also entered the regional market with its own aircraft in 1987, flying them under the Midway Commuter logo with Midway vice president Dick Pfennig in charge. Midway purchased regional Fischer Brothers Aviation of Galion, Ohio as a subsidiary to Midway Airlines. Pfennig made the agreement with Fischer Brothers and had the airplanes in the air just 23 days later from a new Springfield, ILL base.

Using 10 Dornier 228s—homely, noisy but efficient aircraft—the commuter began carrying passengers to MDW from Springfield and Peoria, IL, Green Bay and Madison, WI and Grand Rapids and Traverse City, MI.

Midway ended 1987 surpassing the three-million passengers carried mark by nearly 751,000. Annual departures totaled more than 78,000 and employees numbered about 3,500. Midway Commuter carried 433,000 passengers and Midway Airlines now accounted for 65 percent of total traffic at Midway Airport. Total airline departures numbered 123,844.

Just when most passengers and many of the employees began to wonder how the airline would top 1988, Midway, concurrent with its tenth anniversary, announced the opening of a second hub in Philadelphia an event that was precipitated by the untimely demise of Eastern Airlines and a glut of equipment and gates. Midway outbid USAir and planned to offer by the end of the first quarter of 1990, 70 daily departures from PHL with 16 DC-9-30s acquired from Eastern, though none of the Eastern employees came with the deal.

New destinations from Philly would include 20 cities as well as Toronto and Montreal. Midway also established a marketing agreement with Canadian International Airlines.

Midway Commuter saw ALPA arrive on its property too in 1989, with a separate MEC structure from that of the pilots who flew the jets. One Midway employee said 1989 was "a turning point into the next decade..." Turning point would prove to be an understatement. While the airline had fought its way past other economic difficulties, 1990 would see a fight that would make Daniel in the lion's den look like a walk through the park on a sunny summer afternoon.

Midway management underestimated the tenacity with which USAir would defend its East Coast turf. On October 19, 1990, a year after announcing the opening of a second hub, David Hinson told employees of the airline's intent to sell the Philadelphia hub to USAir. Hinson said "Our problem at Philadelphia is straightforward. At 60 departures per day we are not big enough to create an efficient operation. We need approximately 100...We need capital to buy airplanes, train personnel...financial institutions are unwilling to lend us the money."

Hinson believed the sale to be a necessary one, especially after the Iraqi invasion of Kuwait sent fuel prices soaring. "If we try to continue as we are and fuel prices do not decline, we may place our company in serious financial difficulty." Hinson also admitted at the time that "economic circumstances are changing so fast it's difficult to see too far ahead."

The plan was to reduce the number of employees, ground some of the older DC-9s, eliminate the 737s entirely, but leave the commuter in tact.

The commuter had just begun flying the 29-seat Embraer Brasilias and had a fleet of 28 aircraft of its own serving 18 cities. Hinson's final words during the announcement were that "we need to back up a little and restructure our company in order to have the financial strength to fight another day." Unlike General MacArthur in the Philippines, however, Midway would never again return in such strength.

On October 23, 1990, the airline suspended its regular dividend payment. The next day, Standard & Poor's downgraded its rating on $54 million on Midway's convertible exchangeable preferred stock to single C from triple C. "Given their cash position and balance sheet and the likelihood that they will continue to be adversely affected by high fuel costs in the weeks and months ahead, I, for one, can't see how they can last much beyond the first quarter of 1991," said one New York-based securities analyst. "In retrospect, Midway's decision to buy the Philadelphia hub from Eastern for $100 million would have been a mistake even if there had been no fuel crisis," said the analyst.

As for the possibility of another airline taking over Midway, something rumored for weeks at this point, Hinson said "the airline is not seeking a buyer, but if the move were in the best interests of the stockholders, it would of course, be considered." Hinson added, "If Midway was to be taken over, it would have to be a hostile takeover, because right now we have no intention of selling it."

In October, 1990, ALPA helped formulate a plan to make Midway an attractive to Delta. Delta, however, was in discussion with PanAm. ALPA then approached Northwest. While both the Midway and Northwest MEC, along with ALPA national, began the initial dialogue at Northwest, efforts were swift that would bring Midway and Northwest management together to begin direct negotiations themselves.

But the sand in Midway's hourglass was rapidly running out.

In March 1991, Midway Airlines filed for Chapter 11 bankruptcy protection. By June, the carrier had lost $37.1 million for the year. MD-80s were being repossessed at the gates. The airline had no money to get the lavatories serviced on the Brasilias. The swift turboprops disappeared one by one as pilots on both the jets and the commuter were furloughed. Morale hit rock bottom. An air of depression filled the corridors of the terminal like black smoke. Some days, passengers tried to comfort the employees aware of the state of the airline. Each day employees would arrive at work, wondering if this would indeed be their last. In a desperate 11th hour move, Midway made a deal with Fidelity Bank in Philadelphia to lease 17-old Eastern DC-9s at 1/3 the cost of some of the current aircraft.

It proved to be too little too late. Many ALPA pilots believed those new aircraft were worn out before they arrived.

Midway pilots saw things they'd never encountered with such regularity before—Minimum Equipment List (MEL) stickers all over the cockpits as a lack of parts pushed the capabilities of aircraft and their crews to the limits.

Then, just when many had given up, the deal with the white knight out of the northwest appeared. September 24th, Northwest Airlines agreed to purchase the leasehold interests of the 21 gates Midway Airlines held at MDW for $20 million. The agreement seemed to pave the way toward an agreement to purchase all Midway assets. Some analysts believed, however, that a deal would be struck only for key Midway assets, such as Midway had done with Air Florida.

Jack Hunter, with Rothschild Securities in Chicago said "I don't think (Northwest) would want to assume Midway's debts. Northwest has debts of their own...I don't think they could afford an outright purchase."

Other airline analysts believed a purchase was forthcoming and that this posturing was an attempt on the part of Northwest to catch up with United, American, and Delta.

Southwest and TWA were also rumored to be interested in the carrier. Southwest indeed put forth a $109 million offer for the airline which the bankruptcy court turned down in favor of the $174 million offered by Northwest.

By mid-October 1991, Midway employees began to breath a bit easier. Smiles seemed to return to the terminals as rumors flourished about the results of the Northwest buyout. Very quickly, initial plans were offered by Northwest for training and the overall transition to the red tail look. 3800 of Midway's 4300 employees would have jobs. Taking employees with them was not a part of the Southwest plan.

By early November 1991, however, details and plans from Northwest had slowed to a trickle and anxiety levels again began to rise. Had the deal hit a snag? No one was talking. On November 11th, Midway employees read a *Wall Street Journal* article, Northwest Air Seems Hesitant On Midway Bid. "Northwest Airlines is sending signals that it is having second thoughts about its proposed purchase of certain Midway Airlines Inc. assets."

At first, it was thought as a possible ploy for wringing a bit more out of Midway for a little less money, but they were unaware of the flurry of letters being swapped between Midway and Northwest management.

In the letters, Northwest denied Midway's claim that it ever had a contract to buy Midway's assets because Northwest's Board of Directors had never consummated the deal.

Northwest also claimed Midway management had provided erroneous revenue data to the Department of Transportation, which Northwest had been using to organize its original bid for Midway. A letter from Northwest's Senior Vice President and General Counsel, Richard B. Hirst, said "Midway admitted to Northwest that its DOT data are unreliable."

This information only became available, Hirst continued, after Northwest had paid Midway the $20 million for the gates. Northwest was also concerned about incurring "substantial environmental liabilities."

Then, came the final paragraph. "Upon review of the transaction, Northwest's Board of Directors voted last night not to approve the Midway transaction. Accordingly, the proposed transaction between Northwest and Midway for the acquisition of the remainder of Midway's assets is terminated."

At midnight on November 13, 1991, Midway Airlines ceased operations.

On December 24th, federal bankruptcy judge John Squires approved the return of $68,000 from the Christmas party fund to employees since they'd all lost as much as three weeks pay, as well as any unused vacation time, when the airline shut down. Each employee received about $17. The judge also approved the paying of bonuses to some 57 Midway employees who hung around to shut the airline down after November 13th. President Thomas Schick, for instance, picked up $50,000 while chief financial officer, Alfred Altschul and customer service boss Lois Gallo each received $30,000.

In a letter dated February 24, 1992, U.S. Department of Transportation's Inspector General A Mary Sterling told Congressman James L. Oberstar (D-MN) that although there were certainly differences in the software used to generate passengers at Midway and Northwest, "nothing came to our attention...to indicate that Midway deliberately filed misrepresented sample data."

Nearly 1000 Midway pilots read these words in the newspaper as they waited in line to collect their unemployment checks.

8

The Pilot and the Personal Computer

Throughout this book, I've made it pretty clear that I believe a pilot looking for a job today needs not only basic computer skills, but a good working knowledge of the Internet, as well as some of the proprietary online services such as CompuServe and America Online in order to effectively seek out and use the vast wealth of information available. Here are a few ideas to keep in mind.

To make your computer system work effectively, you'll need a fast telephone modem—56kbs is the maximum at this time—and a good Internet Service Provider (ISP) to connect you to the Internet. Check with your local cable television operator as well, since many are offering cable modems that allow even better access to the Internet.

You'll also need a good software package that includes a word processor—like MS Word—in it to be able to maintain and update your resumé. No one takes their resumé to someone to store and update anymore. There are certainly organizations that can offer you advice on your resumé in addition to what we spoke about in Chapter 4, but you'll want to store it on your own computer so it can be easily updated and printed out to send with applications and cover letters. The reason I suggest MS Word is that it has come to be the accepted standard on computers these days. Personally, I always liked WordPerfect better, but found myself changing because everyone else did. The advantage of using MS Word to store your resumé in, is that you can now e-mail someone a copy that they can open on their machines and print out a perfect copy.

Computerized Logbooks

One of the great tasks computers accomplish well is managing numbers, as in your flight times. I'm a proponent of computerized logbooks (see accompanying article) because they take most of the drudge work out of filling out applications and keeping track of your total time in all of the dozens of various categories airlines and corporations seem to want to see your time broken down into.

One that I have used for years and found easy to use is Aerolog for Windows from Polaris Microsystems (800-336-1204). The benefit to me— once all of the data was loaded into the software—is that every time I want to update my resumé, I just open the software, start up a report I organized once and in just a few seconds, I have updated totals. But a computerized logbook is much more than simply one report, because it offers you the flexibility to break down your times in any fashion imaginable. Aerolog contains its own copy of an FAA Form 8710 that you'll need to fill out before a checkride. Click this and in five seconds you have all the numbers you need to transfer to the form. Since I'm flying for a Part 135 charter operator, my logbook also tracks my flight and duty times to keep the Feds happy should they want to know whether I'm in compliance. (Fig. 8-1.)

The other major convenience of a computerized logbook is that you don't need to reenter the same data more than once. For example, when I add a new flight by typing in N45ML, my logbook already knows that this aircraft is a C-550. It also knows that I have organized a special tag to

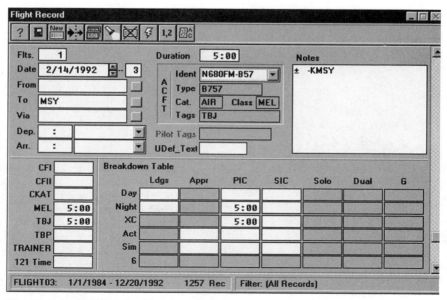

Figure 8-1. AEROLOG—log entry page.

track my jet time separately from my other flight time. Then, when I ask it to tell me how much jet time I have, it searches by the tag for jets and displays the result. But best of all, if someone calls me up and says how much Citation time did you have in 1997, I can put them on hold and be back with them in less than a minute with the answer.

When it comes to computers, the word crash has almost as dire a consequence as it does when you're flying an airplane. While this is definitely a hazard when using an electronic logbook, it won't cause chaos if you backup the program's data files. I make two backups of my logbook after every flight. It takes about 30 seconds to accomplish this task and has saved me three or four times already, such as when the hard drive on my system simply disintegrated a year ago. After the new hard drive was installed, I reinstalled the program, restored the data from the backup file, and was back in business in no time at all. Always, always back up your computer's data files. That includes your resumé and cover letters.

Online Message Boards

A few years ago, a discussion of online message boards focused on aviation would probably have been limited to AVSIG (Aviation Special Interest Group) on CompuServe. But as with just about everything else that relates to computers and the Internet, last week's news is old news online. (Fig. 8-2.)

For those of you new to the Internet and message boards, you can now find them on America Online and on a number of aviation Web sites—many of which are highlighted in the Web site section later in this chapter. Over the years, I've come to rely on CompuServe's AVSIG as the granddaddy of all aviation sites, a kind of virtual water cooler where aviation people hang out, meet new people, and talk about just about anything aviationlike. It is, however, only one of four or five aviation-related boards on CompuServe. There are also a dozen subboards to the AVSIG alone that focus on subjects such as commercial airlines chat, career chat, corporate jobs, training issues, and just plain news items that people want to jaw about. A hot discussion item as I write this, is what will happen at Federal Express—will they strike or back down? The discussions are heated at times, but always interesting and always informative, because the talk comes from the people right there in the trenches.

What also makes any of the boards on CompuServe valuable—and most other message boards, as well—is the ability to post a message in search of advice about your career decisions or to find someone who just interviewed with the company you received a call from. What you need to join the foray, is a CompuServe account—currently $9.95 per month—

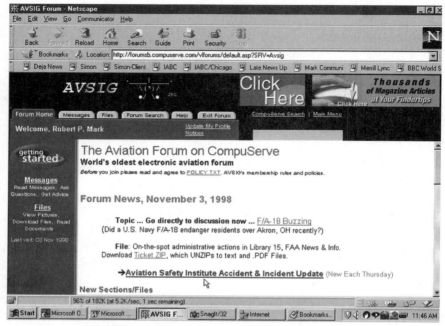

Figure 8-2. AVSIG OPENING SCREEN ON CompuServe.

and Internet access. You simply make the connection to your ISP, open your CompuServe software, and go for it.

America Online is the other major online service. In fact, America Online owns CompuServe, but simply runs it as an alternative forum. CompuServe has always taken a rather formal, perhaps more business-focused look at issues, while AOL tends to be a bit more like the Wild West of online services. Now this does not mean that the information on AOL is any less credible, it simply means you may need to work a bit harder to verify the information you're receiving there. I have found that during the writing of this book, I found more pilots of various categories—student, airline, corporate, etc.—that were willing to talk to me from AOL than there were from CompuServe. This could just mean, of course, that the people on CompuServe did not care to respond. But it also might mean that the people on AOL are just a bit friendlier if you're new to online services. (Fig. 8-3.)

One important difference between CompuServe and America Online that needs mentioning is that most of the people on CompuServe use their real name. On AOL, you can use an alias. Mine on AOL is: C550PILOT@AOL.COM. You can look up an AOL alias and, in some

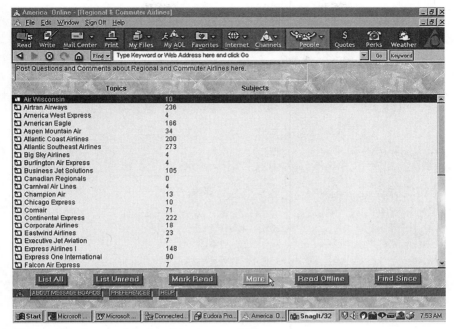

Figure 8-3. Regional Airline Directory on AOL.

cases, learn the real identity of the writer with a little extra effort. The least expensive entry to AOL is currently $4.95 per month.

When you do sign up with AOL, be sure to visit writer and pilot Greg Brown's two flying sites on the system. They're chock full of the kind of information and resources you'll need to make friends online and search for that new job. After you sign on, go to the aviation section by typing in the Keyword: aviation. From there, click on commercial aviation, then career center, and then to "Greg Brown's Career Forum." His other page is for pilots new to aviation and is called "New Pilots, Students, and Wannabe Forum."

Online Message Boards—
A Taste

If you've never visited any of the online forums on CompuServe, AOL, or anywhere else, or perhaps have never been online before, you simply can't imagine what happens. Here's a taste of a forum from AOL that was downloaded from Greg Brown's Aviation Careers Forum. The subject this night was law enforcement flying, so not only will you see a bit of what online chatting looks like, you'll learn something about flying for the Drug Enforcement Agency (DEA). These are not the real IDs of the people in this session, however, except for "Paperjet," who is Greg Brown.

PaperJet: Welcome to "Law Enforcement Flying" on the Aviation Careers Forum.

PaperJet: Our guest is "Tom," DEA special agent and pilot.

Tom: Glad to be here.

MDiprosper: Ok, what do you fly? And how long have you been with law enforcement?

Tom: I am currently flying the MD-500&BO-105 helo along with the Merlin 3B &4C.

Tom: I started my L.E. career in 1971.

Tom: After a seven-year tour as a Naval Aviator with USMC (US Marines).

PaperJet: Tom, is your primary duty intercepting drug runners?

Tom: We don't chase other A/C—that's Customs' job.

PaperJet: So what are your primary flying duties?

Tom: We mainly do aerial surveillance, transport some prisoners/ witnesses and evidence.

Tom: Air-to-ground surveillance.

PaperJet: Is the aerial surveillance electronic?

Tom: We use stabilized binocs and Flirw/video (Infrared)cam

Tom: We have infrared equipment but I can't comment on locations or plans.

PaperJet: Hi, Bill. Go ahead with your question.

BILL: Are the tethered balloons doing any good at the s/w-ern boarder?

Tom: The tethered balloons are operated by U.S. Customs—I have no knowledge of their efficacy.

BILL: Tnx.

PaperJet: Tom, since customs chases the drug runners, what is the purpose of most DEA surveillance?

Tom: Our air-to-ground surveillances are done in conjunction with ongoing criminal investigation.

Tom: Usually undercover operations.

PaperJet: I know your title is Special Agent/Pilot. What percent of your time is spent doing investigating on the ground?

Tom: Up until about six years ago I was doing some undercover work—now most of my time is spent flying and staying current in the various a/c.

PaperJet: Jan, go ahead with your question.

Jan: What background do you have to have to get into this line of work?

Tom: A college degree and preferably some previous L.E. experience, although that is not absolutely required.

PaperJet: Taja, go ahead with your question.

Taja1983: What is the age limit?

Tom: The age limit for entry to DEA is 37.

PaperJet: Bill, go ahead with your question.

BILL: Some of the TV coverage has shown night-time surveillance. Is that done with FLIR?

Tom: That is correct.

BILL: Tnx.

Rich: Do you join the DEA and then hope for a flying position?

Tom: We only have 110 total pilots in DEA—you must enter and qualify as a Special Agent, have two years "on the street" and then make application for entry to the air unit.

Tom: The Special Agent Basic School is in Quantico, Va., and lasts approximately five months done.

Rich: Competition pretty fierce?

Tom: Yes it is.

PaperJet: Taja, go ahead with your question.

Taja1983: How many agents are in line waiting for a flying position?

Tom: Sorry, Taja I don't know—but we have a "pool" of qualified agents waiting.

Taja1983: Thanx...see you later.

Sue: Mach, there are a couple of us ladies out here who were wondering how your flight suits compare to the Naval Flight Officers. And seriously, how long did you fly in military prior to DEA?? <G!>

Tom: We wear green flight suits on helo duty and they look much better on us than NFOs!

Tom: I flew seven years in USMC.

PaperJet: Hi, GKenn! Go ahead with your question.

GKenn90806: Is a degree required prior to Quantico?

Tom: At present, that is affirmative.

GKenn90806: Two or four year?

Tom: Four year.

PaperJet: Triplebus, go ahead with your question.

Triplebus: Do DEA use txp's during intercepts and what is atc's role?

Tom: We do not do intercepts—that's U.S. Customs.

Triplebus: Gotcha.

PaperJet: Tom, what's a "typical" day like on the job? Is it as exciting as it sounds?

Tom: I have to say that the flying is very exciting and rewarding—up until last year we were doing a lot of flying in South America. Very challenging.

PaperJet: GKenn, go ahead with your question.

GKenn90806: What type of aircraft?

Tom: In South America I flew a CASA 212, in the early part a C-47.

PaperJet: Tom, are we making any headway against drugs in SA and Central America?

Tom: Now we are getting political.

PaperJet: Something like the C-47 used for surveillance? Or is it a gunship?

Tom: No gunship—we flew support missions for National Police of country supported.

PaperJet: Interesting! Sounds like good use of your Marine background. Scary very often?

Tom: It was great fun and exciting also very challenging.

PaperJet: Tom, are the requirements similar for most law enforcement jobs?

Tom: I would say that probably they are—stringent background investigation.

PaperJet: How could someone prepare if that's what they'd like to do for a career?

Tom: A lot of our recent recruitment has come from state and local police agencies.

Tom: Although that does not by any means limit anyone who should be interested.

PaperJet: Is a law enforcement degree beneficial?

Tom: I would have to say yes, although I have never been directly involved in recruiting.

PaperJet: How does one apply to the DEA?

Tom: We have offices in all major cities—just call and ask for the employment package (Special Agent)—it has all the forms and info.

PaperJet: I have one last question. I hear from lots of retiring mil. helo. pilots. Much demand for rotor?

Tom: It is about 50-50 in that regard.

PaperJet: Dual ratings desirable?

Tom: Yes—most of our pilots are dual rated but that is not a require-ment.

FRED: Tom...do y'all still do night no light ops into outlying airports in our area.....(please don't arrest me for asking..lol)

Tom: Most of our night ops to outlying airports would be training/proficiency events.

PaperJet: SkyJock1, go ahead with your question.

John: Sorry I was late...What is the typical number of hours (helo) to be accepted into the DEA?

Tom: Just having a helo rating alone would not qualify you for entry to DEA—our helo pilots must possess a Comm helo and helo instr rating + 250 hours + DEA checkride to perform operational DEA missions.

PaperJet: Welcome to "Law Enforcement Flying" on the Aviation Careers Forum.

PaperJet: Our guest is "Tom," DEA special agent and pilot.

PaperJet: Any more questions for our guest?

TIM: Tom, re Fred's earlier question, do you folks do night training ops lights out?

Tom: No we don't do that—some surveillances require it and we have waivers.

TIM: Thanks, Tom.

PaperJet: Well, Tom, thank you for joining us tonight!!!!

Tom: You are very welcome—I enjoyed it and hope that it helped!!

PaperJet: Is it OK for folks to email you with questions they come up with later?

Tom: Sure—no problem.

PaperJet: Thanks and have a great evening!PaperJet: How about a round of applause for Tom!

Tom: Same to you.

These forums take place in real time, so you'll sign on and watch all of this text scroll across your screen as you listen in—figuratively speaking. I've found these kind of interactive forums to be of great benefit, because you can ask questions right as it is all happening—no matter where you are in the country.

Stay Sharp for the Simulator Ride on your PC

No doubt about it, getting the call for an interview is the best news you can ever receive—short of hearing you've gotten the job, that is. The key to moving from the interview stage to the hired stage involves a number of steps, not the least of which is often a simulator ride. Again and again, companies report that one of the major reasons pilots are not hired is that they are not able to walk the talk when it comes to flying instruments.

Sometimes it's simply nerves that do a pilot in, but most of the time pilots just haven't practiced enough, relying on the last few approaches they've flown to get them through the test. When asked about taking some simulator time before the sim check, most pilots will tell you they will, or even that they should, but let's face it, buying simulator time is not always cheap. And then you have to make the trip to the airport or somewhere else if no simulator is available close by.

Yadah, yadah, yadah!

If you want to pass a simulator check ride, you have to know more than the basics. You must be able to flawlessly display them under stress.

When you add to the fact that your simulator ride might be in a Boeing 747, any rusty procedures will simply be amplified.

But now, thanks to a new Advisory Circular—AC-61-126—a pilot can use a personal computer to stay sharp on IFR procedures. And, if you're just now working on your instrument rating, you can even log up to 10 hours of the time you spent on a PC toward your instrument ticket provided the conditions of instructor endorsement, curriculum content, and supporting study materials are met.

Everyone knows that an airplane is a noisy, uncomfortable classroom. That's why simulators have thrived in the past for teaching the basics of flying on instruments. But most simulators are expensive and sometimes out of reach for some flight schools and certainly too expensive for an individual.

Michael Thelander, director of marketing for ASA (Aviation Supplies & Academics—www.asa2fly.com/asa), makers of one PCATD (Personal Computer Aviation Training Device) software package, "On Top," said "There has always been a market for these devices for individuals. Long before PCATD approval by the FAA, thousands of pilots used the basic software to give them a sense of situational awareness, procedural understanding, and the ability to think ahead of the airplane before the next approach. The PCATD is a hardware/software/syllabus combination of the same basic package. While an individual may not need the entire training syllabus, the value is still there."

Thelander discussed an issue often voiced by pilot applicants, especially unsuccessful ones. "It's a given that simulators do not fly like real aircraft. Without seat-of-the-pants sensations and in the absence of sound and inner-ear cues, they're different. Students seriously planning for a professional pilot career should look at training facilities that start simulation training early, at the instrument student level. They should avoid an all-aircraft curriculum if they know a simulated checkride may be around the next corner." A simulator of any kind supports the training you do in an aircraft and helps prevent the information overload that often occurs when students try to accomplish everything in an airplane. (Fig. 8-4.)

Whether or not a pilot can log the time on a PCATD is really not important because, there simply is no substitute for having shot 12 NDB approaches to 1 Left at Dulles before you hop into the 747 simulator for a sim check. The procedures you need to understand to fly an NDB, or a full VOR or a localizer approach, will be so set in your mind, that you'll easily be able spend more time worrying about the really important stuff, like how to fly the 747!

Despite my sense of humor here, you should know that seldom does a potential employer expect you to smoothly fly a large-aircraft simulator when you've been flying much smaller aircraft. But they do expect you to be

Figure 8-4. ASA's On-Top PCATD cockpit.

able to keep the aircraft right side up and to notice trends off of target air-speeds, altitudes, and headings. That means your scan must be fast and accurate. That only comes with practice. And a PCATD is an inexpensive method of maintaining that proficiency, something that is often difficult, even for those of us who fly for a living. Thelander added that "Many airline pilots use the basic software at home in a nonloggable version because they never get a chance to hand-fly the aircraft any more and also because they know that when the chips are down, they need to be prepared."

Some PCATDs—like ASAs—come in integrated modules that begin at the least expensive level with a simple software purchase that costs about $395. The most expensive version of ASA's PCATD cost $2,995, but includes the software, an avionics panel that includes realistic nav/com units, a control yoke, rudder pedals, and a six-lever power control quadrant that with throttles, mixture, and prop controls and even gear and flap switches. The software emulates seven different single-engine and one twin-engine aircraft. A demo version of the ASA software can be downloaded right to your PC for evaluation at: www.asa2fly.com/asa

Another popular PCATD is made by Jeppesen, the people who brought you Jepp charts and other major aviation training systems for almost

70 years. The Jepp system is designed only to work with a fully function-
ing console that looks remarkably similar to a Cessna 172 cockpit included
in the basic price. The unit plugs directly into the parallel port on your PC.
A full-featured version of the Jeppesen FS 200 Flight Simulator costs
$3099.00. The least expensive editions will ring up $925 on the cash regis-
ter. More information on the options Jepp offers on their software can be
found at: www.jeppesensanderson.com. We'll also talk about how the
PCATDs fly in the next section.

But one final word on PCATDs—don't confuse price with value. At
first glance, the basic cost or the cost of options can make one of these
devices seem rather expensive. But when you consider it as a part of your
overall career plan, it might turn out to be a relatively inexpensive investment in your career. If you get the job you want because you aced the simulator portion of the interview, a few hundred or perhaps even a
thousand dollars, will seem a cheap price to pay. If that logic is lost on
you, try amortizing the cost over the 25 or 30 years you may be flying professionally. Let's see now, if I spent $1000, or less, divided by 25. Yep, I'd
spend $40 a year—a mere $3.33 per month—for the job of my dreams.

Flying a PCATD

Flying a personal computer aviation training device (PCATD) is not as
easy as flying an airplane. Like its big-brother siblings, Flight Safety's
Level C & D simulators, or even the nonmotion Frasca machines, a
PCATD requires a pilot's constant attention for almost the entire session
since these devices lack the "feel" of a real aircraft. That's why you need
regular simulator sessions. Try and pass a checkride in a simulator never
having flown one before and the results could be enlightening—to say
the least—and devastating to your career at worst. The question, however, is whether or not a PCATD can give you the real-world experience
and simulator feel you need to feel confident going into a ride.

In my opinion as a 20-year instrument instructor and line pilot in a turbojet aircraft, any IFR pilot—whether they're planning on a prehire simulator check or simply want to remain current—should own a PCATD.
The experience is that real and that valuable. Much like any other kind of
computer simulation system, there are a number of add-ons that sit next
to your monitor, such as a throttle quadrant, rudder pedals, or a radio
stack to make your experience more realistic. The more familiar you
become with the PCATD, the more realism you'll most likely want to add.
For evaluation purposes, however, I simply used ASA's basic "On Top IFR
Proficiency Simulator," right out of the box and found the computer gen-
erated panel to be more than adequate for my needs.

I installed the simulator on a 200 MHz. Pentium machine equipped with 128 meg of RAM, a 17-inch monitor, a Kensington track ball in place of the standard mouse, and a Microsoft 3D Sidewinder Pro joystick. I had my system up and running about five minutes after I loaded the software from the CD-ROM—and with not one call to technical support necessary. I did learn that my PC lacked the video adapters necessary to run all aspects of the PCATD adequately, so the graphics outside my cockpit were rather mundane. I'm told that will improve when I update my system with a 16-bit video card.

The quality of the graphic presentation of the instrument panel on the PC monitor is astounding, however, actually much better than I'd expected. The experience was so real, that I quickly lost track of the fact that the PCATD did not look like the inside of an airplane and was actually sitting on the desk in my office.

What makes the PCATD so valuable is simply the variety of experiences it offers, from your choice of just about any IFR airport in the country, to any of six aircraft types to fly, to a wide range of weather and malfunction situations that you can add to your flight, many at random for added realism. And let's not forget that the PCATD flies like a real simulator, no small challenge in itself.

Since my nonrevenue flying is often in a Piper Arrow, I chose that aircraft as the default for my system. I next chose the cockpit layout with an HSI (Horizontal Situation Indicator), RMI (Radio Magnetic Indicator), and even a panel-mounted GPS (Global Positioning System) to keep me on my toes. But operation of a PCATD makes a few assumptions about the pilot. For instance, it assumes you already know how to use an RMI, an HSI, or an autopilot. If you don't, you'll need to work with an instructor for a bit to understand what is happening. The system does, however, offer you instruction on programming the GPS, since almost all of these are a little different.

A PCATD, like the more sophisticated simulators, offers an overhead map for you to keep track of your progress across the ground throughout the flight. You can toggle out of the instrument panel and over to the map at any point and never lose your place, although, as in a larger simulator or an airplane, the distraction of fussing with switches and buttons can affect your instrument scan for the few seconds the sim needs to lose a hundred feet of altitude.

After setting up the cockpit and aircraft defaults of my system, I set the weather at 400 foot overcast with a 2 mile visibility and took off for a little airwork. Steep turns were just as easy—or as tough depending upon your proficiency—as in other sims I've flown. I even tried a few stalls and found the PCATD to be quite realistic to both the airplane and the more sophisticated, full motion simulators. This airwork was even more challenging because it was performed in the clouds.

I decided the best test of my PCATD was to take it out for a little cross-country work. I planned a flight from PWK (Chicago-PalWaukee) to RFD (Rockford, IL) to shoot some approaches. After setting the weather parameters for takeoff, I added in a few variables, like the clouds being near minimums on all the approaches and only a slight northeast wind. The variability factor the PCATD offers meant that during some letdowns, the clouds might be below minimums calling for a missed approach, while at other times I could land or make a low approach and return to the system — but I'd never know for certain when it was going to occur. I also set the forecast weather to deteriorate some during the flight, but again, I would not know exactly when. The one personal limitation I set for myself was to begin the session and hand-fly it continuously to a final landing at RFD, provided the weather remained above minimums. I made certain I had enough fuel on board to get back to Chicago if I needed it.

The takeoff — as well as much of the Arrow's performance throughout the rest of the flight — was agonizingly slow since I made the aircraft weight equivalent to maximum while I set the outside air temperature to 96 degrees. During the en route phase, I tried some more steep turns, slow flight, and stalls for proficiency. I continued on airways to RFD where I cleared myself to the NDB for the full ILS 1 approach. If you haven't played with the NDB in your airplane or a simulator recently, this is the place to shake out the cobwebs. Many prehire simulator checks like to add in an NDB approach as a starting point simply because they know that a sharp pilot who performs well here, will most likely do well on the ILS or VOR approaches.

I saw the runway at minimums on the first approach, but executed a low approach and flew the published miss out for a few turns around the holding pattern before shooting a B/C 19. At minimums on the back course, I saw nothing, so I went around again and headed back for an NDB 1 approach. I was pretty sure I would see nothing at minimums, but the variability of the simulator's cloud layering gave me just enough of a glimpse of the runway that I knew it was there, but not enough to land. I missed again and vectored myself to the RFD VOR (thanks to a quick toggle to the overhead map for a heading) and set up for a full ILS 7. Thankfully, I landed to a full stop on this one. Total time from takeoff to landing was an hour and twenty-five minutes. By the time I applied the parking brake at RFD and took a deep breath, my hands were sweaty. It had been quite a realistic workout!

A major benefit of the PCATD is the system's evaluation function. During lunch a little later, I sat back down in front of my computer and punched up the instant replay. There, I was able to see just what my holding patterns, as well as all of my approaches, looked like from not only the overhead perspective, but also in a profile view. The PCATD even displayed my ability—or lack of it at times—to stay on the glide slope during my two ILS approaches.

My only word of caution to an IFR pilot using one of these devices is not to get too dependent on the fancy gadgets installed, like the autopilot and flight director, or even two nav systems. Let the simulator do its job by failing some things along the way. Better to find out how well you cope from the comfort of your desk than watching things fall apart when you're in the soup. You'll also never stay in touch with the "feel" of the simulator unless you hand-fly it quite a bit.

Do you need a PCATD? Any pilot who doesn't have one installed in their desktop computer is missing out on one of the most exciting and least expensive methods of staying IFR current, whether it is to prepare for a prehire simulator check or just to be ready for the real world of IFR flying.

Virtual Airlines

Into every pilot's life a little fun must fall—if you're going to remain sane during your hunt for just the right flying job—and here it is. Your opportunity to become an airline pilot without a written exam, without sweating through an interview or even worrying about which way to enter the hold when you miss the approach on the ILS 1 to Dulles in the simulator. Joining a virtual airline will deliver a great deal of fun and enjoyment, with very little aggravation. And best of all, joining a virtual airline will cost you nothing—virtually. But here to explain more about what a virtual airline is, how it works, and why you should consider not filling out one more application for the rest of the afternoon until you're tried one, is Sean Reilly, the dean of the virtual airline set.

Virtual Airlines 101: A Crash Course

By Sean "Crash" Reilly of WestWind Airlines

In the later part of the evening and occasionally into the wee hours of the morning, a hearty group of individuals—most of them seemingly rational, grown men—sit perched in front of computer monitors with sweaty palms tightly clenching flight yokes. Distant cries of, "Honey, come to bed" have long since fallen on deaf ears as, with razor-sharp concentration, these airmen skillfully guide their aircraft down glide slopes to airports across the world. The late night silence is shattered by loud screeches of rubber on runway immediately followed by the deafening whine of reverse engine thrusters and finally by sighs of relief from the flight deck. The equipment, crew and thousands of virtual passengers have safely arrived at their destinations. Just another routine day in the life of a pilot flying for a virtual airline.

Are you one of those people whose eyes are constantly cast skyward, watching aircraft pass overhead? Ever dreamed of flying the heavy iron or wondered what it would be like to be a pilot for an airline? Want to develop a greater understanding of, and/or appreciation for, real-world aviation? Looking for more structure from your flight simming than booting up to the trusty Cessna at Meigs and then having to figure out where you want to fly? If you answered yes to any of these questions, I highly recommend you consider pursuing a "virtual career" with a virtual airline.

Virtual Airlines (VAs) are not a new concept—they've been around for several years. Though I'm not certain of their exact origins, I believe their birthplace can be tracked to Flight Sim forums of online services like AOL or CompuServe. VAs were started by a small group of individuals with a passion for flight simulation, who decided to create the virtual equivalent to a real-world airline. They created a structure where flight sim enthusiasts could fly specified routes in aircraft other than the Cessna 182. With the introduction of a software program called Flight Shop by the now defunct BAO Software, it was possible for these enthusiasts to create custom designed aircraft of all types and colors. This further stimulated the growth of the virtual airline industry. Now it was possible to fly a much wider variety of aircraft not included in the default Flight Sim program.

While VAs based online grew in popularity and number, it really wasn't until the boom of the Internet that—if you'll pardon the pun—VAs really took off. As a conservative estimate, there are 100+ virtual airlines operating either on the Net or online today. They employ (virtually speaking) thousands—probably tens of thousands—of pilots of varying skill levels and aviation knowledge. These numbers will only increase as Microsoft continues to improve on its versions of Flight Simulator, as computer hardware improves, and as outstanding add-on software and programs designed to enhance the virtual flying experience (like Squawk Box and Pro-Controller which enable you to fly with real-time ATC!) are introduced. The future of flight simulation, and of the virtual airline industry, is extremely bright.

To back up a step, to give you a better sense of how a virtual airline operates, let's compare the structure of a typical VA to that of a real-world commercial airline. Let's say, one day you wake up and decide you want to pursue a career in aviation. The first thing you'll do is get over to your local airport and sign up for flying lessons. You'll begin in a small, single-engine aircraft and eventually work your way up to more complex, multiengine aircraft. During the process, you will earn various ratings. After you've accumulated the necessary hours and secured the required ratings (and shelled out tens of thousands of dollars in the process) you'd race down the personnel office of your favorite airline, fill out an application and—with some luck—be issued a polyester uniform and be hired on as an entry-level pilot. Most entry-level pilots begin by flying twin turboprops.

Switching to the virtual world, this is the jump-in point for most virtual airlines. Most assume you have secured the necessary training and accumulated the minimum number of hours to step into a twin turboprop aircraft such as a Beech 1900D or Brazilia. Your first step, once you decide which VA you want to fly for, is to log on to their Web site and follow the links to the pilot application form. Fill out the form and wait for a letter of acceptance. Once accepted, many VAs will require you to take basic training in their entry-level aircraft. Training varies from one VA to another. If you are looking to increase your knowledge of, and appreciation for flying, I highly recommend you select a VA with the most realistic training program possible. A few VAs, including the one I fly for (www.flywestwind.com) , offer training programs crafted by real-world certified flight instructors (CFIs). Some even go so far as to provide downloadable, custom-designed training center scenery and training adventure files! Our training file introduces some "hairy" weather conditions that makes for quite a landing challenge.

Much of what you learn in a virtual training program will mirror what you would actually learn in a nonvirtual program. The more realistic the program, the better. If you survive the training exercises and return the training aircraft in the same condition you received it in (wheels round on all sides, no scratches on the bottom, wings attached and props unbent) you are issued virtual wings and your career begins. Congratulations and welcome aboard! An important caveat for those who may be wondering—there is no charge to fly for any of the VAs I'm familiar with. Even the virtual jetfuel is free!

Okay back to the real-world for a moment—when you fly for an airline, you fly aircraft painted in your airline's colors. In the virtual world, VAs offer a livery of aircraft—ranging from small turboprops all the way up to heavies like the Boeing 747 and 777—painted in their unique colors. When choosing a VA, I recommend you not only consider the fancy paint job of the fleet, but also how well each aircraft actually flies. Better VAs offer both attractive color schemes and highly accurate flight models of every aircraft in their fleet. Some VAs also offer pilot operating handbooks (POHs) for each aircraft. In case you are wondering, VA aircraft can be downloaded, free of charge, from each VA's respective Web site. The process is not complicated nor is the process of loading those aircraft into your Flight Simulator program. Most aircraft come with readme.txt files that walk you through the process. Some VAs, including ours, offer self-installing aircraft files for maximum pilot convenience. A real plus.

In the real world, commercial airlines fly predetermined regional, domestic and/or international routes. Larger airlines fly to destinations throughout the world using a wide variety of aircraft while smaller airlines may fly only regional or domestic routes with a limited fleet. It goes without saying that the types of aircraft each airline files is determined by the nature of the routes they fly. The same applies to the virtual world. Some VAs concentrate solely on regional or domestic routes with few aircraft types while others maintain diverse fleets of aircraft that fly a multitude of routes to destinations throughout the entire virtual world. When looking for a VA, consider the type of flying you most enjoy (short hops, long hauls, or a combination) and the type of aircraft you like to fly. VAs with small fleets concentrating on regional routes can be every bit as fun to fly for as larger VAs with diverse fleets flying a full-blown assortment of routes. It is a matter of personal taste. There is even one VA I know of that flies only classic (vintage) airliners. Quite specialized and clever.

One area of confusion, for those unfamiliar with the VA concept, that should be touched on is where the flying actually takes place. VA Web sites are similar to FBOs at your local airport. This is "home base" where you come to sign up, check out (or download) aircraft, scenery, adventure files, and other add-ons. In addition, training materials, POHs, etc., can be read and printed out from the Web site. You can also communicate with other members of the VA from the Web site. Some VA sites offer bulletin boards or forums for their members. All flying is done using a Flight Simulator program (99.9% are of the Microsoft variety). To be clear, you don't fly from within the Web site of a VA. The world you fly in is determined by the Flight Sim software you are using. (Fig. 8-5.)

A word about career advancement. In the real-world, most entry level pilots flying twin turbos want to rapidly log as many hours as they can in order to advance to aircraft of more substance—the latter promises a pilot greater challenge and a far more attractive salary. Career advancement in most VAs works much the same way. You begin your career in smaller aircraft, build hours (which you should log in your Flight Sim log book), and advance in aircraft type-ratings over time. Most VAs require you to, on regular intervals, report your hours to management. This is usually done via an online PIREP (pilot

Figure 8-5. Westwind Airlines Virtual Fleet.

report) form at your VA's Web site, or by e-mail. Most VAs have an "honor system" for flight time reporting. A few require you to submit your Flight Sim log book from time to time. As you accumulate hours, you are promoted in aircraft type and gain increased bragging rights. Sorry to say, however, that in the virtual world, salaries are the same for small turboprops as they are for triple sevens—zip, nada, zero. But consider the bright side—you didn't need to spend tens of thousands of dollars for flight training and you don't need to wear those polyester uniforms.

A few other things to consider as you search for a virtual airline—first off, please take time to look at many VAs before signing on with one. Download and test-fly various aircraft from at least a few VAs to see if they fly as good as they look. Look critically at the management structure/team of a VA—is it well-organized? Do the managers seem knowledgeable? Consider the longevity of the airline. Many VAs disappear as fast as they appear, which is a real frustration for pilots who then have to sign on with another VA and, in many cases, begin their career at the bottom again. Take a good look at the training program, the route structure, the minimum requirements to maintain active pilot status. Are they to your liking? You might also look critically at the Web site of any VA you are considering. Look for depth and organization of content. Does the site offer interesting and informative information? Does it offer a "pilot help" section? Has it been updated recently or is the content dated? Does the site contain a pilot and/or visitor comments page? If it does, what are people saying there? Does the VA promote and support real-world aviation? Ours is a proud sponsor of GA Team 2000's "Stop Dreaming/Start Flying" campaign to increase real-world general aviation pilot starts. Has the site and/or airline received any awards and, if so, for what and by whom? Another great resource to consider are the flight sim magazines—like this one. Has the VA received any favorable press? Does the VA offer specialty divisions such as a Cargo or Charter Division? Does the VA host special events such as fly-ins? While you might not want to choose one VA over another for any single thing listed above, you should consider the bigger picture—a combination of the above—as an indicator of how well-run and stable a VA is, and how seriously the management team takes their operation. Just because it is a virtual operation doesn't mean it shouldn't be a professional one.

Another thought regarding career advancement in the virtual world—if you would like to do more than just fly for a virtual airline, after accumulating some hours and experience with your VA, inquire about a management position within the organization. The single most important ingredient of any successful virtual airline—large or small—is its management team. It takes more than a few talented people to run a successful virtual airline—it takes many, all working in sync towards a common goal. A virtual airline is only as good as the sum of its leaders. VAs are always in need of creative and talented individuals to fill positions such as hub managers, aircraft and scenery designers, personnel directors, route creators, training directors, special project directors, Web designers, and others. Though the pay isn't great—"virtually nothing," in fact—a management career in a virtual airline can be highly gratifying.

Perhaps the most enjoyable thing I have found about flying for a virtual airline is the creativity and camaraderie that exists between our members—and the flight sim community as a whole. While our airmen come from literally all over the world, are of different ages, cultural backgrounds, etc., we all have one thing very much in common—a passion for flying, be it simulated, real-world or both. The sense of "community" that exists within the structure of a successful virtual airline, coupled with the common interest we share, is second-to-none. Consider a career with a virtual airline—it's a blast!

An additional list of Virtual Airline and Flight Sim resources is available in Appendix C

Sean "Crash" Reilly is the "Virtual" Executive VP of Marketing & New Biz Development and co-founder of WestWind Airlines. He can be reached by e-mail at sean.r@ix.netcom.com or via the WestWind Airlines web site at http://www.flywestwind.com.

Web Sites No Pilot Can Do Without

One of the reasons you bought this book was as a resource. Since I've been talking about how important the new electronic community is to your job search, this review of some unique aviation Web sites should help you on your way toward finding the ideal job. If you have not had the fun of whiling away an hour here and there surfing aviation Web sites, this list— as well as the additional lists stored on most sites themselves—should prove valuable. But be careful. Sometimes you can start clicking on one link after another and lose track of where you are—or where you were— before you know it.

This section includes just a few of the Web sites I've looked at that have value. But they are hardly all of the sites available. While running a search on the Alta Vista Search engine, I typed in the keywords "aviation" and "jobs" and had 170 Web site addresses returned to me. Let me know if you find some additional aviation sites that prove valuable and I'll include it in the next edition of this guide. E-mail the site address to me at: rob@mark-comm.com.

You can also track me down via this book's Web site: www.ufly.com/pilotcareer

A word of caution, too—Web sites come and Web sites go. Many contain information that is difficult to independently verify since a Web site might be run from someone's basement or an office atop the World Trade Center and you'd never know which it was. Information found on the Web should normally be checked with similar or contrasting viewpoints from other sources before you make any career decisions. Prior to sending money or offering a credit card to any company on the Internet, call their telephone number and talk to a live person. Some, of course, don't offer a phone option. But a reputable company should quickly answer any e-mails sent their way. If not, pass them by. This message posted on one Web site I visited explained Web site concerns perfectly_ "Disclaimer: This database may not be complete or up-to-date. Do not rely on it in any way."

http://www.smilinjack.com/—Smilin Jack Site—Interesting little page produced by a guy who liked the Smilin' Jack cartoon from years ago. It contains one of the largest lists of worldwide airline Web sites around— about 150—from all points on the earth. Also an extensive list of weather related Web sites and even a long list of airport Web sites from the United States, Europe, and the Far East. Great place to add research info before the interview.

http://AIR-online.com/contents.shtml—Aviation Internet Resources— Since all pilots love to venture forth with an opinion on just about everything, you'll find this site of immense help. Here you'll find—in addition to a very large list of aviation news—at linked sites like Yahoo's aviation news site—a chance to interact with other aviation professionals and enthusiasts about the airline industry specifically, airports, and aircraft related items. Best of all this site includes searchable Web databases of airlines, airports, and aviation related Web sites. Includes http://AIR-online.com/goto/Yahoo.shtml

http://www.propilot.com/—ProPilot Site—OK, I'll tell you that this site grabbed me at first because the first photo is the cockpit of an airplane I spent many hours in—an EMB-120 Brasilia. But I also found this a place with a wide range of divergent opinions on issues, from a stunning attack on pay-for-training, to an active chat board for students, to a real-time chat room for pro pilots and even a link to the TRACONs at MIA, ORD, and DFW that you can listen to the ATC chatter from via Real-Audio.

http://dir.yahoo.com/Science/Aviation_and_Aeronautics/—Yahoo Aviation—I like Yahoo because this site is a great jumping off point to dozens more information sources on aviation organizations, history, aerospace companies, and women's aviation sites.

http://www.atlanticcoast.com/docs/index.html—Atlantic Coast Airlines—This is an example of an easy-to-read, well-organized site. ACA is always on the lookout for pilots. Here you'd probably find enough basic info to get through an interview and sound rather well informed. A good bet might also be to take some of this information and ask a few more questions online.

http://www.ufly.com/—The Student Pilot Network—For the aspiring pilot or for those who are just beginning to think about a professional career or for those who are well on their way. This network page offers a host of valuable resources and should be one you bookmark and return to often. On one visit I found a great link to an aviation scholarship text that had just been published, as well as another to Greg Brown's *Turbine Pilot* book for those of you who are flying piston airplanes around wondering how a turbine engine even works. There's also a searchable database of flight schools and downloadable interviews with professional pilots in various corners of the industry.

http://www.ftmag.com/homepage2.htm—AOPA *Flight Training* Magazine—If a good pilot is always learning, they need a magazine focused on just those concerns. *Flight Training* does that. In FT mags' own words, they are "Aviation's How-To, Back-To-Basics magazine for new pilots, their instructors and those who own and operate flight training schools. Many experienced pilots also find *Flight Training* magazine's monthly issues provide an excellent up-to-date review of aviation training information and piloting techniques.

http://interactive.wsj.com/pj/personal.cgi—The *Wall Street Journal* Interactive Edition—If you're in search of work as a pilot, you already know I recommend the *Wall Street Journal* as an important source of up-to-date information about the airline or other company you might be interviewing with. But for just $29.95 extra per year, WSJ print subscribers can build their own page—I've made mine the home page I log into each day—and track anything they find of interest by adding as many keywords to their profile. Then, when you log on each day, the journal will bring you stories that match your keywords. Want to know what the Midwest Express, Continental, and Coca-Cola are doing before your airline or corporate interview? Find out here. Then print out the piece and save it to your folder for future reference. This site is a must-have!!

http://www.pilotslounge.com/—The Pilots Lounge—This site features aviation job listings, an aviation message board, and an online pilot supply store.

http://www.pilotshop-usa.com/index.html—The Pilot Shop—A Midwest-based pilot supply shop that offers great products at great prices as well as an online ordering form.

http://www.wso.net/iads/wwwboard/wwwboard.html—The Crashpads.com Message Board—Crashpad is the term used by pilots who need a place to live close to their airline base. Kind of a nifty little resource that offers—in addition to a list of crashpads all over the United States—a nice message board, a great list of aviation Web sites and even a link to stock quote—to help answer that often asked question at an interview.

http://www.geocities.com/CapeCanaveral/4285/index.html—Air Nemo—Good source of airline sites, as well as a breakdown of just about every two and three-letter company code around the world. Finally, you'll be able to learn what "EIN" stands for—Aer Lingus actually. Also a good set of statistical aircraft information, as well as aviation news site links. Some of the graphics here do get a little gaudy!!

http://www.avweb.com/—AV Web site—If you can get past the awful colors and the banner ads spread all over the place, there's actually a lot of good information here. A number of searchable databases, as well as some well-written, informative aviation columns. I missed the FAA Administrator's talk at the recent AOPA convention, but was able to download the speech and replay it on your PC via Real Audio from here.

http://www.landings.com/—The Landings—There's just lots and lots of stuff here—regulations, airworthiness directives, NTSB briefs, the Airman's Information Manual, and a listing of aircraft country prefixes. Now when you see an aircraft whose tail number begins with YV, you'll know they're from Venezuela. Also includes a solid list of aviation search engines.

http://www.airnav.com/—Air Nav—provides free detailed aeronautical information on airports and navigational aids in the USA. Type ORD in the navaid database and you'll learn more than you probably ever wanted to about the VOR/DME that serves the world's busiest airport.

http://www.faa.gov/—The Federal Aviation Administration site—Besides the career database that pulled up a recently announced GS-14 pilot position (salary $66,138 to $85,978) the site also houses considerable information about some of FAA's major components—Air Traffic Service, Regulation and Certification, Airports and Security. Also includes links to tons of FAA statistical information.

http://www.mcgraw-hill.com/aviation/index.html—McGraw-Hill Aviation with links to Aviation Week, the World Aviation Directory, and Business and Commercial Aviation magazines.

http://www.itn.net/cgi/get?itn/cb/aow/index:XX-AIRLINES—Internet Travel Network site—Normally, I wouldn't worry about such an airline passenger focused site, but this one contains a great list of aviation newsgroups that you can subscribe to and stay abreast of insider's opinions.

http://www.af.mil/—Air Force Link—This is the official site of the U.S. Air Force and a jumping off place to tons of information about places you might see if you join the Air Force, like OTS—Officer Training School.

http://www.nbaa.org/—National Business Aircraft Association—This site does not post job information, but if your goal is a career as a corporate pilot, this is a source of information about what is happening in this segment of the industry. The information, however, is somewhat limited if you are not an NBAA member.

http://www.aviation.org/—The Aviation Safety Connection site—This is a site I would not pass up. Focused on human factors issues in flight safety, this site offers an informative online newsletter—Cockpit Leadership, as well as access to an online discussion forum about safety issues. It also includes analysis of NTSB (National Transportation Safety Board) and ASRS (Aviation Safety Reporting System)reports.

http://www.aopa.org/—Aircraft Owners and Pilots Association—Although AOPA is not necessarily a career focused organization, the site does offer terrific message boards for aviators of all experience levels, a well-stocked library, as well as impressive database access. If you're not an AOPA member—join. It's the best $39 per year you'll ever spend.

http://www.atwonline.com—Air Transport World magazine—ATW is a business magazine focused on airline issues from the majors to the regionals, but all from a management standpoint. Not a great deal of original content here, but it does offer you a chance to search back issues of the magazine for information.

http://aviationweekly.com/—Aviation Weekly—This is an actual radio show format brought to you over the Internet, so you'll need Real Audio and a fast modem to make it all happen. But you can visit on Friday nights and spend a couple of hours with two wild and crazy guys who are plane nuts. They love everything about aviation and will talk about anything. Listeners are invited to call in. Visit the site and pick out an archived show to listen to when you have time. It's worth it.

http://www.bts.gov/oai/—The U.S. DOT's Office of Airline Information—More statistical info than you'll ever have time to digest. Choose from such show stoppers as, On-Time Statistics, FAA Statistical Handbook of Aviation, or Sources of Air Carrier Data. Seriously worth it.

http://www.air-transport.org/—Air Transport Association—This is the trade organization that represents most of the major airlines in the United States. You'll find policy speeches here, as well as much airline statistical information.

http://www.2010.atmos.uiuc.edu/(gh)/guides/mtr/home.rxml—University of Illinois Online Meteorology Guides—These are pretty nifty online workbooks about all facets of weather, including a great section on severe storms, weather forecasting, cloud types, and air mass explanations.

http://www.beapilot.com/—Stop Dreaming, Start Flying—This is the official site of the new Learn to Fly program sponsored by most of the major aviation organizations in the country. Great information source for those new to flying. The site includes great articles, such as Susan Paul's "Twenty Life Lessons I Learned From Becoming a Pilot," that makes great reading no matter what your experience level.

http://www.corporatepilot.com—Corporate Pilot. Com—Designed to help business and corporate aviation flight departments and pilots find each other for full-time employment or just more contract work. Some actual job listings posted here.

http://www.AviationWeb.com/index.htm—Aviation Web—This site offers a large searchable database of aviation training facilities around the U.S. Search for schools by name, state, or even zipcode.

http://www.airforce.com/—The U.S. Air Force Career Forum—This site is organized very much like an actual air force base and offers career information to various people in addition to pilots. Archived version of online career seminar held in late 1998 is worth reviewing. A little light on information, but worth reviewing if you're considering the Air Force as an option.

http://www.awgnet.com/bca/—Business and Commercial Aviation magazine—Some interesting stories from each month's issue, as well as regular columnists.

http://www.bizjet.com/iban/default.html—Business Air News—regular business aviation focused news stories, as well as a searchable database of back issues.

http://www.aerospace.bombardier.com/—Bombardier Aerospace—Company site that includes lots of information about Bombardier regional and corporate aircraft in both photos and news stories.

http://www.jeppesen.com/—Jeppesen corporate site—Information on lots of Jeppesen training materials, not the least of which is info on the FS-200 flight simulator reviewed elsewhere in this book.

http://www.asa2fly.com/asa—Aviation Supply & Academics—This is the site from which you can download a demo copy of the company's PCATD software, as well as view a catalog of many other useful aviation books and supplies.

http://www.aviationjobsonline.com/—Aviation Jobs Online—Tough to review this site because you must pay the fee to gain access to the actual job listings. Might be worth trying for a month or two to learn more. Does include a busy message board system for free.

http://findapilot.com/—Find A Pilot—Posts lots of job openings and also offers pilots an opportunity to post their resumé online. Resumés are only on a list however and are not searchable through any special parameter by a potential employer.

http://www.boeing.com/—The Boeing Corporate site—Information on Boeing products and lots and lots of great downloadable photos as well as a great screensaver of Boeing aircraft.

http://www.airbus.com/—The Airbus Corporate site—Airbus aircraft family site with information and photos.

http://www.nationjob.com/aviation—Nation Job's Aviation—Various kinds of aviation jobs from major companies here, but only a few for pilots. But you only need one, right?

http://www.flyingjobs.com/—ATP (Airline Transport Professionals)—Totally commercial, but a good source of the information you may need to get through some of the written knowledge exams. Also includes short explanations of their various certificate programs such as the multiengine rating or CFI training.

http://www.weather.com/aviation/—The Weather Channel Aviation site—The title says it all. It's a gotta have.

http://www.iccweb.com/federal/fedjobs2.htm—Federal Jobs list site—This is a pay for information site about federal jobs. May contain information on pilot positions, but was not verified at press time.

http://www.webring.org/ringworld/rec/aviation.html—The Web Ring—There's just no way to describe The Ring, except to say you must try it. This is a continual link from one aviation site to another. There were simply too many linked together to review. Let me know—via e-mail—which ones you find really useful.

http://www.aeroimages.com/—Aero Images company—You want aircraft photos, these people have photos—commercial, corporate, military, you name it. You can download any of the images for free. Copies that do not have the Aero Images logo printed on them will cost you—but not much. I found my long sought-after Dornier 228 photo here. This is a must bookmark site.

http://www.airnet.com/index.asp—Air Net Corporate site—If you enjoyed the story (elsewhere in this book) visit the site and learn about up-to-date hiring information for pilots.

http://www.microwings.com/—Micro Wings, the International Association of Aerospace Simulations—Not particularly attractive graphically, but if you've turned into a flight sim crazie while you search for that dream flying job, you need to bookmark this site. You'll find software reviews, links to other sites—like the flywestwind.com mentioned elsewhere—and even links to software add-ons for your simulator. Take a break from your job search for just a little while and have some fun.

http://www.polarisms.com—Polaris Microsystems—This is the corporate site for the company that developed and sells the electronic pilot logbook—Aerolog for Windows. The site includes a downloadable demo of their electronic logbook program for evaluation.

http://www.ntsb.gov/—National Transportation Safety Board—Official site of the NTSB and simply loaded with information of interest to pilots. Search a database of some 41,000 aircraft accidents reports to read and learn from. Also a plenty of aviation statistics that might be valuable during an interview.

http://www.wiai.org/—Women in Aviation International—Women in Aviation is dedicated to the encouragement and advancement of women in all aviation career fields and interests. Women in Aviation provides year-round resources to assist women in aviation and to encourage young women to consider aviation as a career.

http://www.trade-a-plane.com/—Trade A Plane—This is the yellow tabloid of just about anything anyone could ever want in aviation, from aircraft parts to pilot jobs. It contains a searchable classified site for employment that will set you back $2.95 per month, or $29 if you purchase a year's worth.

http://www.pilotwait.com/—Professional Pilots Wait Time site—Ever wonder where to eat when you're on a layover? This site lists dozens of restaurants and hotels that have been reviewed by pilots for level of comfort and service as well as the relative cost.

http://www.raa.org/—The Regional Airline Association site—This site include photos and stats on regional airline aircraft, as well as the latest details about how the various companies in this segment of the industry are performing.

http://www.aerolink.com/—Aerolink.com—In their own words, "Aerolink is an aviation specific farm. No eye candy, no page counters and no propaganda. Nothing to crash your browser. Instead, just links. Probably more links than any other aviation Web site." At last count, Aerolink listed over 7000 aviation-related sites.

Employment: (Reviewed elsewhere in the book)

http://www.aeps.com/aeps/aepshm.html—Airline Employee Placement Service

http://www.airapps.com//index.html—Air Inc

http://www.fltops.com/—Fltops.com site

http://www.pilotswanted.com/—Berliner/Schafer Aviation Consulting Group

http://www.upas.com/—Universal Pilot Application Service

http://www.aeps.com/—Airline Employee Personnel Service

http://www.avcrew.com—AvCrew.com site

Official Union Web Sites

http://www.alpa.org/—The National Air Line Pilots Association site

http://www.alliedpilots.org/—The Allied Pilots Association (American Airlines Pilots)

http://www.fedexpilots.org/—Federal Express Pilots Association

http://www.iacp.org/—Independent Association of Continental Pilots

http://www.ipapilot.org/—Independent Pilots Association (UPS Pilots)

http://www.dalpa-lax.org/—Delta's ALPA site for Los Angeles base

http://www.awalpa.org/—America West's ALPA site

http://www.usairwayspilots.org/homepage.htm—USAirways ALPA site

Simulator Training

http://www.simulator.com/—Simcom Training Centers site

http://www.flightsafety.com/—Flight Safety site with searchable database for aircraft type and training locations

http://www.simuflite.com/—Simuflight Training International site

Aeronautical Academia

http://cid.unomaha.edu/~unoai/uaa.html—The University Aviation Association

http://www.aero.und.edu/Academics/Aviation/index.html—University of North Dakota, Aviation Department

http://www.ftmag.com/collegiatedirectory.htm—*Flight Training* magazine's Collegiate Flight Directory

http://www.aviation.uiuc.edu/—University of Illinois at Champaign, Institute of Aviation

http://www.tech.purdue.edu/at/—Purdue University, Aviation Technology

http://raptor.db.erau.edu/—Embry-Riddle University
http://comairacademy.com/—Comair Aviation Academy
http://www.spartanaero.com/—Spartan School of Aeronautics
http://www.amerflyers.com/—American Flyers

So there you have a glimpse at what's happening electronically in aviation. But again, this is just that a glimpse. Spend some time and you're bound to find other tools for your PC that will be of value to you in your job search. Be sure and share them with the rest of us.

The End ... or Just the Beginning?

So here you are at the end or, hopefully, the beginning (Fig. 8-4). You've had a glimpse of what lies in store for you if you decide that a career as a professional pilot is where you want to be. I've spoken about some of the jobs, as well as how to find them. I've spoken about the ratings you'll need and how to pick them up. I've discussed the schools available and how to finance your career. Best of all, I've spoken with people who have made it—people who conceived a plan and followed through on it until they reached their goal. A plan is very important, but the most important part is you must keep moving towards your goal. Don't let anything or anyone get in the way (Fig. 8-5). If someone says no, realize that means no ... today. Tomorrow it could be a maybe. If you try something and you fail along the way, realize that you really didn't *fail*. You learned something. You learned what doesn't work. If you learned something, you didn't fail. Remember, you never fail until you stop trying.

Here are some final words from Lt. Michael Fick from my interview with him about the Air Force. These thoughts sum up a great many of the feelings of the other pilots I've spoken to. I asked him how much he really enjoyed what he was doing. He said, "You can't beat getting paid to fly. There's no other fun like it." Notice he didn't use the word "work" in that sentence. Flying is hardly like work (Fig. 8-6). "The real work comes in those years when you're learning to become a rated pilot."

Computers are a must have if you want to stay on top of the latest information about who is hiring and how to connect with them. I hope you'll keep me abreast of your progress. To make that a little easier, I'd like to again offer my e-mail and Web site address for this book. Please do make it a point to let me know what portions of this book worked well for you and which ones might be improved. If for no other reason, e-mail me when you get that next job.

This book's Web site URL:www.ufly.com/pilotcareer

My e-mail address is: rob@mark-comm.com

Figure 8-6. 727 at STL.

EMB-120 Brasilia, Flight Crew Standardization Manual

Training and Checkrides

In an effort to provide more information and a better understanding of what to expect and what is expected of you during a checkride, we have put together the following information:

FAR Part 135 requires the following tests and checks. The following is a brief summary of each.

Federal Aviation Regulation for 135.293—Initial and Recurrent Pilot Testing Requirements.

This is for both PIC and SIC pilots and is required once a year. There are two parts to this test, a written or an oral test on the pilot's knowledge in the following areas:

FARs 61, 91, and 135, company operations specs. and the Operations Manual.

Aircraft, powerplant, major components and systems performance and operating limitations, standard and emergency procedures, contents of the approved Flight Manual.

Weight and balance limitations.

Use of navigation aids.

ATC procedures.

Meteorology.

Consists of a competency flight check given by the Administrator or check pilot in that class and type of aircraft.

The competency check may include any of the maneuvers and procedures currently required for the original issuance of the particular pilot certificate as required for the operations authorized and appropriate to the category, class, and type aircraft involved.

The instrument proficiency check required by 135.297 may be substituted for the competency check.

135.299 Pilot-in-Command: Line Checks: Routes and Airports

A. This check is required once a year.

   The flight check shall:

   Be given by an approved check pilot or by the Administrator.

   Consist of at least one flight over one route segment, and

   Include takeoffs and landings at one or more representative airports. In addition, for a pilot authorized to conduct IFR operations, at least one flight shall be flown over a Civil Airway, an approved Off-Airway route, or a portion of either of them.

135.301 Crewmember: Tests and Checks, Grace Provisions, Training to Accepted Standard

If a crewmember who is required to take a test or a flight check, completes the test or flight check in the calendar month before or after the calendar month in which it is required, that crewmember is considered to have completed the test or check in the calendar month in which it is required.

B. If a pilot being checked under this subpart fails any of the required maneuvers, the person giving the check may give additional training to the pilot during the course of the check and may require the pilot being checked to repeat any other maneuvers that are necessary to determine the pilot's proficiency. If the pilot being checked is unable to demonstrate satisfactory performance to the person who is conducting the check, the certificate holder may not use the pilot, nor may the pilot serve as a flight crewmember in operations under this part until the pilot has satisfactorily completed the check.

135.297 Instrument Proficiency Check Requirements

This check is for PlC only and is required every six months.

You are required to satisfactorily demonstrate at least one instrument approach procedure using an ILS, a VOR, and an NDB facility. These approaches must include at least one straight-in approach, one circling approach in conjunction with a VOR or an NDB, and one missed approach. Each approach demonstrated must be conducted to the published minimum for that procedure.

The instrument proficiency check consists of an oral or written equipment test and a flight check under simulated or actual IFR conditions.

The flight check includes navigation by instruments, recovery from simulated emergencies, and standard instrument approaches involving navigational facilities which that pilot is to be authorized to use.

The instrument proficiency check must, for a pilot-in-command of an airplane under 135-234A, include the procedures and maneuvers for an airline transport pilot certificate in that particular type of airplane.

If the pilot-in-command is assigned to pilot more than one type of aircraft that pilot must take the instrument proficiency check in each type of aircraft to which that pilot is assigned in rotation.

Note: All flight maneuvers involved in flight training and/or check-rides will be performed as outlined in the POH and prior to taking the aircraft. All maneuvers will be discussed with the pilot-in-training or flight check.

Flight Check

Evaluation will be based on your knowledge and understanding of the airplane, systems, components, and knowledge of its weight and balance limitations.

Preflight

Equipment Exam (oral or written)

Subjects requiring practical knowledge of the aircraft, its powerplants, systems, components, and operational performance factors.

Normal, abnormal, and emergency procedures and limitations related thereto.

Evaluation will be based on the accuracy of your explanation of the operational purpose of each item and the thoroughness of your inspection.

Preflight Inspection

You may be required to conduct a Visual inspection of the exterior and interior of the airplane, locating each item in the preflight checklist contained in the normal procedures section of the AFM, and briefly explaining the purpose of inspecting it.

You should demonstrate the use of the prestart checklist, appropriate control system checks, starting procedures, radio, and electronic equipment checks.

Taxiing

Evaluation will be based on safety, proficiency in handling the airplane, consideration for other aircraft personnel on ramps, taxiways, proper use of power, brakes, flight controls, and nose wheel steering.

Takeoffs

You may be required to properly demonstrate any or all of the following types of takeoffs to determine that you are competent on performing takeoffs under normal and emergency conditions and under various meteorological conditions, and that you can safely reject a takeoff when appropriate.

Normal—normal takeoff, which begins when the airplane is taxied into position in the runway to be used.

Instrument—(lower than standard takeoff)

Crosswind—(Up to the max limiting component)

Powerplant failure—(V1 cut) simulated failure of the most critical powerplant.

At a point after V1 and before V2 that, in the judgment of the examiner, is appropriate under the prevailing conditions.

Rejects—A normal takeoff which is rejected after reaching a reasonable speed considering runway length, surface conditions, wind direction and speed, brake heat energy, and any other factors that may adversely affect safety.

Instrument Procedures

Area Departure and Area Arrival

Objective is to determine that you can adhere to ATC departure and arrival clearances, including assigned radials, and proper use of naV1gation facilities.

Holding

You may be required to hold at various fixes during the checkride as determined by ATC or the examiner.

Evaluation will be based on entry procedure, compliance with the holding direction/radial, timing, and staying within the holding airspace.

You will be expected to maintain altitude within + or −100 feet of the assigned altitude and holding airspeed shall be maintained within + or −10 knots of the predetermined airspeed.

You may be required to perform the following:

At least one normal ILS approach with all engines operating.

At least one manually controlled ILS approach with a simulated failure of one powerplant. The simulated failure should occur before initiating the final approach course and continue to touch down or through the missed approach procedure.

At least two nonprecision approach procedures that are representative of the nonprecision approach procedures you are likely to use.

Acceptable Performance Guidelines

Altitude shall be maintained within + or −100 feet of prescribed altitude during initial approach, and within −0 to +50 feet of MDA or DH as appropriate.

Circling Approaches

You may be required to perform at least one circling approach under the following conditions:

The portion of the circling approach to the authorized minimum circling approach altitude should be made under simulated instrument conditions.

The approach should be made to the authorized minimum circling approach altitude followed by a change in heading and the necessary maneuvering required for maneuvering to the landing runway.

The circling approach should be performed without excessive maneuvering and without exceeding the normal operating limits of the airplane. The angle of bank should not exceed 30 degrees.

Acceptable Performance Guidelines

Altitude shall be maintained within −0 to +100 feet of MDA until the airplane is in a position from which a normal approach to a landing can be accomplished, and preferably until turning final.

Missed Approaches

You may be required to perform at least two missed approaches with at least one from ILS approach.

A simulated powerplant failure may be required during any of the missed approaches.

Descent below the MDA or DH prior to initiation of the missed approach procedure should be disqualifying.

In-Flight Maneuvers

Steep Turns

You may be required to perform at least one steep turn in each direction. Each steep turn should involve a bank of 45 degrees with a heading change of at least 180 degrees but not more than 360 degrees.

Acceptable Performance Guidelines

You will be expected to maintain altitude within + or −100 feet of the entry altitude and a bank angle of 45 degrees + or −5 degrees, and recover within + or −10 degrees of the assigned headings. Airspeed shall be controlled within + or −10 knots entry airspeed.

Approaches to Stalls

For the purpose of maneuver, the required approach to a stall is reached when there is a perceptible buffet or other response to the initial stall entry. At least three approaches to stalls can be required as follows:

One in the takeoff configuration

One in the clean configuration

One in the landing configuration

At the examiner's discretion, the performance of an approach to a stall may be required in one of the above configurations while in a turn with a bank angle between 15 degrees and 30 degrees.

Specific flight characteristics may be required:

Slow flight

Unusual attitudes

ETC

Powerplant Failures

In addition to the specific requirements for maneuvers with simulated power-plant failures, the examiner may require a simulated powerplant failure at any time during the checkride.

Landings and Approaches to Landings

You may be required to demonstrate the following landings:

Normal landing.

Landing in sequence from ILS.

Crosswind landing, if practical under existing meteorological, airport, and traffic conditions.

Maneuvering to a landing with simulated powerplant failure.

Landing under simulated circling approach conditions.

A rejected landing (including a normal missed approach procedure) that is rejected approximately 50 feet over the runway threshold. This maneuver may be combined with instrument, circling, or missed approach procedures; but instrument conditions need not be simulated below 100 feet above the runway.

A zero-flap visual approach to a point where in the judgment of the examiner, a landing to a full stop on the appropriate runway could be made.

V1. Normal and Abnormal Procedures

During the checkride you should fully understand and readily be able to look up and apply proper abnormal procedure as the examiner may simulate these conditions at any time during the ride.

Manual ground start

Ground start failure

Fuel system failure and low fuel management

Electrical system failures

Hydraulic system failures

Pitot static system failure

Compass system failure

Runaway trim

Windshield heat failures

Engine operation with abnormal conditions

Precautionary engine shutdown

Air start

Aborted air start

Single engine landing

Fire drills

Decompression

Emergency Procedures

You will be required to have an adequate knowledge of, and the ability to perform, emergency procedures appropriate to the situation, per flight manual procedure.

Normal Takeoff

The flying pilot will advance the power levers smoothly to 75% torque, and will call "Set Takeoff Power" by 50 KIAS. The nonflying pilot will acknowledge "Takeoff Power Set," and trim the power levers to the take-off setting.

Nonflying pilot calls "80 knots, engines normal, panel clear."

Nonflying pilot calls V1 then "Rotate at Vr." Flying pilot will rotate smoothly to 10 degrees pitch attitude. When safely airborne, the nonfly-ing pilot will call "Positive Rate," and the flying pilot will respond "Gear Up." The nonflying pilot will acknowledge and retract the gear.

At 400 feet AGL and V2 +20, the nonflying pilot will call "400 feet," the flying pilot will call "Flaps Up." The nonflying pilot will retract the flaps and acknowledge.

At 1000 feet AGL the NFP will call 1000 feet." The FP will call "Set Climb Power, After Takeoff Checks," and the NFP will acknowledge.

The standard climb rate will be reduced from 10 degrees to a max of 7 degrees pitch out of 1000' AGL. Target speed is 170 KIAS.

Rejected Takeoff (Practice)

For training, practice rejected takeoffs at a reasonable speed (60 kts) and using less than maximum braking, with consideration for the available runway, surface conditions, wind, etc. To emphasize the importance of using immediate full braking, use the following procedure when per-forming a rejected takeoff during simulated engine failure.

The instructor pulls the power lever of an engine (or other simulated emergency).

The trainee calls, Abort, and pulls the power levers to ground idle, while applying brakes lightly, simultaneously calling Maximum braking.

The trainee maintains aircraft steering with the rudder pedals. If neces-sary (in the left seat) the trainee may use the steering handle while the instructor holds the controls neutral.

If necessary, retard the power levers to full reverse.

Bring the aircraft to a complete stop, or clear the runway and come to a complete stop, and set the park brake.

After completing the relevant checks the instructor will terminate the simulated abnormality or emergency.

Note: The instructor may simulate other malfunctions to initiate an aborted takeoff.

Lower Than Standard Takeoff

To train for reduced visibility takeoffs, the following procedure will be used. A view limiting device will be placed over the window to simulate lower visibility requirements.

Align the aircraft with the center line of the runway. Allow enough forward motion of aircraft to straighten the nosewheel.

Adjust heading bug to the runway centerline, select altitude preselect, and check the magnetic compass. Recheck the standby attitude indicator against the EADI.

Adjust the reduced V1sibility deV1ce for RVR requirements.

Complete the takeoff checklist. While holding brakes, advance power to 40% torque.

Release brakes and continue with power advancement to 75% Tq. Call Set Takeoff Power. IP will set final power setting and call "Power Set." During initial takeoff roll, the PF will maintain visual contact with runway centerline through reduce visibility screen and maintain aircraft alignment with rudder pedals.

When the IP calls Vr, rotate to 10 degrees pitch up.

This procedure may take additional runway time so adV1se ATC accordingly. Proper execution of procedure will have the aircraft at or near centerline for takeoff run, a smooth transition to instruments at rotation and heading control within 5 degrees after takeoff.

Engine Failure at V1

This is a normal takeoff until the point of engine failure/fire, therefore, procedures are identical with normal takeoff through the point of gear retraction.

Flying pilot maintains directional control with rudder and rotates smoothly to approximately 10 degrees nose up. Gear is retracted with positive rate of climb.

Flying pilot establishes climb at V2. Flying pilot Calls "Engine Failure, Check Feather." (If not feathered or fire exists, perform Engine Fire memory items.)

At 400' AGL, the flying pilot levels off, accelerates to V2 +20 KIAS then commands "Flaps Up."

Establish a climb at VYSE and set maximum continuous power. Perform Engine Fire or Precautionary Engine Shutdown Checklist.

"After Takeoff" checklist is completed after the Precautionary Engine Shutdown checklist or Engine Fire checklist, as dictated by the failure condition.

Engine Failure Practice at Attitude

Before demonstrating engine failures on takeoff, the trainee should practice the procedures at altitude in the takeoff configuration. Generally, five thousand feet above the ground is considered a safe altitude. Once in takeoff configuration and at V1, the instructor simulates an engine failure by reducing the power lever to flight idle. The trainee should maintain heading, correcting initial roll and yaw with as much control deflection as needed. The aircraft should be stabilized at V2 while climbing at least 400 feet, then leveled off and allowed to accelerate to VYSE. The maneuver is complete after flap retraction and the trainee stabilizes the aircraft in a climb at VYSE. The trainee should make all callouts required on the Engine Failure/Fire on takeoff checklist.

Simulated Engine Failure on Takeoff at or Above V1

During initial training prior to the first V1 engine failure, the instructor will advise the trainee that he will be simulating an engine failure at V1.

Observe normal takeoff procedures until V1 is reached.

The instructor calls V1 then pulls the power lever to flight idle. The trainee then calls for maximum power. "Maximum power" indicates maximum continuous takeoff power is set.

At Vr the trainee rotates to 10 degrees pitch attitude.

When a positive rate of climb is established the trainee calls, "Gear Up," "Engine Failure, Check Feather." If the instructor simulates an autofeather failure or fire, the trainee will call for the appropriate actions on the Engine Failure/Fire on Takeoff checklist.

Establish in a climb V2 speed until reaching 400 ft. AGL minimum, then level off and accelerate to V2 +20 KIAS and call "Flaps Up."

Establish a climb at V2 +20 KIAS. The trainee then calls for the immediate action items if not previously accomplished and then the appropriate checklist.

At 1000' AGL, the trainee calls for the "After Takeoff Checklist."

Airwork

Familiarization

Handling characteristics are explored by turns at normal and low cruise speeds with and without the use of the flight director.

Demonstrate the yaw effect produced by power changes. The trainee should become familiar with the rudder trim requirements needed as torque is increased or decreased.

Extend flaps at maximum flap speeds and note the effect on pitch and trim requirements. Retract the flaps noting the pitch change required to maintain altitude.

Maneuvering at Minimum Control Airspeed

These maneuvers demonstrate the degree of controllability available while in close proximity to the prestall buffet. They provide the opportunity to practice control techniques that are most beneficial in the low-speed regimes encountered during takeoffs, landings, and powerplant failure emergency situations.

Maneuvering at minimum speed is practiced in both cruise and landing configurations, and will consist of straight flight, turns, climbs, and descents. Flight at minimum speed will be practiced within several knots of computed stall shaker speeds.

Acceptable performance parameters are:

Airspeed plus 5−0

Altitude plus or minus 100′

No unintentional stalls

Approaches to Stalls

These maneuvers afford familiarization with the airplane handling characteristics in the initial stall buffet region and provide training in stall recognition and proper recovery. Approaches to stalls should be practiced at a minimum altitude of 5000 feet above ground level. Approaches to stalls will be practiced in the takeoff, departure, and landing configurations. Approaches to stalls will be induced by smoothly increasing the angle of attack until the first indication that is usually the stick shaker. (Do not trim below 120 kts.) Recovery shall be initiated at the stick shaker. When recovering, the PF should advance power while calling max power, then place both hands on the flight controls. During recovery the instructor will simulate max power when max power is called for. This will avoid overtorquing the engines. Recovery should be made by holding pitch attitude at 7 degrees pitch max with coordinated flight control usage and zero change of altitude. Stalls will generally be accomplished in a series. Each stall will be considered complete as the airspeed accelerates to 140 knots.

At the instructor's discretion, at least one stall will be demonstrated with a bank angle of 15–30 degrees.

I. Takeoff Configuration
   A. Flaps <200 KIAS down

B. Gear <200 KIAS 15 degrees

C. Condition Levers—Maximum RPM

D. Power 10% Tq.

E. Bank 0–30 Degrees Max.

F. Altitude—Maintain

At the stall indication (stick shaker and stall horn):

A. Call for "Max Takeoff Power"

B. Level the Wings

C. Reduce Pitch Attitude to 7 Degrees Max.

D. Check Flaps 15 Degrees

E. At a Positive Rate of Climb (indicated in training by an increase in airspeed in level flight)

Call "Gear Up."

F. At V2 +20 KIAS, Call "Flaps Up."

G. At 140 KIAS call "Climb Power and After Takeoff Checks."

Acceptable Performance

Torque and T6 limits not exceeded.

No altitude loss on stall recovery.

Smooth entry and recovery technique.

No secondary stall.

Departure Configuration (Clean Stall)

A. Gear Up

B. Flaps Up

C. Condition Levers <200 KIAS Max. RPM

D. Power 10% Tq.

E. Brake 0–30 Degrees Maximum

F. Altitude Maintain

At the stall indication (stick shaker and stall horn):

A. Call for Max Takeoff Power

B. Level the Wings

C. Reduce Pitch Attitude to 7 Degrees Max.

D. Check Flaps Up

E. Check Gear Up

F. At 140 KIAS call Climb Power (and) "After Takeoff Checks"

Acceptable Performance

1. Torque and T6 limits not exceeded.

2. No altitude loss or secondary stall.

3. Proper control of aircraft throughout maneuver.

Landing Configuration

A. Gear <200 KIAS Down

B. Flaps <200 KIAS 15 Degrees

C. Condition Levers Max. RPM

D. Flaps (150 KIAS 25 Degrees)

E. Flaps (135 KIAS 45 Degrees)

F. Power 10% Tq.

G. Bank 0–30 Degrees Max.

H. Altitude Maintain

At the stall indication (stick shaker and stall horn):

A. Call for "Max Takeoff Power, Flaps 15"

B. Level the Wings

C. Reduce Pitch Altitude to 7 Degrees

D. At a Positive Rate of Climb (indicated in training by an increase in airspeed in Level Flight), call "Gear Up."

E. At V2 +20 KIAS call "Flaps Up"

F. At 140 KIAS call "Climb Power (and) After Takeoff Checks"

Acceptable Performance

1. Torque and T6 limits not exceeded.

2. Little or no altitude loss.

3. Smooth entry and recovery technique.

4. No secondary stall.

Steep Turns

This maneuver affords practice in controlling the airplane with greater-than-normal bank angles. Turns are in level flight with bank angles of 45 degrees continuing for 180 degrees or 360 degrees. Use 160 KIAS when practicing steep turns. Make two consecutive 360 degree turns, one in each direction. Do not pause at straight and level between turns.

Note: For 160 kt in level flight, approximately 35% TQ is required, and add approximately 8% during turn. Overcontrolling pitch attitude during steep turns can produce a buffet and stick shaker at speeds of 160 knots at training weights.

Acceptable Performance

1. Airspeed plus or minus 10 KIAS.

2. Altitude plus or minus 100'

3. Bank Angle + or −5 degrees

4. Recovery within 10 degrees of assigned heading.

Emergency Procedures Training

General

A captain trainee is trained for and expected to recognize, analyze, and correct malfunctions, abnormals, or emergencies while operating the aircraft. The captain must be able to maintain aircraft control and superV1se the flight crew through their duties while making correct, timely decisions. If the aircraft isn't in a critical phase of flight (takeoff, short final, go-around, etc.) the captain monitors and coordinates the other crewmembers while maintaining aircraft control, thereby assuring effective use of crew resources. Under the above concept, the flight crew carries out its emergency checklist actions using the Challenge and Response concept unless coordination is not required with a second crewmember. As always, the captain must ensure that crewmembers' actions are timely and correct.

First officer trainees are trained for and expected to perform actions per their condensed emergency checklists, expanded emergency checklists, or any appropriate emergency actions as directed by the captain.

For training purposes, the instructor sets up an emergency scenario. The trainee (captain or first officer) performs duties through simulation as the instructor leads him or her through the scenario. The instructor assumes the role of the other crewmember to maintain procedural continuity for the trainee.

Engine Shutdown and Relight

Engines will be secured as the emergency dictates. The Precautionary Engine Shutdown Checklist will be used. The Engine Air Start Checklist will be used to restart an engine.

Note: The minimum altitude to shut an engine down in flight for training/testing purposes is 5000 feet above the terrain.

Handling with One Engine Inoperative

With an engine in simulated feather, practice turns at normal single engine cruise speed and approach speeds.

Engine Fire Inflight

When an engine fire occurs in flight, aircraft control is the primary responsibility of the crew. To simulate an engine fire for training purposes, the instructor will press the Fire Warning Test and advise "FIRE ON THE LEFT/RIGHT ENGINE." The flying pilot will call "Condition Levers 100%, Set Max Power." If the airplane is in a configuration other than clean (except on an approach and depending on circumstances), the crew will proceed to clean it up per stall recovery procedure. Following this, the flying pilot will proceed with the Engine Fire Memory Items:

A. Power Lever (affected engine)—Flight Idle

B. Condition Lever—Feather and Check. If no Feathering occurs:

C. Electric Feather Switch—ON and Check

D. Condition Lever—Fuel Cut-Off

E. Fire Handle—Squeeze and Pull

F. Fuel Pumps—OFF

G. Agent A—Discharge

Engine Fire Inflight

Following completion of the memory items the flying pilot calls for the Engine Fire Inflight Checklist.

Emergency Descent

The instructor will call "rapid depressurization." The trainee will carry out the immediate action items:

Oxygen masks—Don, depending on altitude

The instructor will simulate an altitude above 10,000 ft. if not already above 10,000 ft.

Communications—Establish

Passenger oxygen—As required

Emergency Descent—Perform

The trainee will reduce the power levers to flight idle and call "condition levers maximum RPM." At 200 kts. call "gear down," roll into a 30-degree bank turn, and lower the nose to maintain the lowest of VMO or 200 kts. After established in a descent, the captain trainee should advise the PNF to advise ATC and squawk 7700. The trainee will then call for the Emergency Descent checklist. The instructor will advise the level-off altitude if below 10,000 ft.

Normal, Abnormal, Alternate, and Emergency Procedures

During training and proficiency checks, the pilot must be able to demonstrate the proper use of, and procedures for, as many systems and situations as the person conducting the training/checks finds necessary to determine adequate knowledge of and ability to perform such. During training sessions, all appropriate procedures/maneuvers must be covered. These normal, abnormal, and alternate procedures are detailed and explained in the Normal, Emergency, Abnormal, and individual systems sections of the Airplane Flight Manual.

Examples of such systems failures include (but are not limited to):

Engine: Fuel:

Low Oil Pressure   Low Pressure Light

Chip Detector Light   Fuel Pump Failure

EEC Light   Low Fuel Temperature

Electrical: Flight Control:

Generator Failures   Trim Runaway

Battery Overheat   Elevator/Aileron Jamming

Inverter Failures   Rudder Speed Switch Failure

Jammed Rudder

Flap Failure/Asymmetry

Hydraulics: Pressurization:

Loss of Green or Blue System   Decompression

Main Pump Loss   Duct Leaks

Landing Gear Failure Bleed Overheat

Anti-Skid Failure   Park Failures

NORMAL, ABNORMAL, ALTERNATE, AND EMERGENCY PROCEDURES

Navigation   Ice Protection

EADI   Windshield Overheat

EHSI   Leading Edge Deice Failures

RMI

LOC/GS

Judgment

Throughout the maneuvers, the flight crew must demonstrate judgment commensurate with a high level of safety. In determining whether such judgment has been demonstrated, consideration will be given to adherence to approved procedures and action in situations requiring a decision where there is not a prescribed or recommended procedure.

Captains are expected to maintain ATP standards at all times. First Officers are Captains-in-training and Captains are expected to assist them in the development of Captain qualifications. First Officers are not required or expected to perform to ATP standards; however, their performance should be to the point that they are the obvious master of the aircraft and the outcome of the maneuver or procedure is never in doubt. Captains shall know and recognize their own capabilities and limits and should never allow a first officer to continue beyond the point where that Captain cannot safely recover the aircraft.

Approach and Arrival Procedures

Area Arrival

The flight crew shall adhere to ATC clearances and properly use available navigation facilities and equipment.

General

Both pilots will review approach plates for:

Type of approach

Track/course—navigation aids/frequencies

Procedure turn/FAF/glide slope intercept altitudes

Decision altitude/minimum descent altitude

Determination of missed approach point, if applicable

Missed approach procedure

Callout Through

18,000 feet—set and crosscheck altimeters

10,000 feet—sterile cockpit

1000 feet prior to any assigned altitude

500 feet prior to assigned altitude

Verify aircraft at correct altitude(s) when established on the final approach segment. Altitude can be verified at intersections, FAF, and markers.

During flight director ILS approaches, confirm proper mode and G/S capture.

Call "1000" feet above DH or MDA Both pilots crosscheck instrument indications for presence of warning flags.

Call "500" feet above DH or MDA and each 100' increment thereafter.

Call any significant deviations from profile altitudes, airspeeds, or descent rates throughout the approach and touchdown or missed approach.

Call "Runway in Sight" when appropriate or call "DH" or "MDA" and "MAP" if runway environment is not in sight.

The flying pilot will execute a missed approach if:

At DH/MAP and runway not in sight

Landing is not authorized per FAR 91.116(c)

Holding Patterns

Holding may be accomplished utilizing any radio navigational aid (VOR station, VOR intersections, ADF tracks, localizers at compass locators, etc.). Important to any holding entry procedures are: proper heading, altitude, airspeed (slow down approximately 3 minutes before arrival at holding fix), proper entry into the pattern, and correct holding procedures within the prescribed airspace. Holding procedures shall be accomplished according to the criteria established in the Airman's Information Manual (AIM).

Normal Holding Configuration

Gear Up

Flaps Up

Np 85%

Bleeds Auto

Packs Low

160 KIAS (175 KIAS—Maximum)

Approach and Landing

Within 30 n.m. or fifteen minutes to landing, the flying pilot will call for the "Approach and Descent Checks." The cockpit is considered sterile below 10,000 feet MSL, and both pilots must be extremely vigilant upon arrivals in the terminal areas. Crews must remember that the max allowable airspeed below 10,000 feet is 250 KIAS, 250 KIAS within Class B, and 200 KIAS in the Class C & D and under a Class B floor. The only cockpit to cabin contact in the critical phase of flight is for airplane operational duties (i.e., notification of Flight Attendant on Approach and Descent Checks, etc.)

Instrument Approaches

Instrument approaches are accomplished using different types of radio aids. Each pilot must be familiar with the pictorial profiles for each approach type. The profiles depict a nearly ideal situation, a "textbook" approach. Often vectoring, the pilot's own navigation or a particular situation inserts the aircraft into the profile in other than an ideal position. The pilot must bring the aircraft to the correct configuration and airspeed for its position on the approach.

Each type of approach in the "real" environment requires study to determine where the gear and approach flaps should be extended, where the approach airspeed should be established, where timing commences and what time interval should be used.

As soon as the landing runway is known, all navigation radios that can be utilized will be tuned and identified.

Instrument Approaches

Monitor speed and rate of sink closely. Regardless of whether the approach is being made with raw information or the integrated instrument system, all instruments and indications must be continually cross-checked.

Note: Vref during instrument approaches refers to Vref for 25 degrees flaps.

ILS Approaches

Plan to intercept the localizer approximately 5NM from the outer marker at glide slope interception altitude and 150-160 KIAS. Extend the flaps to

15 degrees. At 1 dot below the glide slope, lower the gear, initiate the Landing Checklist, and select 25 degrees flaps. Stabilize the aircraft on the localizer and glide slope at Vref +10 KIAS.

Single Engine ILS

A single engine ILS is configured as a normal ILS. Airspeed should still be maintained at Vref +10 KIAS.

Nonprecision Approaches

Nonprecision approaches are similar to ILS approaches, except that the nonprecision approach lacks the electronic glide slope cues for vertical guidance. After the procedure turn or approximately 3 to 5 miles from the FAF extend the flaps to 15 degrees. A mile from the FAF or prior to intercepting the final approach segment, lower the gear, initiate the Initial Landing Checklist, and extend the flaps to 25 degrees. At FAF begin descent at Vref +10 KIAS and at a rate that will have the aircraft at MDA in time for a normal glide path to the runway. This normally requires a descent rate of 1000 fpm, but will need to be adjusted depending upon distance from the FAF to runway, altitude to be lost, winds aloft, etc. Descend from MDA on a normal glide path.

Single-Engine Nonprecision Approaches

The single engine nonprecision approach configuration is flaps 15 degrees and landing gear down. Single-engine nonprecision approaches require a high degree of planning and should be performed only if no safer procedure exists. The final approach segment and maneuvering will be flown gear down, flaps 15 degrees, and 130 KIAS. When landing is assured, command "Final Checklist, flaps 25 degrees." Descend from MDA on a normal glide path.

Missed Approach— Both Engines

If a missed approach must be made:

- Set go-around torque, flaps 15 degrees
- Rotate aircraft to attain V2 +10 KIAS (10 degrees max)
- Retract the gear with a positive rate of climb
- Proceed as with normal takeoff
- Notify ATC

Execute a missed approach when:

- Visual reference with the runway is insufficient to complete the landing
- A safe landing is not possible, or
- Instructed to do so

Single-Engine Missed Approach

If a missed approach must be made:

- Apply maximum power (110%/816 degrees C), flaps 15 degrees
- Rotate aircraft to attain V2 (10 degrees max.)
- Retract the gear with a positive rate of climb
- Proceed as with engine failure takeoff
- Notify ATC

Category I Approaches

Category I approaches are predicated on the use of raw data, flight directors, or approach coupler. All altitude references are made to the barometric altimeter.

Flight Guidance and Autopilot Integration

The modes of the flight guidance systems and autopilot should be coordinated for the purpose of continual cross-checking. During maneuvering to the final approach course, the HEADING knob sets a reference bug and selects the heading for the autopilot. When on an intercept, the modes are set through NAV to APPR for automatic localizer/glide slope capture. As the localizer is captured, the flight guidance and autopilot fly to and track the localizer. As the glide slope is captured, the flight guidance and the autopilot will fly the ILS. At not less than 200 feet AGL the autopilot must be disengaged.

In the event of a missed approach or rejected landing, the flight guidance system should be switched to the go-around (GA) or SYNC as soon as practical.

Caution: Flight guidance and autopilot commands must be cross-checked with the localizer and glide slope displays.

Coupled Approach

The autopilot is capable of flying an approach course when selected to the appropriate mode with the navigation information set properly. Additionally, the autopilot can control aircraft pitch when established on ILS glide slope. During a coupled approach, the flying pilot operates/monitors the autopilot, selects altitudes until glideslope interception, and controls the airspeed.

During maneuvering to the final approach course, select heading mode of the flight guidance and the autopilot. Heading follows the heading marker. Pitch commands are manual or autopilot hold.

During intercept set modes to approach. The flight director and autopilot will transition from heading to VOR/Localizer course automatically. Pitch will remain manual or autopilot altitude hold.

As glide slope is intercepted, autopilot altitude hold disengages and pitch is controlled to fly the glide slope.

If proper visual cues are called out, the flying pilot disconnects the autopilot and makes normal descent to a landing within touchdown zone.

If "Decision Height" is called or the aircraft is not in a position from which a normal approach to the runway can be made, the flying pilot calls "Missed Approach," and executes a go-around.

If the flying pilot elects to continue the approach, the nonflying pilot continues to monitor his flight instruments until touchdown, giving speed and sink information and warning against deviations from the normal descent and speed profile.

On nonprecision approaches utilizing the autopilot, the minimum altitude with autopilot engaged is 50 ft. below the MDA. (FAR 135.93)

Landing from an Instrument Approach

No descent or operation may be conducted below the MDA, or no approach may be continued below the authorized DH, unless:

The aircraft is continuously in a position from which a descent to a landing on the intended runway can be made at a normal rate of descent using normal maneuvers, and that descent rate will allow touchdown to occur within the touchdown zone of that runway.

The flight visibility is not less than the visibility prescribed for the approach procedure, and

One of the following visual references for the intended runway is distinctly visible and identifiable to the pilot.

The approach light system, except that the pilot may not descend below 100 feet above the touchdown zone elevation using the approach lights as a reference unless the red terminating bars or the red side row bars are also distinctly visible and identifiable.

The threshold

The threshold markings

The threshold lights

The runway end-identifier lights

The visual approach slope indicator

The touchdown zone lights

The runway or runway markings

The runway lights

On a straight-in nonprecision approach with a visual descent point (VDP), the aircraft is at the VDP (unless delaying descent until reaching the VDP would require abnormal procedures or excessive rates of descent to land).

Pilots conducting instrument approaches should utilize visual cues as they become available during the approach. At DH or MDA, the pilot must be satisfied that the total pattern of visual cues provide sufficient guidance to continue the approach and landing, and if not, a missed approach is executed. If the approach is continued, it is based on a "see to land" concept, and it is imperative that the required visual reference be continuously maintained.

VFR Traffic Pattern

The basic approach pattern is fundamental to all approaches. No matter how or where the approach pattern is joined, the aircraft should be flown according to the basic pattern from that point to touchdown.

Downwind Leg

Complete Approach checklist prior to entering the traffic pattern. Enter at 1,500 feet above the terrain, (if possible) with airspeed of 150–160 KIAS. Extend flaps to 15 degrees. Opposite the approach end of runway, extend landing gear, initiate Landing Checklist and begin normal descent at 130 KIAS. At approximately 45 degrees to the runway threshold, turn to base leg.

Base Leg

Make a normal descent, 500 to 700 fpm, condition levers to max rpm.

Final Approach Leg

As turn is made to final, extend flaps to 25 degrees. The airplane should be on a normal glide path with a touchdown target 1,000 feet beyond the runway threshold. Proper glide slope angles (2-$\frac{1}{2}$ to 3 degrees) and sink rate (500–700 fpm) must be established early in the approach to avoid high sink rates close to the ground. The airspeed should be stabilized at Vref +10 KIAS 500 feet above the field.

Landings

General

Landings may be broken down into normal, crosswind, rejected, single-engine, zero flap, and windmilling propeller.

Normal Landings

A normal landing is made with flaps 25 degrees by maintaining glide path to touchdown target. At 200 feet AGL, reduce power to arrive in the flare at Vref +5 KIAS. At approximately 50 feet, apply back pressure to flare aircraft. Pitch should be adjusted to allow the main gear to contact the runway in the touchdown zone at Vref.

Note: 45-degree flap landings can be made to lower landing speeds and shorten landing distance when:

The aircraft is being flown by visual reference

Flaps are selected to 45 degrees prior to 500 feet above TOZE, and Vref 45 degrees is computed and used for landing.

Flaps will not be extended to 45 degrees if flap movement will destabilize the approach or cause excessive pitch changes close to the ground.

Maintain directional control to touchdown and stay in the center of the runway. Lower nose smoothly to runway.

Note: GND IDLE/MIN REVERSE is an effective means of slowing the aircraft after touchdown. Limit use of REVERSE THRUST below 60 KIAS to reduce blade erosion and FOD.

CAUTION: Avoid using REVERSE in areas of standing water. The NFP guards the control column after nose wheel contact to prevent pitch-up. Avoid excessive pinning of the nose wheel that could cause lifting action on the mains.

Brakes (including any differential braking necessary) may be applied any time after touchdown. Rudder steering can be used for directional control. Use nosewheel tiller steering only at lower speeds.

Crosswind Landing

The maximum allowable crosswind component including gusts is 25 knots. (If braking is less than good, reference appropriate crosswind limits.)

On final approach, establish a crab angle to maintain aircraft on the extended centerline of runway. Maintain crab angle until just before touchdown, then use rudder to align aircraft with the runway.

Keep wings as level as possible. Use only enough bank to maintain zero drift, touching down on the upwind wheels. Do not hold off downwind wheels. Prompt runway contact will greatly aid in stabilization on rollout. After touchdown and during rollout, aileron correction should only be that necessary to maintain wings level.

Rejected Landing

Whenever a landing is abandoned or rejected, the pilot should:

Apply go-around power immediately. Advance the throttles smoothly and progressively.

Rotate to a nose-up altitude, so as to attain a climb as soon as possible. Call for flaps 15 degrees.

CAUTION: At very low altitudes, flap retraction will be delayed sufficiently to ensure terrain clearance should the aircraft settle during wing flap retraction.

With positive rate of climb, call "Gear Up."

When landing is rejected early in the approach, the handling of power, flap, and gear may be altered slightly to obtain greater speed and terrain clearance margins.

Approach and Landing— One Engine Inoperative

When landing with one engine inoperative, approach and landing techniques should be kept as similar as possible to that of a two engine approach and landing.

CAUTION: After touchdown, GND IDLE or REVERSE could cause control problems. Apply reverse thrust slowly if used to slow down during landing roll.

No Flap Approach and Landing

The pilot should always carefully evaluate runway length and condition, wind, weather, and the instrument approach available when selecting a runway for a no (zero) flap approach and landing. Maneuver the aircraft maneuvering speed. Gear is extended as for a normal approach. Plan to establish a long final approach on a normal glide slope as early as possible. Use every approach aid available.

Having the proper power, airspeed, and sink rate established early in the approach will minimize any large corrections near the ground.

Reduce power approaching runway threshold. Touch down with little or no flare. Use brakes and REVERSE but expect an extended landing roll.

CAUTION: Maximum 5 degrees Attitude Nose Up during flare and touchdown.

Windmilling Propeller
Approach and Landing

The pilot should always carefully evaluate the length and runway condition, the wind, the weather, and the instrument approach available when
selecting a runway/airport for landing with a windmilling propeller.
Consider the drag of the windmilling prop when planning the approach.
Final approach will be flown at 15 degrees flaps. Fly the aircraft onto the
runway with flaps 15 degrees.

Appendix B
Interview Questions

The following questions were compiled from multiple interviews:

1. Describe a time where you were subordinate but assertive.
2. Describe a time where you disagreed with a company policy.
3. Describe a time where you were disappointed as a leader (CFI or captain).
4. Have you ever failed a checkride and why?
5. Why do you think you never failed a checkride?
6. What was the most challenging part of your career and why?
7. Are you happy about your college grades?
8. Do you think you could have done better?
9. Have you applied anywhere else?
10. How come you aren't receiving job offers now?
11. Have you ever had a difficult time with a crewmember?
12. Have you ever changed a company policy or procedure?
13. What will you do if you do not get hired by UAL?
14. Where have you been applying?
15. Are you willing to accept employment as a CFI again just to keep yourself current?
16. Tell us a little about yourself.
17. How did you become interested in being a commercial pilot?
18. How did you get to where you are today?
19. Why do you want to work for United?
20. Tell me about your flight training.
21. Tell us about the biggest work decision you have ever had to make.

22. What was the most difficult part of your flight training?

23. Tell me about a time you felt a company policy was personally distasteful.

24. Tell us about a situation where you broke company policy.

25. Give me an example of an situation in which a company policy was unfair to you. How did you cope with this problem?

26. Tell me about a time you were fairly criticized.

27. Tell me about a time that you were unfairly criticized?

28. Tell us about a time that you became involved with a problem faced by a peer or subordinate.

29. What was the most difficult situation you have experienced in establishing a rapport with a Crewmember.

30. Have you ever flown with anyone who talks too little or too much?

31. Describe the perfect work environment.

32. What is the best decision you have ever made?

33. What was the worst decision that you ever made?

34. What leadership role are you most proud of?

35. What is you greatest failure as a leader? Student?

36. When have you gone above and beyond the call of duty?

37. Have you ever diffused a situation between yourself and a crewmember or between two other crewmembers?

38. Have you ever flown with a crewmember who was not satisfied with your performance?

39. Tell us about a time your flying was criticized other than on a checkride?

40. What was the most negative event in your life?

41. What was the most positive event in your life?

42. Has a subordinate ever questioned you?

43. Who have you respected for their leadership qualities? What were these qualities?

44. What are the most important qualities of a practical leadership philosophy?

45. Give an example of a time that you were part of the problem and not the solution.

46. Do you think of Flight Engineer as a leadership position?

47. Have you ever had to diffuse an argument between a supervisor and another Crewmember?

48. Did you ever disagree with a decision made by a supervisor?

49. Have you ever worked with someone you disliked? How did it affect your job performance?

50. Tell me about a major success you've had.

51. Has anyone ever disliked you? How did you handle it?

52. Name a time that you reversed a decision.

53. Tell us about your college years.

54. Tell us about a conflict in the cockpit and how you handled it.

55. Tell us about a person you have worked with that was hostile. How did you deal with that person?

56. Tell us about a company policy that was completely outrageous.

57. Share with us a situation where you had to ignore company policy/procedures/checklist and had to improvise.

58. Tell us about a time where you made up your mind then changed it?

59. What projects have you accomplished in the past year?

60. How do you make decisions?

61. Have you ever disobeyed company decisions?

62. Have you ever made a bad decision?

63. Have you ever been described as "hard headed?" By whom?

64. In what kind of social situations do you "freeze up?"

65. How skillful do you think you are in sizing up people?

66. What has been the most difficult situation that you have experienced in trying to establish rapport with a crewmember?

67. Tell me what you know about wake turbulence.

68. Tell me about a time where you disagreed with a captain.

69. Tell us about your flying career.

70. We would be interested to know how you became interested in becoming a pilot and what steps you have taken to achieve this goal.

71. Tell us about the biggest work decision you had to make.

72. Would you describe yourself as being more logical or intuitive in making decisions?

73. Have you been tempted to break company policy to get the job done?

74. Give a brief summary of all the leadership positions you have held.

75. Have you ever had to take over the leadership role unexpectedly?

76. Give me an example of when you feel that you failed in a leadership position.

77. Why should we hire you over the other candidates?

78. Describe a time that you made multiple decisions with little information.

79. When did you arrive in DEN?

80. Where are you staying?

81. What type of fog is that outside?

82. Which hydraulic system on the 737 uses engine driven pumps?

83. What items are located on this system?

84. Read a weather report for two specified times.

85. What is the trend of the weather?

86. Can you land at the airport with the reported weather?

87. What kind of weather conditions would you need to be able to land?

88. What would the weather need to be at your filed alternate (both precision and nonprecision requirements)?

89. Explain the ILS DME RWY 8R approach to Denver Stapleton A/P:

90. If you are on the 220 radial 13 miles out at 10,000 and ATC gives you a heading of 270, what would you do?

91. If you are given a descent to 8,000 and a heading of 020 to intercept the localizer cleared for the approach, when will you start to descend?

92. If your dump rate is 1,500 lbs/min, how long will it take to dump 21,000 lbs?

93. Draw a holding pattern on an en route chart.

94. What is the radio call upon entering holding?

95. What is the max holding airspeed for a turboprop?

96. What is outbound timing for holding at 14,000?

97. How would you enter holding on the missed?

98. How soon would you slow down?

99. Give the definition of MEA and MOCA.

100. What is Class A airspace, what is the max speed in Class B, and max speed below lateral limits?

101. What is the standard day temperature in C at 20,000?

102. Calculate numerous time/airspeed/distance problems.

103. Determine the visual descent point on an approach.

104. Why do you have to continually monitor an ADF but not an ILS?

105. What is V1 and how is it used?

106. What is the critical engine in the Beech King Air and why?

107. What is P-factor?

108. Why do you need 5 degrees into the good engine after an engine failure?

109. What kind of A/C and D/C power do the T-44 generators produce?

110. Read a weather line.

111. Numerous time/distance to descend problems.

112. Heading versus radial questions.

113. What are the FAA weather minimums?

114. What are the FAA fuel requirements?

115. Numerous time to fuel dump problems.

116. Define visual descent point. What kind of obstacle clearance do they provide?

117. How is a MSA defined?

118. Can a controller vector you below MSA?

119. What is the formula for hydroplaning? What are the considerations?

120. What is the difference between type 1 and type 2 de-icing fluid, between de-ice and anti-ice fluid?

121. What is an inversion layer and what is associated with it?

122. What weather is associated with a cold front/warm front?

123. What kind of front is associated with a snowstorm?

124. Do snow flurries indicate a stable or unstable air mass?

125. What is the significance of the temperature/dew-point spread?

126. What is advection fog and what causes it?

127. What is an ILS hold line and when is it activated?

128. What is the criteria to continue an approach at the MAP?

129. What does it mean if you hear "PAN, PAN, PAN."

130. What is minimum and emergency fuel?

131. What would you do if you received a windshear report inside the FAF? What if a Cessna 152 reported an airspeed gain of 25 kts?

132. What are outside signs of windshear?

133. What would the weather brief say that would indicate windshear?

134. What are the considerations for landing in a wet or slick runway?

135. What would you do if you went NORDO in IMC right after takeoff, what altitude would you fly?

136. What if the weather is above takeoff rain but below landing mins?

137. You have a sick passenger and the nearest landing minimums are two hours away. What do you do next?

138. What are the considerations for takeoff/landing behind a "heavy jet?"

139. Define accelerate/stop and accelerate/go.

140. What is a balanced field takeoff?

141. Where is the final approach fix for a precision approach?

142. What is virga and what is it associated with?

143. What if the weather goes below mins while you are outside/inside the FAF?

144. If you want to track 330 degrees, with a 30-kt crosswind, what crab angle would you need?

145. If you are cleared for the ILS, are you cleared for the localizer?

146. What is mountainous versus nonmountainous terrain?

147. Calculate numerous time to climb problems.

148. What is the touchdown zone and the requirements?

149. Explain some lost comm scenarios.

150. Talk through a normal approach/landing from 100', 3-4 miles out.

151. What do you do with an engine failure after you rotate?

152. What is Mach buffet?

153. What are the conditions for use of anti-ice?

154. What is a Northeaster, an Alberta Clipper, and an occluded front?

155. What is your fuel planning if the weather is below minimums?

156. What is clear air turbulence and where would you find it?

157. What is the MAP for LOC35 Den (old approach)?

158. In what state is the highest mountain, and how high is it?

159. What would you do for rapid decompression? What altitude would you descend to if you were over the Rockies?

160. What does the arrow in the plan view indicate?

161. How close would you fly to a thunderstorm?

162. What is the transponder code for hijack?

163. When do you need to file an alternate?

164. What are the alternate minimums?

165. How do you check the VORs?

166. What factors are involved in figuring V1 and Vr?

167. What do you do if your engine fails before at V1?

168. Tower issues windshear alert as you are #2 for takeoff. Explain if this is significant to you.

169. How long does a windshear alert last?

170. What does it mean to climb out at Vx or Vy?

171. What is the lowest altitude you would fly on a route with no published altitude?

172. When would you slow down to enter holding?

173. Which leg of the hold will be shorter/longer if you have a headwind/tailwind in the inbound leg?

174. When can you descend below the minimums?

175. Why is nitrogen used to inflate most aviation tires?

176. Where do you see yourself in five years?

177. What was the lowest grade you received in college?

178. Why do you believe you received this grade?

179. Have you ever flown with someone you really didn't like?

180. You are flying as captain and discover that your aircraft is going to be grounded by a mechanical discrepancy. There is a spare aircraft available but it will take approximately 30 minutes to get the aircraft to the gate. How will you prepare your crew for the aircraft swap? What will you tell the passengers? What will you ask the flight attendants to do?

181. What qualities make a good captain?

182. What would your friends say is your biggest weakness?

183. Have you ever been arrested?

184. Have you ever been the subject of an FAA accident/incident investigation?

185. How many moving violations have you been issued since you got your driver's license?

186. Would you rather fly with a good friend or someone who is extremely competent but difficult to get along with?

187. When do you have to declare a departure alternate?

188. Can your departure alternate also be your destination airport?

189. What other airlines have you applied with?

190. How did you choose these airlines?

191. How big of a factor is salary in your decision-making process?

192. How far apart are runway centerline lights?

193. How far apart are runway edge lights?

194. Alternating white and yellow runway edge lights are an indication of what?

195. What has been the scariest thing you have experienced as a pilot?

196. What are your three strongest personality traits?

197. What would you do if you couldn't fly?

198. What was the biggest sacrifice you made to become an airline pilot?

Virtual Airline Internet Resource Guide

by Sean "Crash" Reilly

There are hundreds of great, and not-so-great, Web sites containing Flight Sim and Virtual Airline-related content on the Net. Here are just a few of the better Web sites that I recommend to all flight simmers—especially those who fly for a virtual airline—to enhance their virtual flying experience. Enjoy!

Virtual Airline Listings/Links: Here are a few places to go if you're searching for a virtual airline to fly for. It's easier to provide online resources containing direct links to the 1001 VAs on the Net than it is to list URLs for all of them here. Good hunting!

- **VADU Active VA Listing:** http://www.vadu.com/vai/active_va.htm
- **Virtual Airline Directory:** http://www.va-home.com/vad/
- **MicroWINGS VA Directory:**
  http://www.microwings.com/CoolWebs.HTML#VA
- **Guide To FS (Virtual Airlines):**
  http://www.wwguide.com.au/fs/fsairlin.htm
- **Yahoo (Virtual Airlines):**
  http://www.yahoo.com/Recreation/Games/Computer_Games/Genres/Simulation/Flight/Virtual_Airlines/Airlines/
- **Lycos (Virtual Airlines):** http://www.lycos.com/cgi-bin/pursuit?query=virtual+airlines&matchmode=and&cat=lycos&x=29&y=13

VA News and Information: Virtual Airlines have become so popular that news services were developed to track the "latest and greatest" goings-on at many. A great way to keep abreast of what is happening in the VA industry.

- **Virtual Airlines News Flash:**
  http://virtualairlines.com/vanf/printedition.htm
      The on-line virtual airline news service that has been around the longest. Includes on-line print edition (updated daily) plus Real Audio™ daily news updates and weekly shows! Also contains an archive of past Virtual Airline news stories.
- **Virtual Airline World News:** http://www.va-home.com/vaw/
      A very comprehensive and up-to-date news service. Very well done.
- **Virtual Pilot Publishing News Service:**
  http://www.vpmag.com/news/index.html
      Another site with up-to-date virtual airline and Flight Sim-related news content.
- **Virtual Airlines Press Service (VAPS):**
  http://www.vadu.com/vaps/vaps_now.htm
      A real-time industry news press service. Nonedited. From VA spokespersons' mouths to your eyes. Good archive section.

Free Add-On Panels: If you're going to fly the advanced aircraft in any VA's fleet, you're going to need better panels than the default ones in Flight Simulator.

- **Eric's Freeware Panels:** http://www.flightsim.com/efpanels/
      Eric Ernst, a real-world commercial pilot is arguably one of the best panel designers out there today. Download them free from this Web site.

Navigation Software: If you'd like to navigate with greater ease and accuracy, you should consider using a GPS and/or flight planning software. There are several programs out there. Here are a few.

- **Chris Brett's EFIS:** http://www.formulate.clara.net/
      EFIS is the original moving map system for glass cockpit like those in newer Airbus and Boeing aircraft. Download his shareware EFIS here.
- **Super Flight Planner:**
  http://members.tripod.com/~theflyguy/sfl/sfl.htm
      Still in beta (for free download), a powerful and versatile Windows 95/98/NT application to visually display navaids and plan flights for use with Flight Simulator 95 and 98. It also allows creation of great ATC adventures in a snap.

Airport and Navigation Data/Information: Here are additional real-world navigation resources that will help you get to where you want to go in the virtual world.

- **How Far Is It?** http://www/indo.com/distance/
  This site allows you to plug in "From" and "To" information and then determine the distance from one point to another. It also provides you with specific latitude/longitude and elevation information which can be used in GPS and other navigation.

Weather: Those serious about flight simming like to fly in weather conditions other than the default Flight Sim weather. Below are three weather-related links. The first two provide weather conditions that you can then manually plug in to your Flight Sim's weather dialog. The third is a freeware software program that actually will import real-world weather condition data into your Sim.

- **Aviation Weather Center:**
  http://www.awc-kc.noaa.gov/awc/Aviation_Weather_Center.html
  Aviation weather galore including winds aloft forecasts, National Weather Service information, AIRMETS and SIGMETS. Everything but the NY-METS<g>

- **NOAA's National Weather Service:**
  http://weather.noaa.gov/weather/ccus.html
  Additional very detailed weather information for your flight planning. Searchable by region and specific areas. The server is a bit slow (government Web site) but it's worth the wait.

- **Real Weather5.0:** http://www.avsim.com/mike/weather/index.html
  Freeware program that allows you use weather data that is downloaded from the Internet in your actual Flight Simulator flights. The program will work in FS98 and FS6.0 and it is a Windows 95-based program.

Flying with Real-time ATC: If you want to enhance your virtual flying by one-hundred fold, you absolutely must check out the following links. The programs below are hugely responsible for the continued growth and success of the VA industry.

- **Squawk-Box:** http://www.flash.net/~n5pyk/sb/dl.html
  An outstanding Flight Sim (freeware) add-on software program that enables you to receive real-time air traffic control for the entire duration of your flight. Now you can communicate with ATC—live—just like in the nonvirtual world.

- **Pro Controller:** http://www.netxn.com/~jgrooms/atc/index.html

  If you'd rather do the flight directing than the flying, this is the program that will enbable you to direct other aircraft in the virtual skies. Now you can be a virtual Ground, Tower, TRACON, and/or Center controller!

- **Simulated ATC Organization (SATCO):** http://www.satco.org/

  The official organization of "virtual" air traffic controllers (many of whom are real-world air traffic controllers) who will direct you through all phases of your Squawk Box flight—from push back to shut down. They will provide clearance, vectors, and even put you on SIDs and STARs! These are the guys on the Pro Controller end of things.

Sites with Great "Links" Pages: A few additional sites that contain links to all kids of flight sim-related sites.

- **VADU:** http://www.vadu.com/

  A full-service FBO of the VA industry. They provide flight sim enthusiasts useful information in a way that allows them to get the most out of their experiences. Lots of links.

- **Pre-Flight Checklist:**
  http://www.geocities.com/TimesSquare/Arcade/6721/links.html

- **Mike's Flight Sim Homepage:**
  http://www.geocities.com/CapeCanaveral/7090/

Essential Web Sites: The following Web sites contain more information related to flight simming that you can view in a single sitting. Bookmark and surf both. They are outstanding resources.

- **Avsim.com:** http://www.avsim.com/

  Flight Sim and industry news, product reviews, free downloads, online surveys, links to the some of the very best flight sim sights, flight simmers guides, help files and tutorials, patches and fixes, a listing of many VAs, and much, much more. Plan to spend considerable time surfing the site, and loving every minute of it.

- **Flightsim.com:** http://www.flightsim.com

  Another "must visit" site. This site has been built over several years and is jam packed with everything related to simming, much like the Avsim sight. It also hosts online forms for many of the virtual airlines which makes it unique.

Reprinted by permission of *Full Throttle* magazine, 5376 52nd Street SE, Grand Rapids, MI 49512, 800-821-1707, www.ftmagazine.com

AirNet Systems, Inc. Pilot Salary System

Jet Captain		
	Annual	Hourly
1	38,000	15.22
2	38,000	15.22
3	38,000	15.22
4	38,000	15.22
5	40,000	16.03
6	42,000	16.83
7	44,000	17.83
8	46,000	18.43
9	48,000	19.23
10	50,000	20.03
11	52,000	20.83
12	54,000	21.63
13	56,000	22.44
14	58,000	23.24
15	60,000	24.04
16	60,500	24.24
17	61,000	24.44
18	61,500	24.64
19	62,000	24.84
20	62,500	25.04

Jet SIC		
	Annual	Hourly
1	24,000	9.82
2	26,800	10.74
3	28,200	11.30
4	31,300	12.54
5	33,300	13.34
6	33,900	13.58
7	34,300	13.74

Prop Captain		
	Annual	Hourly
1	18,000	7.21
2	21,000	8.41
3	23,000	9.21
4	24,500	9.82
5	26,000	10.42

Courtesy AirNet, 1998

Flight Operations Abbreviations and Glossary

Courtesy: fltops.com

A&P Airframe and Powerplant Mechanic's license.

ALPA Airline Pilots Association.

AME Airline Medical Examiner.

APA Allied Pilots Association (American Airlines).

ATP Airline Transport Pilot certificate.

ATPw Airline Transport Pilot written test passed.

Block-to-Block Aircraft flight time logging method (from departure block to destination block to restrain movement).

Class I Medical Certificate (also called First Class Medical certificate).

Class II Medical Certificate Minimum certificate to pilot commercial aircraft.

Comm Commercial pilot certificate.

Domicile Location designated as pilot's home or crew base for scheduling purposes.

DOT Department of Transportation.

Duty Rig Duty rigs are special contract provisions that influence how an airline schedules and compensates pilots. For example, the contract will specify the pilot's duty time beginning before aircraft departure and after the aircraft is blocked at the end of the day. If, for example, an airline has a 1:2 duty rig, this assures the pilot will receive not less than one hour of flight time and pay for every two hours of duty.

FAA Federal Aviation Administration.

FAR Federal Aviation Regulation.

FBO Fixed Base Operator.

FE Flight Engineer's certificate, or Flight Engineer.

FEw Passed Flight Engineer's exam.

FO First Officer, co-pilot.

FPA Fedex Pilots Association.

IACP Independent Association of Continental Pilots.

IPA Independent Pilots Association (UPS).

Jump Seat Noncrew cockpit observer seat. Used to identify reciprocal airline pilot travel privileges.

Line Pilot, Line Holder Full-time pilot employee who is assigned a specific "line of flying," or schedule of pilot trips for each month.

LEC Local Executive Council of ALPA local governing body.

LOFT Line-oriented flight training. Pilot training that is based on actual airport routes for the airline and as used in simulator training.

Major Airline fltops.com designates the Big 13 domestic airlines as those global or majors with more than $1 billion in annual revenue generated from flight operations.

MEC Master Executive Council is the central governing body for ALPA at an airline.

ME Multiengine rating is regulated and issued by FAA and keyed to medical certificate.

Multi Multiengine hours logged.

Multiengine aircraft Two or more aircraft engines.

PFE Professional Flight Engineer is a separate designation apart from piloting.

PIC Pilot in Command hours logged.

Reserve A Reserve pilot who has not been designated a predetermined amount of flying for each month.

SASE Self-addressed, stamped envelope.

Seniority Pilot tenure based on date of hire.

SO Second Officer.

SWAPA Southwest Airline Pilots Association.

TT Total flying time logged.

Turboprop Aircraft powered by turbine engine to turn propeller.

Type Ratings FAA ratings of pilot proficiency by aircraft type.

UPAS Universal Pilot Application Service.

Index

About the Author

Robert P. Mark is an airline transport-rated pilot with over 5100 hours of flight time logged in some 37 different aircraft. Mark, a former Midway Airlines pilot who currently flies a Cessna Citation for a Chicago-area charter company, has interviewed with numerous airlines and corporate flying department managers during his aviation career. He holds current instrument and multiengine flight instructor certificates.

Mark is the author of *The Joy of Flying*, second and third editions, as well as *Becoming a Professional Pilot*, both published by McGraw-Hill.

Robert Mark hosts the Aviation Careers Forum on America Online (AOL).

He has also written numerous articles that have been published in *Airline Pilot, Career Pilot, Flying, Aviation International News, Flying Careers, Professional Pilot,* and *AOPA Pilot Magazines.* His aviation work has also appeared in the *Chicago Tribune Sunday Magazine.*